Philosophy of Engineering and Technology

Volume 24

The Philosophy of Engineering and Technology book series provides the multifaceted and rapidly growing discipline of philosophy of technology with a central overarching and integrative platform. Specifically it publishes edited volumes and monographs in: the phenomenology, anthropology and socio-politics of technology and engineering the emergent fields of the ontology and epistemology of artifacts, design, knowledge bases, and instrumentation engineering ethics and the ethics of specific technologies ranging from nuclear technologies to the converging nano-, bio-, information and cognitive technologies written from philosophical and practitioners' perspectives and authored by philosophers and practitioners. The series also welcomes proposals that bring these fields together or advance philosophy of engineering and technology in other integrative ways. Proposals should include: A short synopsis of the work or the introduction chapter. The proposed Table of Contents The CV of the lead author(s). If available: one sample chapter. We aim to make a first decision within 1 month of submission. In case of a positive first decision the work will be provisionally contracted: the final decision about publication will depend upon the result of the anonymous peer review of the complete manuscript. We aim to have the complete work peer-reviewed within 3 months of submission. The series discourages the submission of manuscripts that contain reprints of previous published material and/or manuscripts that are below 150 pages / 75,000 words. For inquiries and submission of proposals authors can contact the editor-in-chief Pieter Vermaas via: p.e.vermaas@tudelft.nl, or contact one of the associate editors.

More information about this series at http://www.springer.com/series/8657

Belén Laspra • José Antonio López Cerezo
Editors

Spanish Philosophy of Technology

Contemporary Work from the Spanish Speaking Community

Springer

Editors
Belén Laspra
Center for Political Studies
University of Michigan
Ann Arbor, MI, USA

José Antonio López Cerezo
Department of Philosophy
University of Oviedo
Oviedo, Asturias, Spain

ISSN 1879-7202 ISSN 1879-7210 (electronic)
Philosophy of Engineering and Technology
ISBN 978-3-319-71957-3 ISBN 978-3-319-71958-0 (eBook)
https://doi.org/10.1007/978-3-319-71958-0

Library of Congress Control Number: 2017963992

Printed on acid-free paper

This Springer imprint is published by Springer Nature
The registered company is Springer International Publishing AG
The registered company address is: Gewerbestrasse 11, 6330 Cham, Switzerland

Introduction: Thinking Through Technology in Spanish

With around 470 million native speakers, Spanish is the second language in the world in terms of the number of native speakers and already the second language used on Twitter, with a sustained growth on the Internet of 800% over the last decade. It provides cultural identity to the people of 21 countries where it is an official language as well as to over 40 million Spanish speakers in the USA. The Spanish-speaking world is a young, dynamic world with great potential for the future. The boom in the population and the growth of the economy in these countries in recent decades have been accompanied by intense development in the sciences and the arts, an upsurge that includes the philosophy of technology. In Spain, Mexico, Colombia, Argentina, or the USA, philosophers who write – at least part of their work – in Spanish seek inspiration in the literature written in their own language with the aim of understanding the meaning of technology in the present-day world.

A salient reference in this respect is the Spanish philosopher José Ortega y Gasset (1883–1955). In his *Meditation on Technics* (1939) and other works, he argues that human nature is not something given by existence but created by individuals who project themselves beyond mere organic needs. This is the *homo faber* who involves himself through technics both in spiritual creativity and material fabrication, thus interpreting and transforming nature as his new artificial home. It is also a one-way trip through evolution: we do not live with technics, we live inside technics. The great threat of modern technology is its ubiquity and extraordinary power to fulfil any projected future, which moved Ortega y Gasset to express the vertigo of human existence due to awareness of its own lack of limits, "(…) for finding himself in principle capable of almost everything makes it all the harder for him to know what he actually is (…) That is why our time, being the most intensely technical, is also the emptiest in all human history" (p. 319, English translation). Some years before this publication, at the outbreak of the Spanish Civil War in 1936, Ortega left Spain to take up exile in France, the Netherlands, Argentina, and Portugal. He was not to return to his country until 1945, coinciding with the end of the Second World War. A new technological world emerged from that global war: a world of lost innocence and empty consumerism driven by seemingly autonomous technology. The recent history of humankind can also be seen as the struggle of the will to create meaning amidst the new technological reality and to understand and manage ever faster technological change worldwide that is increasingly more difficult to control.

The aim of this volume is to bring together a representative collection of work characteristic of contemporary philosophy of technology in Spanish-speaking countries. In spite of Ortega y Gasset's pioneering work, Spanish philosophers of technology have had little impact and influence outside the boundaries of their own language. Thought in Spanish, but written in English, the contributions of this volume aim to fill the lack of visibility of a particular way of thinking through technology – a particular way of living inside technics. The volume addresses different topics of the philosophy of technology, ranging from ontological and epistemological aspects of technology, through ethical, political, and regulatory issues, as well as issues regarding development and innovation, to philosophical queries concerning new technological frontiers such as cloud computing or nanotechnologies, in addition to five case studies that focus on water supply, power plants, engineering education, running shoes, and the fight against drugs. Many divisions and different classifications of contributions could have been made. The one finally chosen only intends to provide a certain order for the readership.

Authors from different fields exemplify some of the main features of philosophy of technology in Spanish-speaking countries. As to the Spanish side of the oceanic divide, Miguel Ángel Quintanilla and Javier Echeverría are major references in the current landscape of philosophy of technology, both in Spain and Spanish-speaking countries. José Luis Luján is a leading Spanish author in research on technological risk and has published several papers in key international journals. This is a trend in quality research which has also been followed by Jesús Vega, who has focused his research on the epistemology of artifacts. Former head of the Spanish National Research Council (1988–1991), Emilio Muñoz is an authority on Spanish science policy who, over the last 15 years, has reoriented his interests toward the philosophy of technology. Eulalia Pérez Sedeño is also a reference in the Spanish-speaking world for her work focused on feminist perspectives on science and technology. The volume also includes other younger scholars, such as Marta González, Javier Bustamante, and José Manuel de Cózar, who have already forged a solid track record of philosophical research on the new technological frontiers. The volume likewise provides a representation of research in the philosophy of technology in American Spanish-speaking countries. Jorge Núñez (Cuba) is a key reference in research on the role of technology in local development, as well as a promoter of what has been called Latin American social thought on science and technology. The research into social issues concerning innovation carried out by Judith Sutz and Rodrigo Arocena (Uruguay) has achieved international recognition. These authors are also advocates of the use of science and technology to promote social inclusion in Latin America. It is a social commitment also present in the philosophical analysis of Mónica Lozano (Colombia) on the fight against drugs in her own country. From Spanish origin, Inmaculada de Melo has developed her career in the USA, where she has got recognition in the field of ethical studies of reproductive technologies. Jorge Linares (Mexico) and Carlos Augusto Osorio Marulanda (Colombia) are eminent scholars who focus their work on ethical issues and the social appropriation of technology, just as Gloria Baigorrotegui (Chile) does from the viewpoint of her own case studies on large technological infrastructures. Finally, Diego Lawler and Diego Parente, both from Argentina, have researched the ontological and epistemological aspects of technology, and their work has already achieved an international scope. These are some of the 31 authors, the coordinators of their respective chapters, who have written the 19 valuable contributions which make up the present volume. Each chapter begins with an abstract, and a full list with short bios can be found at the end of the book. Our gratitude to them all for their good work. Our recognition as well to many other authors that, for one reason or another, could not be included in this volume, especially to the Mexican philosopher León Olivé – an international reference in the field who could not attend our invitation to contribute and unfortunately died shortly before completing the edition of this volume.

Twenty-five years ago, the first volume addressing the philosophy of technology in Spanish (*Philosophy of Technology in Spanish Speaking Countries,* edited by Carl Mitcham) was published by Springer in 1993. Since then, a new generation of Spanish-speaking philosophers and researchers concerned with technology has emerged. In the second decade of this century, some of the pioneering authors (Miguel A. Quintanilla and Judith Sutz were included in the 1993 volume) and many novel authors struggle to make their voices heard in a world dominated by English formats. This volume provides a renewed opportunity for bringing their views and ideas to the international readership, in a humble attempt to contribute to a dialogue that justifies the universal condition that is commonly attributed to culture. The initiative is also a humble recognition of Carl Mitcham's earlier publication and generous personal efforts in providing Spanish philosophy of technology with some international visibility.

Acknowledgments

Acknowledgments to Paul Barnes for his invaluable support in the correction of English language throughout the book.

Oviedo, Asturias, Spain — José Antonio López Cerezo
Ann Arbor, MI, USA — Belén Laspra

References

Mitcham, C. (Ed.). (1993). *Philosophy of technology in Spanish speaking countries*. Dordrecht: Springer.

Ortega y Gasset, J. (1939). *Meditación de la técnica*, Madrid: Revista de Occidente, 1977. English translation: Thoughts on technology. In C. Mitcham & R. Mackay (Eds.), *Philosophy and technology: Readings in the philosophical problems of technology* (pp. 290–313). New York: Free Press, 1972.

Contents

Part I Ontological and Epistemological Aspects of Technology

1 **Atomism, Artefacts, and Affordances** 3
Jesús Vega Encabo and María Muñoz Serrano

2 **Praxeology Approaches Technology: The Ontology and Epistemology of Our Technological Practices** 17
Diego Lawler

3 **Synthetic Life: Organisms, Machines, and the Nature of Synthetic Biology Products** 31
Diego Parente

Part II Ethical, Political and Regulatory Issues

4 **Valuing Reprogenetic Technologies: Bringing Insights from the Philosophy of Technology to Bioethics** 45
Inmaculada de Melo-Martín

5 **Regulatory Science: Between Technology and Society** 59
José Luis Luján and Oliver Todt

6 **Practices and Knowledge: Philosophy of Biomedicine, Governance and Citizen Participation** 73
Esther Ortega Arjonilla, Silvia García Dauder, Nuria Gregori Flor, and Eulalia Pérez Sedeño

7 **The Culture of Risk: STS Citizens Facing the Challenge of Engagement** 87
José Antonio López Cerezo and Belén Laspra

Part III Development and Innovation

8 Engaging Technologies: Criteria for an Alternative Model of Technological Development 103
Miguel Ángel Quintanilla Fisac

9 Re-thinking Innovation as a Lever for Development Taking into Account Inequality 125
Rodrigo Arocena and Judith Sutz

10 University, Technology and Development: Reflections from the South 139
Jorge Núñez Jover and Galia Figueroa Alfonso

Part IV New Technological Frontiers

11 The Philosophy of Technoworlds and Technopersons 153
Javier Echeverría

12 Ethical and Political Delusion in the Model of Cloud Computing 165
Javier Bustamante Donas

13 The Promises of Synthetic Biology: New Bioartefacts and Their Ethical and Societal Consequences 179
Jorge Enrique Linares Salgado

14 Matters of Concern Regarding Nanotechnology 195
José Manuel de Cózar-Escalante and Andrés Manuel Núñez-Castro

Part V Case Studies

15 Philosophy of Activism and Community Management in Water Systems 211
Carlos Augusto Osorio Marulanda

16 Localities Facing the Construction of Fossil-Fuel Power Plants. Two Experiences to Address the Hostile Face Electricity Infrastructures 223
Gloria Baigorrotegui and María Teresa Santander Gana

17 The Challenge of Transversal Education Through Teaching Ethics in Engineering: From Hubris to Hybrid 239
Emilio Muñoz, Ana García-Laso, and Domingo Alfonso Martín-Sánchez

18 Technique and Technology in the Practice of Distance Running 251
Marta I. González García

19 Public Participation in Science and Technology and Social Conflict: The Case of Aerial Spraying with Glyphosate in the Fight Against Drugs in Colombia 267
Mónica Lozano

Contributors

Rodrigo Arocena holds a degree in mathematics and a Ph.D. in development studies. He has served as rector of the University of the Republic of Uruguay. In the field of social sciences, his main line of research addresses the problem of innovation in underdevelopment, the evolution of higher education in Latin America, the contribution of universities to integral development, and the relationships between research and innovation and social inclusion. He has also worked for more than 20 years in foresight tasks.

Gloria Baigorrotegui holds a degree in industrial civil engineering and a Ph.D. in philosophy of science. She is a member of the Institute for Advanced Studies at the University of Santiago of Chile. Her research focuses on two areas: philosophical and sociological analysis of energy technologies, the environment and social movements, and the relationships between engineering, education, and technology. She also has some research experience in policy areas, including social capital in relation to intellectual capital, energy policy, and government local planning.

Javier Bustamante Donas holds a Ph.D. in philosophy (Complutense University of Madrid), an M.Sc. in science and technology studies (Rensselaer Polytechnic Institute), and an M.Sc. in computer science (Pontifical University of Salamanca). He lectures in moral and political philosophy at the Complutense University. He is the director of the Ibero-American Center for Science, Technology and Society (CICTES) and vice president of the González-Vallés Foundation. He is the author of nearly a hundred publications on cyberculture and Chinese philosophy.

José Manuel de Cózar-Escalante holds a Ph.D. from the University of Valencia (Spain) in evolutionary epistemology and is currently an associate professor at the University of La Laguna (Spain). His areas of research are science and technology studies, philosophy of science and technology, and environmental philosophy. He has published numerous papers on the ethical and social impacts of nanotechnology.

Inmaculada de Melo-Martín is a professor of medical ethics in the Division of Medical Ethics at Weill Cornell Medical College. She holds a Ph.D. in philosophy and an M.Sc. in molecular biology. Her research interests include bioethics and philosophy of science. Most of her work has been on ethical and epistemological issues related to reprogenetics and molecular genetics. She has been particularly concerned with calling attention to the importance of science when making ethical judgments, the importance of ethics when evaluating new scientific and technological developments and proposing public policies, and the importance of attending to the social and political context when assessing science and technology.

Javier Echeverría is an Ikerbasque research professor at the University of the Basque Country (San Sebastián, Spain). He has published 19 books of his own, over 100 papers in scientific journals, and more than 200 articles in collective books. He was the president of the Spanish Society of Logic, Methodology and Philosophy of Science, between 1992 and 2000 and has been a member of the International Academy of the Philosophy of Science since 2006.

Galia Figueroa Alfonso is an assistant professor at the University of Havana and is currently a Ph.D. student at the Center for Development Research (ZEF), University of Bonn.

Silvia García Dauder holds a Ph.D. in psychology and is currently an associate professor of social psychology at the Rey Juan Carlos University, Madrid. She has participated in various research projects on science, technology, and gender and has published several articles on the relationships between psychology and feminism and on pioneering psychologists and social sciences.

Ana García Laso was awarded her degree in telecommunications engineering from the Polytechnic University of Madrid (UPM) in 2010 and accredited as an M.Sc. by the Accreditation Board for Engineering and Technology (ABET) and a Master's Degree in technology for human development and cooperation from the UPM (2012–2014). She is currently a Ph.D. student as well as coordinator and spokesperson for the Social Entrepreneurship, Ethics and Values in Engineering Unit (UESEVI) at the Mining and Energy Engineering School (UPM).

Marta I. González García is an associate professor of philosophy and social studies of science and technology at the University of Oviedo, Spain. She has held research and teaching positions at the Technical University of Budapest, University of Minnesota, Complutense University of Madrid, Carlos III University, and the Institute of Philosophy of the Spanish National Research Council (CSIC). Her research deals with gender and science, scientific culture, and public participation in environmental controversies.

Nuria Gregori Flor holds a Ph.D. in sociocultural anthropology and gender studies. She works as a nurse in the General Health Department at Elche Hospital, Alicante, Spain. She has focused her research on the medical and social management of intersexuality and differences in sexual development (DSD) within the Spanish context. She has participated in numerous projects on intersex issues and has published several articles on narratives around intersexuality/DSD.

Belén Laspra holds a Ph.D. in philosophy from the University of Oviedo (Spain) and a Master's Degree in social studies of science and technology from the same university. She was a Severo Ochoa predoctoral research fellow at the University of Oviedo. She is a member of the Social Studies of Science Research Team at the University of Oviedo. She is currently a postdoctoral research fellow at the Center for Political Studies, University of Michigan. Her research lines focus on the study of scientific culture and its indicators.

Diego Lawler holds a Ph.D. in philosophy from the University of Salamanca (Spain). He works at the Institute of Philosophical Research, Argentinian Society of Philosophical Analysis. Since 2005, he has been a researcher at the National Council for Scientific and Technical Research (Argentina). His lines of research focus mainly on metaphysical and epistemological aspects of technology, theory of action, practical knowledge, and artifacts.

Jorge Enrique Linares Salgado holds a Ph.D. in philosophy from the National Autonomous University of Mexico (UNAM) where he lectures at the Faculty of Philosophy and Humanities. He is a member of the National System of Researchers, of the Mexican Academy of Sciences, and of the International Association of Bioethics. His teaching and research work has focused on the ethical problems of science and technology, bioethics, and contemporary ethics. He is currently the director of the Bioethics Program at UNAM.

José Antonio López Cerezo is a professor of logic and philosophy of science at the University of Oviedo, Spain, coordinator of the STS Network of the Organization of Ibero-American States (OEI), and head of the Social Studies of Science Team at the University of Oviedo. His main academic interest focuses on the study of social and philosophical aspects related to science and technology (scientific culture, citizen participation, risk acceptability), fields in which he has published several books and papers in international scientific journals.

Mónica Lozano is a professor in the Program of Psychology in the Universidad del Rosario, Bogotá, Colombia. Her research centers on the role of science and technology in democratic governance, with a particular focus on the process of social appropriation of science and technology. She is the author of several publications within this field.

José Luis Luján is a professor of logic and philosophy science at the University of the Balearic Islands (Spain). He has been a researcher at the State University of Navarra and the Spanish National Research Council and is member of the CIEMAT Research Unit on Scientific Culture (Madrid, Spain). Author and editor of several books in Spanish and papers in international academic journals, his research focuses on the philosophical analysis of risk assessment and the role of scientific knowledge and values in technology regulation.

Domingo Alfonso Martín-Sánchez holds a Ph.D. in geological engineering from the Polytechnic University of Madrid (UPM). He currently lectures in the Department of Geological Engineering at the Mining and Energy Engineering School of the UPM, where he is vice director of Quality and Social Responsibility. He has coordinated several international cooperation projects for development in Argentina. Author of more than 20 articles related to geology, geophysics, and innovation, he is also coauthor on two industrial patents.

Emilio Muñoz is a technical director of the Social Entrepreneurship, Ethics and Values in Engineering Unit (UESEVI) at the Mining and Energy Engineering School (Polytechnic University of Madrid). He is the coordinator of the CIEMAT Research Unit in Scientific Culture (Madrid, Spain). Among other posts, he has served as president of the Spanish Research Council (CSIC). He is the author of several hundred articles on biochemistry, biotechnology and its social and economic aspects, and scientific and technological policy.

María Muñoz Serrano holds a degree in philosophy from the Autonomous University of Madrid and a Master's Degree in logic and philosophy of science from the University of Valladolid, Spain. She is currently a predoctoral researcher in the Department of Linguistics, Logic and Philosophy of Science at the Autonomous University of Madrid. Her research focuses on the analysis of artifacts and the practices of interaction between intentional agents and artifacts from an ecological and situated perspective. Her approach highlights the importance of materiality and interdependence between artifacts, habits, cultural practices, and knowledge.

Andrés Manuel Núñez-Castro holds a Ph.D. in philosophy of technology and is a member of the Group for Social Research in Nanotechnology at the University of La Laguna, Spain. His recent research focuses on the democratic evaluation of the transfer of nanotech innovations and the social implications of their risks. He has also published several articles addressing the issue of technological risk from different perspectives.

Jorge Núñez Jover holds a Ph.D. in philosophical sciences and an expert qualification in epistemology and philosophy of science. He lectures at the University of Havana, where he is the coordinator of the Department of Social Studies of Science, Technology and Innovation and director of the university's Post-Graduate Program. Social innovation and regional development are among his recent lines of research.

Carlos Augusto Osorio Marulanda holds a Ph.D. in philosophy in the University of Oviedo (Spain). He is an associate professor of the Industrial Engineering School of the Universidad del Valle, Cali, Colombia. His lines of research focus mainly on education for the public participation in science and technology.

Esther Ortega Arjonilla holds a Ph.D. in philosophy of science and is an associate professor in the Sociological Theory Department at the Complutense University of Madrid. Her research interests and main published articles focus on STS, feminist and queer technoscience studies, and the history and sociology of sexuality. She is currently embarking on a new line of research on postcolonial technoscience and its intersection with the categories of "sex, gender" and "race, ethnicity."

Diego Parente is a professor of philosophy at the University of Mar del Plata (Argentina) and researcher at CONICET (Consejo Nacional de Investigaciones Científicas y Técnicas). He works primarily in the areas of philosophy of technology and contemporary philosophy. His recent books in the field focus on the nature of artifacts and the biocultural dimension of technology.

Eulalia Pérez Sedeño is a professor of research in science, technology and gender at the Spanish Research Council's Institute of Philosophy and professor of logic and philosophy of science. She has lectured and researched at the University of Barcelona, the Complutense University of Madrid, the University of Cambridge (UK), and the University of California at Berkeley, among others. She has carried out research on the history of science and scientific institutions, as well as on the philosophy of science, science, technology and society, perception and communication science, and science, technology, and gender.

Miguel Ángel Quintanilla Fisac is the founder of the ECYT Institute and Professor of Logic and Philosophy of Science at the University of Salamanca, Spain. His major lines of research are the philosophy of science and technology, social studies of science and technology, scientific culture, public communication of science, and scientific and technological policies. He has published numerous books and articles on subjects of logic, philosophy of science, epistemology, history of philosophy, moral philosophy and politics, and philosophy of technology, as well as on scientific, technological, and university policy.

María Teresa Santander Gana holds a degree in industrial civil engineering and a Ph.D. in philosophy of science. She is a member of the Department of Industrial Engineering at the University of Santiago of Chile. Her training, studies, and research have focused on the area of science, technology, and society, specializing in national innovation systems, management and innovation methodologies; entrepreneurial culture, and methodologies. She is in charge of technological projects and innovation with the small- and medium-sized enterprises and public institutions, in addition to evaluating and managing them.

Judith Sutz is an academic coordinator of the Sectoral Commission for Scientific Research, University of the Republic (Uruguay). She researches in the field of science, technology, society, and development. Her main lines of research address the problems of development and underdevelopment based on the interrelations between science, technology, and society. She also studies the problems associated with the production and social use of knowledge, promoting interdisciplinarity in the field of social sciences.

Oliver Todt is a professor of philosophy of science at the University of the Balearic Islands (Spain). He has been a researcher at the Institute of Philosophy of the Spanish Research Council. His principal research interests include the role of scientific knowledge in regulatory and policy processes in the governance of science and technology, as well as decision-making processes in technology innovation and regulation. He is the author of more than 50 publications on these topics in Spanish and international academic journals, including 20 in JCR-indexed journals.

Jesús Vega Encabo is a professor of logic and philosophy of science at the Autonomous University of Madrid. His research interests are mainly focused on epistemological issues (testimony, know-how, epistemic normativity, epistemic agency, perceptual knowledge, self-knowledge, and epistemic virtues, among others) and the study of different epistemic cultures. His work moves between the fields of philosophy of mind, philosophy of science, and philosophy of technology. He has recently coordinated several research grants on scientific practices, knowledge, and artifacts.

Part I
Ontological and Epistemological Aspects of Technology

Chapter 1
Atomism, Artefacts, and Affordances

Jesús Vega Encabo and María Muñoz Serrano

Abstract In this chapter, we argue against what we call the atomist assumption that has governed the recent recovery of artefacts in metaphysics. By drawing some parallels between artefacts and concepts, we hold a relational and holistic view on the nature of artefacts. Artefacts acquire their identity conditions only within a constellation of other artefacts and objects. We will also defend that the identity of an artefact is given by the affordances that constrain our agency within a cultural niche or transformative environment. As constitutive elements of these environments, artefacts do not stand isolated; what they are, fixed by what they afford, cannot be specified without referring to other artefacts, (natural) objects, and the (normative) practices that regulate the engagement with them.

1.1 Introduction: The Artefactual Realm

This chapter aims to be a contribution to the ontology of technology. We take it to be an ontology of action, but not one that privileges the category of "technical action" (Vega 2009). "Technical" hardly qualifies actions in a direct way; it is not easy to say what makes an action a *technical* one if it is not its being just skillful or artful. We think that the focus should change towards what we will call a *transformative environment*, that is, the *milieu* where our agency succeeds in tracing transforming-paths and trajectories. In transformative environments, instrumental actions take place when an articulation means/end is present, sometimes through an intermediary object that serves as an *instrument*. Instrumental actions are a basic component in the skillful and *technical* transformation of environments, but not the

J. Vega Encabo (✉) · M. Muñoz Serrano
Department of Linguistics, Modern Languages, Logic and Philosophy of Science, Theory of Literature and Comparative Literature, Faculty of Philosophy and Letters, Autonomous University of Madrid, Campus of Cantoblanco, Crta. Colmenar km. 15,4, 28049 Madrid, Spain
e-mail: jesus.vega@uam.es

B. Laspra, J. A. López Cerezo (eds.), *Spanish Philosophy of Technology*, Philosophy of Engineering and Technology 24,
https://doi.org/10.1007/978-3-319-71958-0_1

only one. Other kinds of actions, with symbolic, expressive or even theoretical dimensions, are crucial in shaping the transformative environments where we live and act. We cannot lose sight of this complexity of actions when developing an ontology of technology.

Transformative environments are habitats that have been substantially modified by the activities of animals that move in them constantly creating new pathways. Such environments are not necessarily the outcome of represented purposes, but constitute nevertheless the background against which animals deploy their skillful activities and build their technological devices and artefacts. The control over the pathways and the habitat they live in is characteristic of their technological engagement with the world. A second characteristic feature of these (technical) transformative environments is that they help in organizing new forces of transformation that crucially affect the agents that act in them. They are *active* environments, even if one does not want to regard them as genuine agents. They are transformative basically because a complex network of practical possibilities dwell within them. These *real* practical possibilities are usually articulated around objects that we call *artefacts*. These are the outcomes of the interventions that shape transformative environments, that is, independent items detached from them that become the objects the agents engage with to organize their own activities.

Artefacts make up a realm whose metaphysical status is not easy to discern. In this chapter, we will address some problems in understanding the nature of artefacts as items that are part of those transformative environments where agents are able to identify possibilities of action. Our aim will be to challenge an assumption that has driven the recovery of artefacts for metaphysics, the *atomist assumption*, and propose an alternative that emphasizes the role of artefacts within complex transformative environments of artefacts and affordances. In Sect. 1.2, we introduce the atomist assumption in the metaphysics of artefacts by drawing a parallel between artefacts and concepts. Section 1.3 argues against atomism and defends that the physical, functional, and intentional dimensions of artefacts, in so far as they are relevant to the metaphysics of the artefactual realm, require a non-atomist, relational, account. In Sect. 1.4, we describe artefacts as nodes that articulate affordances or possibilities of action in a whole transformative environment. Artefacts only stand out against the background of those environments and cultural niches in which agents pick up affordances and answer to their demands. As constitutive elements of a niche, artefacts do not stand isolated; what they are, fixed by what they afford, cannot be specified without referring to other artefacts, (natural) objects, and obviously the (normative) practices that regulate the engagement with them.

1.2 The Atomist Assumption in the Metaphysics of Artefacts

The last decades have seen a recovery of artefacts for metaphysics. This recovery, as Amie L. Thomasson has remarked, resolves an unfortunate situation of neglect "since understanding the ontological status of artifacts is crucial to understanding the objects of concern in the social, human and technological sciences" (Thomasson

2009, p. 192). Part of this recovery has been governed by the idea that the genuine ontological significance of artefacts asks for the identification of real or nominal kinds (Lawler and Vega 2010; Vega and Lawler 2014). Every artefact derives its identity conditions from its membership to an artefactual kind, that is, from those features that unify the members of the kind. These features could be established either by the very nature of the thing or by criteria relative to the use of the corresponding sortal terms. For instance, the question of what a screwdriver is becomes a question about what unifies the different kinds of screwdrivers. This could be related to the criteria that govern the use of the term and then we treat all of them as belonging to the same nominal kind. But there could also be a *real* nature to be uncovered, the analogue of a microstructure that explains the superficial features of the members of the kind. In order to identify those features that delineate the kind, several options are now at our disposal: we can appeal to the sameness of function to fix the identity criteria of kind membership or to the sameness in the productive intentions of the makers. Many other (intermediary) positions on this issue are now on the table.

We are going to defend that this path to the metaphysical recovery of artefacts has been driven by a reasonable, but misguided assumption. We will call it the *atomist assumption* following from which an artefact possesses by itself a set of characteristics or essential properties as an object that is identical to itself, isolated and independent of other objects. The identity conditions of an artefact are defined by its belonging to a kind (nominal or real) that is delineated by membership criteria that do not apply to other kinds of artefacts or entities. We think that this assumption ballasts our understanding of artefactual and technical environments.

Our talk of atomism in the artefactual realm does not refer to physical atomism but to what is discussed in the philosophy of language and mind as *conceptual* or *meaning atomism*. This is particularly important, because it suggests where to look in order to account for the significance of artefacts in metaphysics. It is not just a question of drawing consequences from the metaphysical dependence of artefacts on the concepts that are constitutive aspects of the intentions of their makers (maybe not only of them). This has been the very enlightening strategy followed by Thomasson (2003, 2007, 2009, 2014) in her metaphysics of artefacts. Our starting point and inspiration is an analogy between concepts and artefacts, such as has been proposed by Fernando Broncano (2008, 2009, 2012). The parallels between them are very instructive. They are so close that one is even tempted to claim that artefacts exhibit the *very same* metaphysical features as concepts. Let us introduce two of those parallels.

First, we can regard concepts as "intellectual" units that organize our experience in judgments. In the same way, we can talk about artefacts as "material" units that organize our experience in actions. Both concepts and artefacts intervene in the delineation of a space of possibility and necessity that is the result of exerting what the Kantian tradition has called our spontaneity. They initiate us in a space of rational constraint and freedom that is characteristic of our inhabiting the so-called second nature (Vega 2012). Artefacts articulate rational constraints in the domain of agency. They help us to give birth to procedures to transform reality that are at the same time tools to transform us as agents with our intellectual and practical abilities and competences. Broncano explains it with clarity: "Asserting a state of affairs under a concept or skillfully bringing about a state of affairs are two ways of transcending the causal order of things through the exercise of spontaneity" (Broncano 2012, p. 89; our translation). Artefacts manifest a way of being that is not detachable from the sort of intelligibility that is tied to the exercise of our spontaneity (Vega 2012).

Secondly, concepts and artefacts share several features concerning their respective normativity. Just as the normativity of a concept is strongly tied to the set of inferences to which we are entitled by its possession, so the normativity of an artefact is given by the set of actions to which a competent agent is entitled. But it is not only that. Both of them exhibit the characteristic structure of an achievement. As F. Broncano rightly points out again, "concepts and artefacts shape our horizon of effective possibilities and their use should be efficient" (Broncano 2012, p. 90). The normativity of an achievement is given both by its success conditions and by the metaphysical dependence of success on the exercise of a competence. Therefore, the achievement in the application of concepts captures the normativity characteristic of judgings; the achievement of putting into use artefacts captures the normativity characteristic of our artefactually mediated agency.

Concepts are rightly viewed as the units of thought; artefacts can be regarded as the units of our ability to transform environments. Units? No doubt. Our challenge to the atomist assumption does not want to deny that artefacts are taken as units; what we will deny is that they have any sort of metaphysical independence regarding other units, mainly artefactual. They need to be characterized relationally and as taking part in meaningful "wholes" in our transformative experience of environments.

Let us explore in more detail the atomist assumption. In order to do that, we should maybe remember two well-known semantic principles of Fregean inspiration: the principle of context and the principle of compositionality. Both of them can also be applied to the domain of artefacts. Or so we claim. But doing so does not come without consequences for our metaphysical understanding of artefacts. The principle of context states that the semantic value of an expression is a function of the contexts in which it appears. There could be a corresponding principle in the realm of artefacts. We could define a sort of semantic value understood as "transformative value" that is the specific contribution that the artefact makes to the different contexts in which it appears. The context includes other artefacts, (natural) objects, agents and specific actions; the artefact needs to compose with this complex set in order to have a concrete value as transformative tool.

The principle of context usually goes with a compositionality principle that claims that the meaning of a complex expression is a function of the semantic values of the units together with the rules of composition. True enough, it is not easy to identify the sort of "complex expressions" that are part of the artefactual realm in the same way as we do in language, but we can coherently think that compositionality is exhibited at least at two levels: at the level of the objects themselves and at the level of the actions they entitle the agents to. Artefacts crucially compose with other artefacts; it is this combinatorial nature of the technical elements that explain the growing complexity of our artefactual world. This structure is better understood as a set of constraints and possibilities that prevent/enable certain combinations. The fitting together of units according to certain rules is also essential to the artefactual realm. The compositionality at the level of actions is easier to grasp; actions mediated by artefacts compose in more complex agential expressions.

Atomism could be viewed as an explanation of the widespread application of both principles in the artefactual realm. Let us take a particularly successful strategy in the metaphysics of artefacts that is committed to some version of atomism: the view in which the essence of the artefact is captured by its (proper) function. The artefact, as the unit that can enter into several different combinations and artefactual contexts, is defined by what it is supposed to do and essentially characterized by a functional property (or properties). There are many ways of spelling out functional views of this sort. Consider one proposed recently by P. Kroes (2012). The identity of the artefact is given by the (technical) artefactual kind it belongs to and there is a functional property that determines in turn kind membership. The functional property in question is fully determined by a "largely successful execution of a largely correct design" (p. 118). The functional property *of the artefact* in virtue of belonging to a well-delineated artefactual kind sets the contexts and combinations open for the artefact.

One could argue that nothing in this definition encourages an atomist reading. On the one hand, success in bringing into existence an artefact of the kind requires that other material and practical conditions be satisfied, among which the availability of other artefacts, agents, and technical capacities is crucial. These requirements can even become explicit in the process of design. What is important to realize is that none of these elements contribute to the fixation of the functional property that metaphysically characterize the artefact as such and its particular "for-ness". There is something that the artefact is supposed to do and this corresponds to the functional property that has been selected by design for each artefact. On the other hand, they become the basic artefactual units that define the possibilities not only of combination but also of being included in new contexts of use and articulation of artefactual complexes. And if it is so, the attractiveness of atomism is now easy to understand: each artefact provides a sort of transformative value; this value fully depends on the functional property that defines the kind of artefact it is; when it is adequately combined with other transformative values, it enables a new complex of artifacts and makes up technical environments.

1.3 Beyond Artefactual Atomism

But is the atomist assumption plausible in the artefactual realm? We think it is not. We can sketch the core of this assumption in the following terms:

(Atomism): Artefacts are units whose transformative value within a context is determined by the properties that define what they are independently of any other element (artefact) of the transformative environment.

We argue that this assumption obscures the nature of artefacts as core elements of transformative environments. We start by identifying at least three dimensions that traditional theories appeal to in order to metaphysically characterize artefacts:

(i) *physical structure* and the capacities of the object; (ii) *purely functional features* or *the "for-ness" dimension*, which involves identifying what the artefact is supposed to do; and (iii) *intentional states* of the agents (designers/users), which contribute to setting the function of the artefact. In the rest of this section, we will argue that if we want to understand the role played by artefacts in transformative environments the atomist assumption is to be rejected in any of these dimensions which underlie almost every metaphysical position about them.

1.3.1 The Physical Dimension

Obviously, physical structure is relevant in order to establish the identity of artefacts; additionally, physical structure is a core element when determining their function, normative use and the kind of interaction that agents establish with them. For instance, the sharp edge of a knife is a relevant structural property for a knife to be recognized as such and to be able to successfully satisfy its function; in the same way, a wheel must be (more or less) round in order to be what it is.

However, to think that the physical and structural properties of an artefact can be conceived in isolation is, if not naive, then at least unfruitful. The physical structure of a given artefact is only significant as far as it is combined with other physical properties of the environment that allow the artefact to be properly and successfully used, for instance. Let us come back to the example of the knife: the sharp edge seems to be a structural property, which is essential to identify the knife as such; but a sharp edge only makes sense when it is combined with other properties of the environment (properties of other artefactual or natural objects, or embodied properties of the agent). So, a knife cuts when there is a surface hard enough or when an agent holds strongly enough what is going to be cut, and, obviously, there is a material with certain structural properties that allows it to be cut. In the same way, the circular structure of a wheel is only materially significant in interaction with other properties of the environment, like the flatness and firmness of a surface.

It could be argued that these conditions are merely contextual; this is true in a sense, but deeply wrong in another. True, because at the same time it is the form and matter characteristic of *being a knife* that counts even if they only manifest the corresponding physical and causal capacities within a context and in the right circumstances. But it is wrong to treat these conditions as part of a context so narrowly understood. It is not just a one-time context of appearance what is at stake here. First, a knife is still a knife in contexts where there are not solid surfaces, and a wheel of a car does not cease to be a wheel when the car gets stuck in the mud (and the wheel cannot satisfy its function successfully). When we say that the physical and structural properties of artefacts only make sense in relation with the physical and structural properties of the environment it is placed in, we are pointing to a wide and stable environment where the "history of life" of an artefact takes place: for instance, the wheel, as an artefact, could not be *conceived* in an environment where there were no flat and firm surfaces. In short, the physical description of an object is not just its structural description, that is, the description that identifies how its function depends on its structure (if we take the function as essential for being such an artefact and the function as fully determined by the structure). An adequate physical description of an artefact should mention other artefacts, natural objects, and even the physical conditions of (intentional) agents. Its structural properties cannot be fixed without the material conditions provided by other artefacts, like for instance the dependence that cars exhibit on wheels and roads, even if this sort of dependence is flexible, variable and comes in degrees. The material dimension of a particular artefact is fixed within a network and a background of other material objects and agents, in such a way that the materiality of the artefact can be seen as just an aspect of the environment of the material culture in which it is inscribed.

1.3.2 The Functional Dimension

Secondly, atomist assumptions are more widely spread with respect to the functional dimension of artefacts. Functional descriptions concern basically the "forness" of the artefact. This very fact allows us to see how any functional description points to other artefacts and objects in a very direct way; a wheel is for moving other things or people, it is not just for moving itself; a knife is for cutting something that is not itself, and so on. But there are other more convincing considerations that argue for the fact that any functional description involves a whole environment of artefacts and that any attempt to determine the function of an artefact in isolation systematically fails. Taken in isolation, the determination of the function of an artefact is always threatened by ambiguity. The function of an artefact can only be disambiguated if we also consider the context in which it is inserted. Let us imagine, for example, a wine decanter, an artefact whose use and function are very specific. A wine decanter isolated from its proper context –and from the environment that it creates around it- is not very different from a vase in which we put flowers. What we could call the intrinsic properties of the object are compatible with both functions and there is nothing in the object itself that lean us towards privileging one or other functional description. Thus, we are before an artefact whose nature remains undecided in so far as it is considered without reference to other artefacts and objects. The environment of artefacts and agents becomes essential in the understanding of the functional nature of the artefact.

Once again, a parallel can be drawn between artefacts and concepts and we can thus apply to artefacts the Wittgensteinian lesson about how to treat misunderstandings about meaning. Appealing to the context is required. But the context extends to the activities, practices and techniques that help us to highlight the *point* of using the word. A stable environment where the history of the life of the artefact makes sense is needed, but also a complex physical context, other artefacts (and their functional profile), objects of any kind, agents, and even cultural practices, ways of doing things, regularities and characteristic reactions, regular action-paths, and so on.

Keeping this in mind, let's come back to the example of the wine decanter/vase for flowers. The object can be functionally defined as a decanter when it is put, within a certain environment, in relation to a constellation of elements: other artefacts (glasses, wine bottles, corkscrews, etc.), certain cultural practices (for instance, the ritual of having dinner in a posh restaurant), ways of doing things (such as the proper activities of waiters and sommeliers), etc. This context, this constellation of elements, enables certain actions for an agent who makes a successful identification of the function of the artefact, of the purposes that grow only in this kind of environment of other objects. It is, precisely, by enabling these actions that a non-ambiguous functional description is possible.

The nature of an artefact is inseparable from the framework of intelligibility in which it is inscribed, and this framework is constituted by a changing, but also stable mutual fitting of multiple material artefacts and agents in interaction. This *fit* is normatively regulated. This explains why artefacts seem to exhibit the normative force characteristic of a demand to act, a demand that is only manifested within the particular settings in which artefacts seem unproblematic with regard to their *proper* function.

1.3.3 The Intentional Dimension

We come lastly to the atomist assumptions that come with the intentional dimension of artefacts. One of the most fruitful views in the metaphysics of artefacts is the one that defends that the nature of artefacts is determined by the content of the intentions of their makers, a content that necessarily includes a substantive concept of them (Hilpinen 2004; Thomasson 2003, 2007). The identity of each artefact, and thus the identity of the kind, is given by the intentions of the makers. There is a set of intentional conditions that are sufficient for the identity of the kind. But the question is then whether these intentions can be isolated in such a way that the "conceptual content" that contributes essentially to the delineation of the kind is fixed without referring to the environmental conditions where the artefact ultimately is placed.

On the one hand, a maker's intentions are the product of a long history of interactions within a cultural, social and physical environment. Instead of conceiving intentions as mental states that can be individuated on their own, we believe that it is more promising to understand agents' intentions as a form of coordination with the environment in which the constellation of artefacts, practices and agents makes some intentions and courses of action available.

On the other hand, and in general, the intentional dimension that artefacts paradigmatically exhibit cannot be identified with the intentions involved in the making or in the intentional use of the artefact or even in the intentions that become recognizable in the very artefact itself. It is not possible to isolate a set of intentions (or intentional states) that is *proper* to the artefact. In fact, its very reality as an artefact comes to life only when it genuinely contributes to make available purposes inside a concrete transformative environment where other artefacts are recognizable as such. That is to say, intentions *grow* within an environment that is already meaningful to the agents, and this meaningfulness cannot be severed from the rich set of intentions other artefacts help to articulate.

In brief, the identity and meaning of an artefact is set only within a complex environment of other artefacts that already are a substantial aspect in the agential involvement of the organisms. Any of the dimensions that could be used to formulate a metaphysics of artefacts –physical, functional, or intentional- need to be considered from a relational point of view. If any of them contribute to define the nature of the artefact, it is because they immediately refer to other artefacts and objects the agents engage with in a particular transformative environment. The physical description of an artefact is only significant in metaphysical terms when its physical capabilities are seen in the light of the physical properties and capabilities of other objects, particularly other artefacts. The functional description remains ambiguous if we do not appeal to a whole environment of artefacts, (normative) practices, and agents. Finally, if the artefact acquires any identity conditions from the intentions of the makers and users, these intentions only grow inside a transformative environment where certain possibilities of action become salient for them and where certain of their competences are crucially tied to other artefacts and objects. Atomism does not seem to be the right strategy to delineate identity conditions for artefacts.

1.4 Environments and Affordances

We have argued that the identity of an artefact -its mode of being- is not defined by a property or a set of properties of the object as such, without reference to other elements that constitute what we have called a transformative environment. Recent contributions in the metaphysics of artefacts, we think, have shared an atomist assumption. We propose instead to view artefacts from the perspective of objects that occupy a certain place within a temporally and spatially extended environment. The artefact is like a node within a complex network of agents, artefacts and practices that normatively regulate their interactions.

The identity of these nodes is, at the same time, essentially tied to the possibilities of action offered by the artefacts. They become detached units within a transformative environment once they are perceptually detected as possibilities of action. The artefact manifests then a set of *affordances*. Artefactual environments are also, and in a fundamental sense, environments of affordances. A non-atomist conception of artefacts goes beyond a phenomenological emphasis on the structure of interaction and engagement between a concrete artefact and an agent. The significance of the artefact grows inside a transformative environment where a set of interrelated affordances become perceptually salient. Stable structures of affordances make up transformative environments and organize the interactions with a network of artefacts.

For Gibson, "[t]he *affordances* of the environment are what it *offers* the animal, what it *provides* or *furnishes*, either for good or ill" (Gibson 1979, p. 127). He was interested in bringing to the foreground a particular conception of the relations an organism maintains with its own environment, relations that should not be viewed as mechanistic exchanges but as meaning-generating interactions. Meaning and value are notions that really hold for the environments where the organism acts. The notion of affordance captures this aspect of the relation by identifying the meaning things have for action. The environment is meaningful because it contains *affordances*, opportunities of action that ground the significance of objects, places and features within it. "The meaning or value of a thing consists of what it affords" (Gibson 1982, p. 457).

This complementarity organism-environment and the idea that affordances are primarily *of the environment* (not of the object) are crucial in our approach. Affordances are perceivable possibilities of action for an agent within an environment and they cannot be isolated as properties of the artefact/object as such. They structure the interactions between agents and artefacts and give meaning to particular objects only through a dense network of exchanges. The meaning (on which the identity of the artefact depends) is delimited by the spectrum of possibilities of action and constraints available for an agent in the environment. Some subset of these possibilities and constraints could be attached to a particular artefact, but this involves a learning process in which agents grasp the rules and classificatory conventions –obviously flexible and open to change- that regulate the interactions with the object and with other objects and artefacts (Muñoz-Serrano and Díaz 2013). In short, affordances constitute, so to say, the potential of mediation the artefact opens for an agent. They guide agency through the artefact.

The notion of affordance that we defend can be summarized in the following five features:

1. Affordances can be seen as *practical landmarks* inside a space of artefacts and agents.
2. Affordances supervene on the relations that are established between artefacts and agents and, therefore, they are better viewed as features of a niche or environment and not of the object itself.
3. Affordances are relationally and dynamically determined by the abilities that enable an agent to engage with the artefact.
4. The environment is structured by the set of affordances or practical landmarks.
5. Affordances solicit actions from agents, they do not either determine them or cause them.

This conception of affordances leads to several conclusions about the relation between artefacts, environments and intentional agents:

First, the mutuality that characterizes the nature of affordances implies, on the one hand, that affordances are established by a process of interaction between agents and objects; there are not affordances prior to the interaction. On the other hand, affordances *invite* the performance of certain actions in relation with those objects, as Withagen et al. (2012) have suggested. The existence of affordances is manifest only under certain conditions of intentional interaction; at the same time, intentional interaction is guided by these affordances.

Second, intentions are not hidden aspects of the interaction; they become perceptually accessible through it. It is, precisely, because the agent understands the context of intentional relations where the artefact is inserted, that she grasps certain possibilities of action, which have a special value for her and are intimately related to the meaning of the artefact. We will call them *intentional affordances*: they are grounded on a shared cultural framework in which agents understand the intentions other people exhibit in relation to the artefact or to the world through the mediation of the artefact (Tomasello 1999). Affordances, as we have said, are practical landmarks that require a certain understanding of the intentional relations that people have towards the object.

The concept of intentional affordance helps us to explain a central fact about the nature of artefacts. Any object, including artefacts, offers to an agent many different affordances and it can be used in different ways. Now, not every possibility of action offered by an artefact is constitutive of its identity, of "what the artefact is". Only a small set of affordances points to the privileged use of the artefact, a use that, somehow, gives access to what the artefact is. The notion of intentional affordance highlights the difference between these particularly relevant affordances and the rest: the privileged use of an artefact is determined by the intentional context the artefact is embedded in. Intentional affordances are those that are systematically exploited by other agents and create a normative use of the object. We do not run into the things; other people present them for us. Among other things, learning to perceive the intentional affordances of a given artefact implies *knowing how* to make proper use of the artefact, *recognizing* a given artefact as such, and also *understanding the intentions of other agents* within the artefactual space.

Third, affordances are necessarily interdependent. Intentional affordances depend upon complex networks of agents, objects and intentions. Then, if artefacts are characterized by their intentional affordances, the resulting network is essential to the identity of the artefact. Atomism, therefore, should be false and the artefact could not be defined without reference to other artefacts and the affordances they exhibit within this particular environment or cultural niche. It is not possible to simplify this relation by claiming that intentional affordances have been "put there" by the maker. If so, it would be possible to understand and use a given artefact without understanding the complete intentional context where it is embedded. It would be enough to get access to the intentions of the maker. Could an artefact be meaningful even if it remains isolated from its artefactual environment? Fiebich (2014) suggests that it could be so for some intentional affordances of the artefact. This falls prey again to the atomist assumption. Intentional affordances are not even conceivable if there is no shared intentional space that entails the existence of other artefacts and privileged (culturally mediated) interactions with objects.

Fourth, in so far as intentional affordances invite agents to perform certain actions, they are an essential aspect in shaping their intentions and they cannot be viewed as a mere outcome of them. This means that in our interactions in contexts of material culture, the notion of intention needs to be reformulated. Malafouris points in this direction when he claims: "The artefact should not be construed as the passive content or object of human intentionality but as the concrete substituting instance that brings forth the intentional state" (Malafouris 2008, p. 33).

To sum up, artefacts are necessarily tied to other artefacts through the affordances that they exhibit for agents that share a cultural niche, a transformative environment where certain possibilities of action become real and salient. As we have said, affordances do not primarily belong to each artefact; they do not attach to the object itself as if they were its properties. They are what they are because they can be detected by agents appropriately tuned to an environment full of artefact networks and regulated by normative practices that exhibit both discriminatory, recognitional, and classificatory aspects. Artefacts are just transformations in the space of pragmatic possibilities of action within a culture, as Fernando Broncano has written (Broncano 2012, p. 100). What a particular artefact affords is an invariant within the multiple relations that contribute to build a certain environment or space of artefacts (and agents). An artefact is just an item within a constellation of objects that hold stable relations among them and enable shared cultural practices. We hold that the identity of an artefact is given by the affordances that constrain our agency within a cultural niche. It is because the artefact is not given in isolation that we are able to culturally constrain the possibilities of action that an object affords and so define the object as such an artefact. The metaphysics of the artefactual realm cannot be developed without taking into consideration the nature of the transformative environments we live in, that is, without describing complex networks of artefacts and affordances.

Acknowledgments This chapter has been written with the support of two research grants funded by the Spanish Ministry of Economy and Competitiveness on the nature of artefacts and their role in epistemic contexts (FFI2009-12054; FFI2013-45659-R). We would like to thank Fernando Broncano for very fruitful discussions on these topics and Diego Lawler that has read previous versions of the chapter.

References

Broncano, F. (2008). In media res: Cultura material y artefactos. *Artefactos, 1*, 18–32.

Broncano, F. (2009). *La melancolía del cyborg*. Barcelona: Herder.

Broncano, F. (2012). *La estrategia del simbionte. Cultura material para nuevas humanidades*. Salamanca: Editorial Delirio.

Fiebich, A. (2014). Perceiving affordances and social cognition. In M. Gallotti & J. Michael (Eds.), *Social ontology and social cognition* (pp. 149–166). New York: Springer.

Gibson, J. J. (1979). *The ecological approach to visual perception*. Boston: Houghton Mifflin.

Gibson, J. J. (1982). Notes on affordances. In E. S. Reed & R. Jones (Eds.), *Reasons for realism: The selected essays of James J. Gibson* (pp. 401–418). Hillsdale: Erlbaum.

Hilpinen, R. (2004). Artifact. *The Stanford encyclopedia of philosophy*. http://plato.stanford.edu/archives/fall2004/entries/artifact1. Accessed 11 July 2016.

Kroes, P. (2012). *Technical artefacts: Creations of mind and matter*. Amsterdan: Springer.

Lawler, D., & Vega, J. (2010). Clases artificiales. *Azafea Revista de Filosofía, 12*, 119–147.

Malafouris, L. (2008). At the potter's wheel: An argument for material agency. In C. Knappett & L. Malafouris (Eds.), *Material agency: Towards a non-anthropocentric approach* (pp. 19–36). New York: Springer.

Muñoz-Serrano, M., & Díaz, A. (2013). Affordances: Construcción social y propiedades dinámicas. In M. J. Miranda, J. J. Alemán, & C. Saborido (Eds.), *Filosofías subterraneas* (pp. 63–72). México: Plaza y Valdés.

Thomasson, A. L. (2003). Realism and human kinds. *Philosophy and Phenomenological Research, 67*, 580–609.
Thomasson, A. L. (2007). Artifacts and human concepts. In E. Margolis & S. Laurence (Eds.), *Creations of the mind* (pp. 52–73). Oxford: Oxford University Press.
Thomasson, A. L. (2009). Artifacts in metaphysics. In A. Meijers (Ed.), *Handbook of philosophy of the technological sciences* (pp. 191–212). Amsterdam: Elsevier Science.
Thomasson, A. L. (2014). Public artifacts, intentions, and norms. In M. Franssen, P. Kroes, T. A. C. Reydon, & P. E. Vermaas (Eds.), *Artefact kinds: Ontology and the human-made world* (pp. 45–62). Amsterdam: Springer.
Tomasello, M. (1999). The cultural ecology of young children's interactions with objects and artifacts. In E. Winograd, R. Fivush, & W. Hirst (Eds.), *Ecological approaches to cognition: Essays in honor of Ulric Neisser* (pp. 153–170). London: Lawrence Erlbaum Associates Publishers.
Vega, J. (2009). Estado de la cuestión: Filosofía de la tecnología. *Theoria, 66*, 323–341.
Vega, J. (2012). Técnica, normatividad y sobrenaturaleza. Una ontología para un mundo de artefactos. *Revista Internacional de Tecnologia, Conocimiento y Sociedad, 1*, 11–24.
Vega, J., & Lawler, D. (2014). Creating artifactual kinds. In M. Franssen, P. Kroes, T. A. C. Reydon, & P. E. Vermaas (Eds.), *Artefact kinds: Ontology and the human-made world* (pp. 105–124). Amsterdam: Springer.
Withagen, R., de Poel, H. J., Araujo, D., & Pepping, G.-J. (2012). Affordances can invite behavior: Reconsidering the relationship between affordances and agency. *New Ideas in Psychology, 30*(2), 250–258.

Chapter 2
Praxeology Approaches Technology: The Ontology and Epistemology of Our Technological Practices

Diego Lawler

Abstract In this chapter, I defend a classic praxeological approach to technology: a robust philosophical reflection on our technological practices demands an adequate theory of action. My interest lies in exploring certain ontological and epistemological issues arising from this programmatic statement. First, I characterize the point of view of praxeology and determine the set of basic properties of technical action. Second, I suggest that the application of this viewpoint to the world of the products of technical action par excellence, namely artifacts, encourages the adoption of Wittgenstein's notion of 'family resemblances' to map that world. Third, I suggest that, epistemologically speaking, the praxeological focus on our technological practices promotes what will be called a 'pluralistic view' of the knowledge embodied in such practices.

2.1 Introduction

The artificial world, the world of technological constructions, is the product of our individual and collective action, and expresses our needs, desires, hopes, fears, and imaginary possibilities (Broncano 2012). It is a contingent world inasmuch as it could be completely or partially different from what it is. To accept its contingency is to grasp its conditions of production. The philosophical current of praxeology tells us that there is no relevant philosophy of technology without an adequate theory of action. This is the key insight of praxeology regarding technological practices. The present chapter explores the ontological and epistemological consequences of this insight.

D. Lawler (✉)
Institute of Philosophical Research, Argentinian Society of Philosophical Analysis, National Council for Scientific and Technical Research, Buenos Aires, Argentina
e-mail: diego.lawler@gmail.com

B. Laspra, J. A. López Cerezo (eds.), *Spanish Philosophy of Technology*, Philosophy of Engineering and Technology 24,
https://doi.org/10.1007/978-3-319-71958-0_2

Technological practices constitute worlds in which humans act, experience and experiment life, imbuing it with sense. These worlds are the product of the transformation of reality; the drive to desire, imagine and produce what does not yet exist. Technological practices express a way of life. At the same time, this mode of life affects humans; i.e., their conceptual abilities and practices are shaped and defined by the technological world. In this manner, our technological worlds are imagined, conceived, and realized from within the interior of a way of life that is itself upheld by structures of technological practices.

A way of life has objective and subjective aspects. Its objective dimension includes all material and symbolic practices, in addition to customs, rites and rituals. Its subjective aspect encompasses socially linked consciences that strive to become self-aware of these practices and act, understand and produce meanings from within them in order to inhabit and transform them.

Such a perspective on the world of artifacts suggests the adoption of an anti-essentialist approach, leading to almost completely abandoning of the notion of *classes* to map the world of artifacts. The notion of classes should be partially replaced by the Wittgensteinian notion of 'family resemblance'. The idea is to encourage descriptions of the artificial world that express the robust and natural manner in which our practices of production and use of artificial objects are interwoven. In epistemological terms, a praxeological approach encourages a view that I call 'pluralistic' regarding the nature of knowledge involved in technological practices and its sources.

In the next section, I present a set of basic properties that characterizes the structure of technical actions from a praxeological point of view. This will elicit the intuition that there is no relevant philosophy of technology without an adequate theory of action. Furthermore, it will establish the background for inferring some ontological and epistemological consequences of our technological practices in the third section.

2.2 Praxeology Takes on Technology Practices: The Basic Structure of the Class of Technical Actions

Praxeology is presented as a general theory of human intentional action (Kotarbinski 1965; Bunge 2002), i.e., deliberate and conscious action. It involves an analysis of the concepts essential to the description, evaluation and planning of intentional action, as well as the development of a set of practical recommendations for promoting its rational development. Technical actions –the basic building blocks of technological practices– belong to a special type of intentional human action, that of *productive* actions. The relevance of praxeology for the philosophy of technology is well-nigh obvious. Technical actions are the most valuable form of intervention, modification and intentional control of reality in order to adapt it to human needs and desires (Ortega y Gasset 1939; Quintanilla 2005; Broncano 2000). These actions, qua human productive actions, are guided by a precise description of the object, event or desired result, and by an amount of acquired knowledge. The implementation of these actions is geared to meeting objectives previously assumed as desirable and valuable. Thus, the agent strives to produce an object, event or state of some kind, aided by a body of knowledge, with a previously determined purpose.

In praxeological terms, the following features characterize the basic structure of technical action:

1. The content of the intention of technical actions involves a plan of action.[1] An agent intends to perform a technical action when he or she has the idea of producing something through the realization of something else. The intention is Janus-faced: one side faces the intended purpose (the object of the action), while the other faces that which guarantees the end (the *means* of technical action). Therefore, the content of the intention articulates the relation between intended end and attempted means (Bratman 1987, 2007). The content of the intention of technical action indicates not merely the objectives sought, but also how to achieve them; i.e., it refers to the means and the organization of courses of action that the agent is to establish and coordinate in time and space.
2. Assuming (1), it follows that technical actions are not occasional or isolated intentional actions. On the contrary, they are strategic intentional actions with the property of being 'composable'; i.e., they have the ability to be jointly articulated and form informational complexes involving representations of the necessary operations to carry out, the objectives sought and the possible results of the actions. To put it another way, and in accordance with (1), technical actions are structured plans or systems of actions that agents previously represent (Marx 1980, p. 131).[2] In short, the fact that such actions can be arranged as composites relates to their previous ordering in action plans.
3. Technical actions are instrumental actions of the second order (Broncano 2000); i.e., they are actions used as appropriate means, which are products of previous technical actions. However, what constitutes 'appropriate means' within the context of a technical action? This is a complex issue. In abstract terms, an

[1] This feature, of major importance for the correct understanding of technical action, has its roots in Aristotle (1970) and his medieval followers; e.g., Thomas Aquinas (1988). For these philosophers, intention is, generally speaking, having an intention to do something. Thus, in their reflections on practical reasoning, they combine the idea of intention with a choice to act in a certain time and space. Therefore, to try something is to choose to do something, be willing to do it, and do it when given the chance. Within this framework, intention is a product of a process of deliberation on possible courses of action. It follows that intention implies a strong sense of planning (Austin 1970), since to have an intention is equivalent to having an action plan. Therefore, the content of the intention of an agent is constrained by the plans for the proposed ends. In other words, what is intended in an action plan depends on what an agent plans.

This insight is especially important because it justifies, in a sense, the transformation of the analysis of intention into the study of the formation and structure of action plans. This lays the foundation for the best interpretation of the content of the intentions in technical action.

[2] Recall Marx's dictum: "what distinguishes the worst architect from the best of bees is this, that the architect raises his structure in imagination before he erects it in reality." (1980, p. 198).

'appropriate means' refers to an artificial entity (*AE*) that makes a causal contribution to the transformation of a state of affairs *A* into a state of affairs *B*, where: '*A*' refers to a state of affairs that is unsatisfactory in view of a set of objectives and values (*O*); '*B*' refers to a state of affairs that instantiates *O*; while '*AE*' is an artificial entity that is causally sufficient for *B*. However, *AE* needs to satisfy other conditions to be an effective means to an end; e.g., it should technically feasible; it should be available; the instructions for its operation should be properly set out; the operators should have the skills, and the implicit and explicit knowledge necessary to make it funtion it (Vega 2001, 2010), and so on. From this point of view, determining the necessary and sufficient conditions that a means must satisfy in order to be an effective means within a technical action sums up various aspects related to the modality of technical action.

4. Technical actions are designed actions. The design is to conceive what does not yet exist; i.e., to give an artificial entity an intentional reality that precedes physical reality (Broncano 2006, p. 105) and then to produce it, effectively bringing it into the world. Designs transform our realities. When designing, certain functional properties are realized in material structures in order to solve problems (Simon 1969; Houkes et al. 2002; Kroes 2012).[3] Design involves deliberation and decision-making against a background of physical, technological, economic, aesthetic, ethical, and safety constraints (Kroes 2012, p. 137).[4] In this sense, the activity of design assumes practical accountability before the results of such design.[5]
5. Technical actions are indirectly guided by reliable beliefs; i.e., they are actions prescribed by a systematic and planned set of rules, secured by the best available knowledge. According to (1) and (2), the content of the intention of technical actions involves composable action plans that can be assembled in the manner of a puzzle. These courses of action are not evaluated according to their truth or falsity, but according to their consistency or inconsistency. However, this property is not independent of the truth or falsity of the beliefs on which the plan is based. The consistency of plans is inseparable from the truth or falsity of the beliefs of their planners (Bratman 1987). It follows that a plan is consistent if and only if it does not presuppose false beliefs. However, we could admit degrees of consistency, which would mean reformulating the above criteria: the degree of consistency of a plan depends on the amount of false beliefs that it presupposes. This allows us to capture something that can occur in technological practice: the

[3]According to Kroes (2012, p. 137), "This physical structure has to be put together, synthesized, from parts (components) some of which may already exist, others of which may have to be tailor-made for the specific case at hand. It is for that reason that designing is often characterized as a synthetic activity, or as an activity that uses synthetic methods, as opposed to scientific research that is commonly characterized as an analytic activity, or as an activity that uses analytic methods."

[4]The fact that it is a deliberative and decisional activity means that designing resembles a process of invention rather than discovery (see Kroes 2012, p. 133).

[5]As Broncano claims, "The action of the engineer introduces changes in history and, in a deep sense, it changes the conditions of existence, although he may not be aware of the consequences of his action." (2006, p.110).

circumstance that one or more of the plans belonging to a complex plan may not rest on the best available knowledge. Whatever the case, this line of argument highlights the fact that the consistency of the plans is linked to the truth of the beliefs of the agents involved. Moreover, as the consistency of a plan is what sustains its viability or realizability, the properties of beliefs (truth or falsity) constitute the basis of its successful implementation. Therefore, if a technology action plan is expressed in action rules, representing operations of intervention in reality, a key issue for the effectiveness and efficiency of the plan is the way such rules relate to the best available knowledge (Bunge 1972; Quintanilla 2005; Toribio 1995).

6. Technical actions are oriented towards the transformation and control of states of affairs, events and objects. Consequently, they can be classified into two main branches: on the one hand, actions of production or execution; and, on the other, actions of organization, management and control (Bunge 2002; Quintanilla 2005). The processing and control of reality involve intervention in the course of things to produce, organize or manage (or alternatively to prevent the occurrence of processes, states, events or objects that would not occur spontaneously). At their most basic level, actions involve both kinds of operations to control and transform matter productively (Leroi-Gourhan 1988).
7. Technical actions are aimed at obtaining valuable results both effectively and efficiently. They are actions in which planning results from the consideration of the most appropriate means to achieve the satisfaction of the best goals. There are two basic aspects to the axiology of technical action. On the one hand, the evaluation of the action considered as technical action: how to perform actions with increasing efficiency. 'Efficiency' is a generic name for a set of practical values, technical efficiency, effectiveness, productivity, etc., which constitute the natural evaluative context for these actions (Kotarbinski 1965; Quintanilla 2005; Quintanilla and Lawler 2000). On the other hand, this involves the evaluation of the objectives, which are accurate representations of what is possible, what is desirable, and what should be avoided (Broncano 2000, 2006, 2012). These objectives are a combination of seizing objective practical opportunities for action, which give rise to technical actions themselves, with the adoption of broad interests, values and social purposes that are considered legitimate to pursue (Broncano 2000, 2009; Liz 2001; Echeverria 2003; Vega 2000, 2010). Therefore, the objectives of a set of technical actions may be assessed taking into account their degree of compatibility with the desirable, possible or available technologies, and more or less valuable potentials to promote technological developments in a rich and broad sense.
8. Technical actions are conditioned by technical culture. The technical culture of a particular social group is characterized by three components: representational (knowledge, beliefs and conceptual representations of technologies); practical (rules, operational skills and knowledge); and evaluative (preferences regarding the design, use and production of technologies and technological knowledge) (Quintanilla 1998). Technical culture carries out the function of filtering the universe of the technological and the non-technological, activating or limiting

perceptions that agents have of their respective technical possibilities of action (transformation of reality) (Ortega y Gasset 1939).

9. The agents of technical actions may be individuals, groups of individuals or representatives designated by these. Cases of agency involving more than one individual raise interesting philosophical problems concerning deliberate collective action (Broncano 2000; Ezquerro 1995). In their capacity to collectively transform and control reality, technical actions may involve cases of strong and equal interdependency and cooperation. Nonetheless, they mainly entail cases of little or no cooperation and weak interdependence (Kutz 2000), given that they occur in institutional settings of a hierarchical nature as a result of the social division of labor.[6]

2.3 Ontological Diversity and Epistemic Plurality

2.3.1 Family Resemblances, Machine Realism, and the Relative Abandonment of the Notion of Class

Technological practices produce artifacts, objects that are real pieces of our world (Broncano 2008; Vega 2011). Artifacts are part of human material culture, together with the knowledge and skills associated with their use and their symbolic values and practices which are embedded or incorporated in them (Lawler and Vega 2009; Vega and Lawler 2014).

How do we classify these pieces of reality? What characteristics do they have? Quine (1969) reminds us: "No entity without identity". If something particular does not have an identity, it is difficult to consider it as an entity of some kind. Philosophical analysis has postulated different metaphysical theses on artificial artifacts and classes. In the following paragraphs, I review two ways to defend the view that artifacts have essential features that make them particular entities that can be grouped into classes. This essentialism may or may not involve a commitment to realism; i.e., the thesis that artificial classes (or however the artificial world may be parceled up) are independent of the beliefs and purposes of humans.[7] I will argue that these two options are unsatisfactory and will use the criterion of multiple realizability (MR) as a means to diagnosing some of the problems they entail. Of course, by adopting this strategy I am acknowledging that we cannot understand what a technical artifact is unless we understand MR (Dennett 1987; Simon 1969; Houkes and Meijers 2006; Lawler and Vega 2012). As a conclusion to this section, I will

[6] The usual approach to collective action (Bratman 2007; Gilbert 1990; Searle 1990; Kutz 2000) presents important limitations in accounting for cases of collective agency within the framework of technological actions, as it focuses on cases of equal cooperation and strong interdependence.

[7] Of course, I am leaving aside the thesis that artifacts lack an essence. Eliminativism is a thesis of this kind. According to eliminativism, artifacts qua artifacts do not exist. Those objects to which we refer to as artifacts are merely aggregates of matter organized in a certain way. Of course, there are different strategies to reach this conclusion. See, e.g., Van Inwagen (1990) or Merricks (2003).

suggest we relax Quine's judgment, partially abandon the language of classes, and adopt the Wittgensteinian notion of family resemblances.

Let me consider the first option: Artifacts are identified and grouped into nominal kinds (Wiggins 2001). A defender of this approach argues that:

(1) Artificial kinds are not real kinds; on the contrary, they are conventional kinds. (2) Artificial kinds have nominal 'essences'. (3) A real essence is something that is governed by an 'internal principle of activity' (Wiggins 2001), which, in turn, specifies the persistence of the object; this principle of activity can be explained in terms of a fundamental law of nature. (4) There is no principle of activity that can be formulated and attributed to an artificial class. This is something that becomes clear, inter alia, in the fact that an artifact can undergo sweeping changes and still constitute an entity. To put it plainly, an artificial-kind term does not identify a substantial unity. (5) An artificial-kind term is a sortal term that classifies its members following a functional description that specifies a nominal essence (i.e., a classification according to an abstract idea). This functional description is indifferent to the internal constitution and the materials of the artifacts within any given artifactual kind. Consider the case of clocks: water clocks, sundials, mechanical clocks, electric clocks, etc. These different types of clocks are gathered together by a functional description that is indifferent to a specific constitution.

It thus follows that the satisfaction of MR becomes trivial and philosophically negligible according to this view. It is trivially satisfied because we can group artifacts following a functional description regardless of their material constitution and their causal effective realizations; at a certain level of generality, there is an infinite group of artifacts that could be grouped together. Hence, the questions of how realization occurs and under which conditions it occurs can be avoided circumvented any loss in understanding. In this sense, it is philosophically negligible.

Let us now consider the second option: Artifacts are identified and grouped into real kinds. Carrara and Vermaas (2009) have shown how we may understand the notion of function in order to provide a foothold to the philosophical intuition that the function of an artifact identifies the real essence of a particular artifact kind. The argument proposes a criterion of identity to classify artifacts, which comprises "a combination of the functions (of artifacts), and other features of artifacts" (Carrara and Vermaas 2009, p. 131). The argument refers to this combination as the "function+ of an artifact" where '+' is what is added to the function, whatever that may be. Under this characterization, two artifacts that have different functions (where 'function' is understood as 'function+') belong to different real kinds because they have distinct essences.[8] The notion of function+ supposes an operational principle and a causal-material structure; this is precisely what makes this notion a good candidate to capture a real essence.

However, if we 'inflate' function in this manner, we lose the notion of a contingent relation between what an artifact does, its material structure, and its operational principles. Thus, it could be the case that the real artifactual kind is multiply realizable

[8] While this conclusion opens up a promising avenue to use ontology as a guide to an epistemology (see Carrara and Vermas 2009), I do not have the space here to delve into the issue.

when you have two instances of that class with trivial differences; meaning by 'trivial differences', differences that comprise only superficial material properties. That is to say, differences that do no suppose a different causal structure. Conversely, if these differences entailed different causal structures, then we would be dealing with different, real artifact kinds, given how artifact kinds are finely parceled. Think about different corkscrews and you will easily imagine this trap. This is clearly not satisfactory.

In short, the nominalist approach offers coarse categorizations and makes MR trivial. The realistic approach offers fine-grained categorizations, but the application of MR becomes problematic. It also leads to an unnecessary proliferation of artificial classes. In both options, an ontological characterization of kinds according to functional groupings becomes problematic because we cannot proceed without first establishing the parameters of a good description of the functional organization of a system. As we have seen, if the description is too coarse, we lose grounds for justifying the reality of kinds, seeing as no robust regularity or principle of activity would justify talk of real kinds. Rather, the clusters would have to be formed in terms of a broad functional description. If the parameters are too wide, our classification loses robustness. Artifact kinds are too heterogeneous. On the other hand, if the parameters are finer, then it is possible to identify real kinds of artifacts (from their normal configurations and operating principles), but the principle of multiple realizability would lose its strength and artifacts kinds would multiple unnecessarily.

However, I suggest that when artifacts are conceptualized as part of human material culture, there is no need to postulate special facts –such as underlying essences or privileged functions– to describe what an artifact is. On the contrary, the identity conditions of an artifact are revealed by displaying our practices of dealing with them. These conditions, of course, cannot be fully expressed in terms of statements of the natural sciences, even though artifacts involve a causal structure; nor can they be minted merely in terms of functions dependent on makers or users, even though artifacts are the results of intentional human agency. To determine what an artifact is according to the vocabulary of classes typical of the natural sciences would lead to a science of the artificial too closely contaminated by the problems of natural science. To argue that artifacts are human classes and that they can be understood in terms of the powers conferred on them by their designers or users involves modeling the field of the artificial on the notion of divine creation. Unlike these options, replacing the vocabulary of classes by the notion of 'family resemblances' means resorting to our human practices of dealing with artifacts to characterize what they are (Lawler and Vega 2012). It means thinking about what they are from the concepts we apply to them, the appropriate judgments we make on them, the inferences that we draw, the collation of the genealogies that trace their lineages, and so on.

The notion of 'family resemblances' is a Wittgensteinian notion that can be stated as follows. Something generally falls into the class of Xs not because it satisfies the definition of 'X' and thus shares the features with all other Xs. Rather, Xs fall into this class because they exhibit various overlapping similarities to one another. We should imagine that the Xs are all bonded together like a rope, i.e., by way of numerous overlapping threads, each of which is shorter than the whole, so that no one thread runs from one end to the other. As long as the overlaps are tight and numerous enough, this does not threaten the integrity of the rope: no continuous thread is required. Similarly, as long as the pattern of overlapping similarities is stable enough, Xs need not have one or more essential feature in common. Wittgenstein uses a metaphor to make this plainer, namely that of a "family resemblance" (Wittgenstein 1958). There may be a definite 'Horan' face, in that all of the extended Horan family look alike and can easily be seen to belong to that family. However, a closer look need not reveal there to be one or more specific feature which all those with the 'Horan' face share. Rather, there might be a pool of features – short ears, square chin, piercing eyes, high cheek-bones – of which all of the Horans have one or more, although there is no one of these features which all those with the Horan face have. This enables family members to resemble each other without sharing an essential Horan-feature.

However, the intuition that gives substance to the idea of family resemblances does not end here, but further reveals an interesting issue regarding the status of artificial objects as cultural objects. Although members of a given class can no longer be grouped according to nominal essences or their shared underlying nature in terms of scientific laws, for Wittgenstein the principle of grouping should be our sense of *what is natural and has value*. In a way, our view is that it is more natural to detect and collect certain features more than others to group members against a background of traits. What is more, according to Wittgenstein, this issue is strongly correlated with our "forms of life".

How do we translate this notion of family resemblances to the world of artifacts? Specifically, what is entailed in the suggestion that artifacts should be grouped according to our sense of what is natural? Two issues are relevant here. On the one hand, we need to identify the presence of sufficiently stable patterns of similarities, a common background of features determining the different families of artifacts, even though there might not be common denominators shared by all members.[9] We are no longer concerned with an essence. Moreover, these patterns of similarities are constituted by our sense of what is natural, by what our perspective detects, groups and collects. The issue concerns the normativity involved in our practices of grouping artificial objects. The perspective is thus not just any perspective; i.e., it is not the point of view of any particular individual or social group. If this were the case, sufficiently stable patterns of similarities would not exist. On the contrary, a perspective is forged in and structured by a form of life. Its dependence on the normativity embedded in our forms of life has a dual function. On the one hand, it stabilizes patterns resulting from similarities into sufficiently robust groups; on the other, it indicates that whatever is relevant or stands out against a background of family commonalities is immediately visible in our material culture. This may seem a diffuse or vague principle of classification, but it is just the opposite, seeing as our practices are themselves, as it were, the medium of the conceptual complexity of the

[9] It should be noted that functional considerations and the *evolutionary* history of these cultural objects may be relevant when assessing the naturalness of these groupings.

field of artificial. Following this strategy, artifacts not only retain a principle of identity, but ontology falls into step with our normative considerations (Vega et al. forthcoming).

This anti-essentialist strategy could be implemented to either a radical or a moderate extent. In its radical version, we must completely abandon the notion of kinds. This version would be plausible only if it were possible to show why anti-essentialism is an absolutely necessary starting point for the world of artificial objects. I do not think this argument could satisfactorily be offered (Lawler and Vega 2010). Rather, I would like to suggest a moderate version, one in which the two terms are combined. That is to say, we talk about family resemblances at a higher level, and the term 'kinds' is reserved for a detailed discussion of particular instances of artifacts. This latter strategy appears to be methodological. While it departs from an anti-essentialist intuition, it suggests that there are different explanatory levels depending on the stance we adopt within our technical practices. We could group artifacts in terms of family resemblances or identify kinds that have a certain robustness; e.g., when considering machines as piston engines. In this case, members of this kind exhibit a unified principle of activity that is constrained by sortal-specific laws. As Lowe (2014, p. 24) points out, "[t]he laws in question are, moreover, distinctively laws of engineering, not mere laws of 'natural science' or 'physics'.". Following this suggestion, we would be promoting descriptions that highlight the way in which our practices are interwoven with artificial objects, and their relationship with how we parcel the artificial world.

2.3.2 *Epistemological Pluralism*

The praxeological approach promotes a pluralistic vision of the knowledge embodied in technological practices. What types of knowledge structure these practices? My impression is that the answer to this question requires focusing on technical action and analyzing the knowledge that structures it. Two types of knowledge promptly emerge. On the one hand, scientific knowledge; on the other, engineering knowledge. Let us see how each of these is related to technological practice.

Bunge (1972) has thoroughly explored the relationship between scientific knowledge and technological practice, culminating in his famous thesis of technology as applied science. I will not discuss this thesis here, but will show how scientific knowledge is relevant to technological practice.

Technological theories are the result of applying the method of science to the solution of practical problems. These theories are composed of a set of rules that prescribe courses of action. Scientific knowledge contributes to the effectiveness and efficiency of the rules that structure technological theories, ensuring the presupposition relationship between a statement of a scientific law, a nomopragmatic statement, and an action rule. Here is an example offered by Bunge (1972, p. 69) and included in, among others, Niiniluto (1993, p. 13):

1. Statement of scientific law: "Magnetism disappears when the Curie temperature (770 °C in the case of iron) is exceeded."
2. Statement of law as conditional: "If the temperature of a magnetized iron body exceeds its Curie point (770 °C), then it is demagnetized."
3. Nomopragmatic statement: "If a body of magnetized iron exceeds its Curie point (770 °C), then it is demagnetized."
4. Rules of technological action: R1 – "To demagnetize a body of iron, heat it until it exceeds its Curie point (770 °C); R2 – "To prevent demagnetization of a body of iron, avoid heating it beyond its Curie point (770 °C)."[10]

The technological rules would hence derive from technological theories based on true scientific propositions. Thus, technological rules are supported "from above" (Niiniluto 1993) by the descriptive-predictive statements produced by basic scientific research. In this conception, technological design is the design of objects from the basis of knowledge extracted from the physical world, then transformed into action rules with the purpose of introducing a novelty in the world.

However, this is not the only epistemic source of technical rules. There is not always a basic scientific theory from which to derive rules. However, it is not just a case of having or not having a theory; on the contrary, technological rules can be obtained in a completely different way. Consider the alternative presented by engineering knowledge.

Layton (1971) rejects Bunge's thesis and advocates a conception of laws based on engineering knowledge.[11] This type of knowledge supports technological rules "from below". Niiniluto (1995, p 129) describes it as follows: "the researcher typically employs background theoretical information and constructs a mathematical model, with manipulable and dependent variables, and tries to obtain relevant empirical information by experimentation and computer simulation.". Technological rules are seen as the product of modeling actual practices, through procedures of trial and error and broad experimental practices where the relationships between the relevant design variables of an artifact are investigated in order to find optimal procedures to achieve the desired effects. These relationships or dependencies often give rise to engineering laws, i.e., robust empirical generalizations of the behavior of artificial objects, when these behaviors are invariant for individuals of a particular class or some artificial family or lineage.

Hence, these laws are postulated *a posteriori*; i.e., after experimental interaction with artificial objects. Technological knowledge would thus be the "discovery of fundamental principles, the deduction of these principles and the testing results through the experiment" (Layton 1971, p. 566). This knowledge is essential in engineering design and in the search for minor as well as radical innovations. At the same time, inserting technological laws into the very fabric of technological practices means that cognitive contents not directly derived from scientific laws

---

[10]Technical rules involve a descriptive formulation of an intentional action, which depicts the structure of a particular hypothetical imperative (Toribio 1995).

[11]An exhaustive characterization of this position can be found in Sandrone (2016).

(technological knowledge) can acquire normative features by postulating epistemological and practical virtues (Vega 2010).

Scientific knowledge does not exhaust the types of knowledge found within technological practices. The very same technological practices can produce technological knowledge, which is relevant to the design and use of artifacts. Both types of knowledge are embodied in our best practices of design and construction of artificial worlds.

2.4 Conclusion

In this chapter, I have introduced praxeology as a general theory of human intentional action. In particular, I have applied it to characterize the basic structure of technical action, our most valuable form of transforming reality, imagining and creating artificial worlds. Technical actions are the building blocks of technological practices. I have explored the structure of those actions in order to gain insight into these practices. Remember the praxeological dictum: there is no comprehension of technology without a theory of action. Furthermore, I have focused on artifacts, the product of our technical actions. I have discussed how to identify and classify these pieces of reality. By appealing to the criterion of multiple realization, I have addressed two version of essentialism. As a result, I have advanced the suggestion of partially abandoning essentialism. I have advocated applying the Wittgensteinian notion of *family resemblance* to the realm of artifacts. Of course, this does not mean that we cannot appeal to essentialism in peculiar cases of artifacts and artifact kinds which exhibit a certain robustness. Finally, I have diagnosed two kinds of knowledge exhibited by our technological practices. In this case, I have argued the case for epistemological pluralism.

Acknowledgments I would like to Andrés Vaccari, Fernando Broncano and Jesús Vega for their contributions to these ideas. This chapter has been written with the support of CONICET and the Spanish Ministry of Economy and Competitiveness (FFI2009.12054; FFI2013-45659-R).

References

Aquinas, T. (1988). *Suma de Teología II*. Madrid: Biblioteca de Estudios Cristianos.
Aristotle. (1970). *Ética a Nicómaco*. Madrid: Instituto de Estudios Políticos.
Austin, J. L. (1970). *Philosophical papers*. London: Oxford University Press.
Bratman, M. (1987). *Intentions, plans, and practical reason*. Cambridge, MA: Harvard University Press.
Bratman, M. (2007). *Structures of agency. Essays*. Oxford: Oxford University Press.
Broncano, F. (2000). *Mundos artificiales. Filosofía del cambio tecnológico*. Mexico: Paidós.
Broncano, F. (2006). *Entre ingenieros y ciudadanos*. Barcelona: Montesinos.
Broncano, F. (2008). In media res: cultura material y artefactos. *Artefactos, 1*, 18–32.
Broncano, F. (2009). *La melancolía del Cyborg*. Barcelona: Herder.

Broncano, F. (2012). *La estrategia del simbionte. Cultura material para nuevas humanidades*. Salamanca: Editorial Delirio.
Bunge, M. (1972). Toward a philosophy of technology. In C. Mitcham & R. Mackey (Eds.), *Philosophy and technology* (pp. 62–76). London: Macmillan Publishers.
Bunge, M. (2002). *Ser, saber, hacer*. Buenos Aires: Paidós.
Carrara, M., & Veermas, P. E. (2009). The fine-grained metaphysics of artefactual and biological functional kinds. *Synthese, 169*, 125–143.
Dennett, D. (1987). *The intentional stance*. Nueva York: Basil Blackwell.
Echeverría, J. (2003). *La revolución tecnocientífica*. Madrid: Fondo de Cultura Económica.
Ezquerro, J. (1995). Acciones, planes y tecnología. In F. Broncano (Ed.)., (1995) *Nuevas meditaciones sobre la técnica*. Madrid: Trotta.
Gilbert, M. (1990). Walking Together: A paradigmatic social phenomenon. *Midwest Studies, 15*, 1–14.
Houkes, W., & Meijers, A. W. M. (2006). The ontology of artefacts: The hard problem. *Studies in History and Philosophy of Science, 37*, 118–131.
Houkes, W., Vermass, P. E., Dorst, K., & de Vries, M. J. (2002). Design and use as plans: An action-theoretical account. *Design Studies, 23*, 303–320.
Kotarbinski, T. (1965). *Praxiology*. Oxford: Clarendon Press.
Kroes, P. (2012). *Technical artifacts. Creations of mind and matter*. New York: Springer.
Kutz, C. (2000). Acting together. *Philosophy and Phenomenological Research, 61*(1), 1–31.
Lawler, D., & Vega, J. (Eds.). (2009). *La respuesta a la pregunta. Metafísica, técnica y valores*. Buenos Aires: Editorial Biblos.
Lawler, D., & Vega, J. (2010). Clases artificiales. *Azafea. Revista de Filosofía, 12*, 119–147.
Lawler, D., & Vega, J. (2012). Realizabilidad múltiple y clases de artefactos. *Revista Iberoamericana de Ciencia Tecnología y Sociedad, 7*(19), 167–178.
Layton, E. (1971). Mirror-image twins: The communities of science and technology in 19th-century America. *Technology and Culture, 12*(4), 562–580.
Leroi-Gourhan. (1988). *El hombre y la técnica (Evolución y Técnica I)*. Madrid: Taurus.
Liz, M. (2001). *Un metafísico en Tecnolandia. Realidad, conocimiento y acción bajo nuevos puntos de vista*. Murcia: Universidad de Murcia.
Lowe, E. (2014). How real are artefacts and artefacts kinds? In M. Franssen, P. Kroes, T. A. C. Reydon, & P. E. Vermaas (Eds.), *Artefact kinds: Ontology and the human-made world* (pp. 105–124). Amsterdam: Springer.
Marx, C. (1980). *Capital. Volume 1: A critique of political economy*. London: Penguin Classics.
Merricks, T. (2003). *Objects and persons*. Oxford: Clarendon Press.
Niiniluto, I. (1993). The aim and the structure of applied research. *Erkenntnis, 38*, 1–21.
Niiniluto, I. (1995). Approximation in applied science. In M. Kuokkannen (Ed.), *Structuralism approximation, and idealization. Poznan Studies in the Philosophy of the Sciences and the Humanities, 42*, 127–139.
Ortega y Gasset, J. (1939). *Meditación de la técnica y otros ensayos sobre ciencia y filosofía* (p. 1992). Madrid: Revista de Occidente.
Quine, W. V. (1969). Speaking of objects. In W. V. Quine (Ed.), *Ontological relativity and other essays* (pp. 1–25). New York: Columbia University Press.
Quintanilla, M. A. (1998). Técnica y Cultura. *Teorema, XVII*(3), 49–69.
Quintanilla, M. A. (2005). *Tecnología: un enfoque filosófico*. Mexico: Fondo de Cultura Económica.
Quintanilla, M. A., & Lawler, D. (2000). El concepto de eficiencia técnica. In G. Denegri & G. Martínez (Eds.), *Tópicos actuales en filosofía de la ciencia*. Mar de Plata: Universidad Nacional de Mar del Plata.
Sandrone, D. (2016). *Aportes para una nueva concepción del diseño tecnológico: un estudio filosófico de su naturaleza y su rol en el cambio tecnológico*. Doctoral thesis. Universidad Nacional de Córdoba.
Searle, J. (1990). Collective intentions and actions. In P. Cohen, J. Morgan, & M. Pollack (Eds.), *Intentions in communication*. Cambridge, MA: MIT Press.
Simon, H. (1969). *The sciences of the artificial*. Cambridge, MA: MIT Press.

Toribio. (1995). Semántica de las reglas tecnológicas. In F. Broncano (Ed.), *Nuevas meditaciones sobre la técnica*. Madrid: Trotta.
Van Inwagen, P. (1990). *Material beings*. Ithaca: Cornell University Press.
Vega, J. (2000). La astucia de la razón en la técnica. *Arbor, 657*, 187–205.
Vega, J. (2001). Reglas, medios, habilidades. Debates en torno al análisis de "S sabe cómo hacer X". *Crítica, 33*(98), 3–40.
Vega, J. (2009). Estado de la cuestión: Filosofía de la tecnología. *Theoria, 66*, 323–341.
Vega, J. (2010). *Los saberes de Odiseo*. Buenos Aires: EUDEBA.
Vega, J. (2011). Técnica, normatividad y sobrenaturaleza. Una ontología para un mundo de artefactos. *Revista Internacional de Tecnología, Conocimiento y Sociedad, 1*, 1–15.
Vega, J., & Lawler, D. (2014). Creating artifactual kinds. In M. Franssen, P. Kroes, T. A. C. Reydon, & P. E. Vermaas (Eds.), *Artefact kinds: Ontology and the human-made world* (pp. 105–124). Amsterdam: Springer.
Vega, J., Broncano, F., & Lawler, D. (forthcoming). *The nature of artefacts. Meaning, history, and agency*. Amsterdam: Springer.
Wiggins, D. (2001). *Sameness and substance renewed*. Cambridge: Cambridge University Press.
Wittgenstein, L. (1958). *Philosophical investigation*. Oxford: Basil Blackwell.

Chapter 3
Synthetic Life: Organisms, Machines, and the Nature of Synthetic Biology Products

Diego Parente

Abstract A brief overview of the ways synthetic biology is discussed in contemporary philosophical debate seems to justify the following hypothesis: this debate is initially motivated by moral questions. While these approaches are obviously valuable in our understanding of the ethical issues involved in these techno-scientific practices, it is also true that this bias has eclipsed other important aspects, specifically those related to a strict ontological discussion of what is really new about these entities produced by synthetic biology. In other words, the hidden question is whether these synthetic products are genuinely different – in ontological terms – from other products of previous modes of domestication or intervention in natural dynamics. This paper accordingly aims to examine certain ontological implications of synthetic biology, especially those related to the way it affects traditional dichotomies such as natural/artificial and organism/machine.

3.1 Introduction

A brief overview of the ways synthetic biology is discussed in contemporary philosophical debate seems to justify the following hypothesis: this debate is initially motivated by moral questions. Questions about the types of threat involved in this new technological modality (Schmidt et al. 2009; Hagen et al. 2016), as well as questions regarding its risks and the cautionary principles to take into account (Kaebnick 2014; Lewens 2013; Carlson 2010), lead us to inquire about the conditions of this 'new phenomenon' (if we can truly call it that).

While these approaches are obviously valuable in our understanding of the ethical issues involved in these techno-scientific practices, it is also true that this bias has eclipsed oher important aspects, specifically those related to a strict ontological discussion of what is really novel about these entities produced (or, in some cases, planned) by synthetic biology. In other words, the hidden question is whether these

D. Parente (✉)
University of Mar del Plata, Mar del Plata, Argentina
e-mail: diegocparente@yahoo.com

B. Laspra, J. A. López Cerezo (eds.), *Spanish Philosophy of Technology*, Philosophy of Engineering and Technology 24,
https://doi.org/10.1007/978-3-319-71958-0_3

synthetic products are genuinely different – in ontological terms – from other products of previous modes of domestication or intervention in natural dynamics.

Consequently, this paper aims to examine certain ontological implications of synthetic biology, especially those related to the way it affects traditional dichotomies such as natural/artificial and organism/machine. In this context, we propose, following Preston and Lewens, a notion that we shall call a 'design *continuum*' that encompasses – undoubtedly with different degrees and densities – the material results of diverse historical modalities of human intervention in nature.

3.2 Synthetic Biology as a Form of Intervening in Nature

What exactly is synthetic biology? In what sense may it involve truly radical innovations in the mode of intervention in nature? Does it really have a relevant ontological impact that is philosophically valuable? Although synthetic biology is a relatively emerging field in historical terms, it constitutes a hybrid between biology and engineering aimed at the design and manufacture of biological components and systems that do not exist in the natural world, in addition to the re-design and production of existing biological systems (Schyfter 2012; Schark 2012).

Synthetic biology is different from other engineering and scientific disciplines both in its approach and its choice of object because it uses the advances of biological research, while formulating new rules for engineering purposes. It thus constitutes a hybrid discipline that combines engineering and science to achieve its goal of producing synthetic organisms (Andrianantoandro et al. 2006).

In this respect, the essential difference between synthetic biology and genetic engineering is that, while the latter is basically "a cut-and-paste operation, in which genetic sequences from one organism (…) are inserted into another organism," the objective of synthetic biology is, in contrast, "to construct whole organisms from scratch" (Preston 2013, p. 650). This implies that such entities lack meaningful connections with historical evolutionary processes.

The latter aspect reveals a second important difference: since these synthetized organisms have no 'ancestors', they cannot be properly placed within the framework of an evolutionary causal chain of historically viable organisms. As C. Preston (2008b) rightly points out, this connection with ancestors exists even in the genetically modified organisms of traditional genetic engineering, which alters z, an already known type of organism, either by adding or removing genes. For instance, DNA can be inserted into bacteria belonging to the human genome, or cows can be made to produce human proteins in their milk (Lee 2003). In both cases, bacteria and cows possess a continuous causal chain between the currently manipulated genome and the historical evolutionary process. At each point in this chain, there has been a viable organism.

A third difference is revealed when the scale in which each of these disciplines operates is considered: synthetic biology introduces a genuinely 'engineer's approach', whereas genetic engineering allows only for a 'craftsman's approach'.

While the artisan cobbles together "variously shaped stones found in nature until a usable structure is achieved (…) the engineer employs standardized parts that have themselves been manufactured, and so can plan out her structure" (Preston 2013, p. 651). These parts, known as "BioBricks", are standardized biological parts which are added into a current global project to create an open and expandable catalogue of genetic parts, such as protein coding sequences, which can be combined in different ways to make synthetic organisms.[1]

The aforementioned differences reveal that certain aspects of synthetic biology are novel in relation to previous domestication methods. These new aspects can all be related to the goal of conveying the organic to the sphere of design, which is conceived in terms of planning, diagrammatic representation of artifacts, and standardization of parts to be assembled. Although this desire for rational and modularized control of the design of artificial organisms is not new at the methodological level (it was present at the origin of genetics as a discipline), both current scientific knowledge and its forecasting capacity are.

As to this context of differences, the questions that emerge are: are these novel aspects of synthetic biology sufficient to break with previous design modalities? If so, would this break affect the traditional ontological divide between nature and artificiality? Following certain suggestions from Lewens (2013), we believe there is a design *continuum* between the 'pure' forms of natural evolution at one extreme (where no intentional agency influences or affects the evolution of a lineage) and, at the other, the forms that stem from the current efforts made by synthetic biology to achieve a completely formalized and standardized rational design. This idea is in turn compatible with the recognition of relevant differences in the degrees of control between the diverse historical variants of intervention in nature.

3.3 Three Problems

3.3.1 First Problem: Does Synthetic Biology Really Alter the Traditional Ontological Divide Between the Natural and the Artificial?

Several perspectives dealing with synthetic biology theoretically, with either a dystopian or an emancipatory stance (for instance, Lee 2003; Baker 2008; Church ans Regis 2012; Schyfter 2012), coincide in considering this new, hybrid discipline as a radical transformation that must be addressed precisely because it represents a transcendental novelty with respect to previous forms of manipulating nature.

For example, Baker (2008) argues that current technological modes of intervention in the sphere of living organisms introduce a qualitative difference with respect to

[1] Synthetic biology, as Adrian Mackenzie claims, works under the assumption that biological space is "a flattened space of technical operations" (2010, p. 194).

previous modalities. This breaks with the traditional dichotomy and produces a sort of "ontological catastrophe".[2] According to Baker, a clear divide existed between natural and artificial entities that is now becoming blurred as a result of certain advances in biotechnology. The traditional division between the natural and the artificial was based on the distinction between *intention-independent* objects (organisms, among others) and *intention-dependent* objects (artifacts). The guiding criterion was the degree of independence of a certain entity with respect to our beliefs or purposes.

In this paradigmatic intentionalist perspective put forward by Baker, the division between natural and artificial is justified by the relationship between certain types of entities and certain kinds of human intentional states. Artifacts exist only because we have certain beliefs and purposes regarding them; they possess characteristics functions that we deliberately create and assign to them. On the other hand, the mode of existence of natural objects (or organisms) does not depend on our beliefs or purposes; neither are their components' characteristic functions independent of what we create or want.

Thus, Baker's argument has two interrelated corollaries: (a) this new technological modality renders the natural/artificial divide obsolete; and (b) it is the first time that such a modification has occurred in the history of our species. Faced with this author's perspective, there are two questions we must ask ourselves: first, whether this distinction between intention-dependent and non-intention dependent objects is really as clear and strong as it is presented in her argument; and, second, whether the type of hybrid objects that challenge the aforementioned divide have only arisen through contemporary biotechnology.

Beth Preston's (2008a) objection to this hypothesis of a recently unleashed "ontological catastrophe" begins with the indication of an erroneous assumption: the idea that, while artifacts are things created by humans to serve their purposes, nature is completely independent of our purposes and activities. According to Preston, the distinction between intention-dependent and non-intention-dependent objects does not distinguish neatly between artifacts and natural objects. Neolithic bioartifacts,[3] such as wheat and cows, reveal this. Their structure is produced both by human practices and by factors which are completely independent of human beings and their activities. Since all bioartifacts share this ambiguous nature, the proper starting point to examine the natural/artificial dichotomy is not European modernity, but the beginnings of agriculture. Modern science and technology, including biotechnology and synthetic biology, are only a series of "footnotes to the Neolithic transition from foraging to farming" (Preston 2013, p. 650). Before agriculture, plants and animals lived their lives free from human intervention. Neolithic

[2] In a similar vein, Keekok Lee (2005) is concerned about the "ontological impoverishment" caused by current biotechnological practices. The implementation of the modern scientific/technological programme "leads inexorably to the increasing elimination of that dualised other. In this view, then, the dualism between human and non-human may finally liquidate itself, if science and technology can in principle systematically and at a deep level transform the natural into the artefactual. Such transformation also produces ontological impoverishment, which is an inevitable part of modern anthropocentrism". (2005: 19)

[3] K. Lee (2005) calls these Neolithic creations "biotic artifacts". For a more extensive characterization of bioartifacts in the context of the intentionalism/non-intentionalism debate, see Parente (2014, 2015).

domestication introduced structural changes not only in some specimens of plants and animals (as in certain tamed animals), but also in the species as a whole.

As these bioartifacts evolved under human manipulation and control, domesticated organisms diverged more and more from their wild versions. This interpretation of the phenomenon of Neolithic domestication leads Preston to conclude that, from an ontological point of view, synthetic biology does not produce anything genuinely new. In fact, we crossed "a qualitative ontological divide when we started making biological artifacts upwards of 10,000 years ago" (Preston 2013, p. 651). Certain features that supposedly characterize the radical novelty of synthetic biology were already present in selective breeding carried out by domesticators in the Neolithic. While lacking the genetic knowledge available today, these domesticators managed to intensify biological functioning, thus creating indefinitely malleable biological meta-materials (Preston 2013, p. 654).

Thus, differences between the organisms domesticated in the Neolithic and the current 'products' of synthetic biology are merely "a quantitative matter of how much of their function and structure is under our intentional control" (Preston 2013, p. 658). In other words, the sophisticated degree of control introduced by synthetic biology has had no ontological impact so far, because the great transformation that questioned the natural/artificial divide began – strictly speaking – in the Neolithic transition from gathering and hunting to agriculture.

Accepting the idea that the innovations of synthetic biology do not have a clear ontological impact does not mean, as Preston suggests, rejecting them as irrelevant. As our interpretation of bioartifacts depends partly on our correct understanding of underlying 'authorial' processes, a philosophical theory interested in exploring their nature requires carefully thematizing the different degrees of intentional intervention in such processes. This is where the author's control is made evident. Undoubtedly, it is reasonable to admit that these degrees of intervention imply mere differences of scale, not of substance, but such 'differences of scale' are philosophically valuable to understand the idea of the bioartifact historically and how this idea helps to partially shape our identity. In Section [3], we shall explore the intricacies of these 'differences of scale'.

3.3.2 Second Problem: Is It Valid to Talk About 'Biological' or 'Living' Machines?

According to Preston, the idea that guides current research in synthetic biology is that

> (…) living things can in principle be designed and constructed from the ground up, just like tables and pots. Achieving this depends in part on the capacity to synthesize DNA, a technique already used to some extent in genetic engineering. But since DNA only works within cells that transcribe and replicate it, a thoroughly synthetic biology calls for synthetic cells as well (…) Synthetic cells containing synthetic genomes would be the 'biological machines' or 'living machines' synthetic biology advertises as its ultimate goal. (Preston 2013, p. 650)

The questions we are committed to answering are the following. In what sense is it possible to think that this techno-scientific enterprise creates 'biological machines'? What image of the living and the mechanical underlies this representation? To what extent does it generate conceptual tensions within the project itself?

First, the main goal of synthetic biology related to the creation of 'living machines' implies a crucial tension inasmuch as this new discipline aims to create stable organic systems that do only what bioengineers want. At the same time, however, it manufactures living beings whose essence is constant change and that therefore possess an inherent plasticity; an aspect that clearly challenges the sought ideal of control (Schark 2012, p. 32).

Even if synthetic biology achieved thorough and definitive knowledge of natural processes and used this knowledge to synthesize artificial cells, these mechanisms would not produce something static in nature. The structure channelling vital processes is itself involved in a process of degradation and reconstruction (Schark 2012, p. 32). One of the main features of living things is this inherent plasticity and 'openness' to change without loss of structure. This condition that organisms always display contrasts significantly with the fixed character of a machine's organization. Consequently, in this context, it is very difficult to assimilate the idea of a 'living machine' coherently.[4]

Thus, there is a tension between two poles within the very idea of a 'living machine'. One pole emphasizes the entity's fixed (*mechanistic*) character, ensuring control and predictability over the created product. The other pole stresses the plastic character of the living entity, its openness to constant change, hindering complete control and significantly reducing the system's degree of predictability. This tension is not solved simply by discovering the origin of the entity under analysis. In fact, a product of synthetic biology is like a machine from the point of view of its origin (it has been made by humans and, in this sense, causally depends on them), but, once produced, it behaves like an organism if we acknowledge how plasticity governs its behaviour and development.

In this respect, an appealing possibility is to explore whether the type of purposiveness of each entity tells us something relevant about the coherence of the very concept of 'living machine'. In order to answer this question, let us imagine a scenario in which synthetic biology actually succeeds in creating a living system from scratch (i.e., a self-producing, self-organizing, self-sustaining, and self-regenerating system). As Nicholson (2013) rightly points out, such a system – despite its synthetic origin – would have the capacity

> (…) to act by virtue of its internal organizational dynamics, to act on its own behalf in accordance with its own norms. (…) the distinction between intrinsic and extrinsic forms of

[4] It could be objected –from a Simondonian perspective– that these features do not really apply to genuine 'machines' or complex technical individuals, which behave progressively, through a 'concretization' process, like organisms at the level of self-regulation. However, the focus of this paper is not Simondon's idea of technical objects, but rather the way synthetic biology conceptualizes the machine by means of the notion of 'living machine'.

purposiveness is far better suited than the distinction between natural and artificial origins to capture the respective features of organisms and machines. (Nicholson 2013, p. 674)

If we accordingly follow this suggestion in order to validly and reliably ascribe the character of *organism* or *machine* to an entity, we should proceed according to the following rule: "Do not look at the origin (whether it relies on processes that are dependent or independent of intentional production); rather, assess whether its form of purposiveness is intrinsic or extrinsic." If intrinsic, then it is an organism (or functions as such). If extrinsic, then it is a machine (or functions as such).

The crucial point of this argument is that the primacy of intrinsic normativity – i.e., a bioartifact's capacity to act according to its own norms within the materials with which synthetic biology works – appears as an irreducible ontological aspect and, at the same time, as a sequence of continuity with the rustic organisms domesticated in the Neolithic. In other words, in parallel with what was shown in Section [2.1.], here again differences of scale in the degree of control do not lead to a genuine categorical leap or a pure ontological novelty.

3.3.3 Third Problem: Does Synthetic Biology Imply a Radical Transformation in the Type of Productive Creativity?

Other apologies of synthetic biology point out two features that would represent a critical break from previous modes of intervention in nature. These two features are the type of creativity involved in the process and the type of selection dynamics under which it operates. Thus, compared to genetic engineering, synthetic biology has a greater range of combinable variants ensured by the production of BioBricks and synthetic cells. At the same time, it represents a form of selection superior to 'blind' natural selection, as it allegedly allows the use of a genuinely engineering approach, with no trace of craftsmanship.

Counter to this argument, however, some problems related to certain assumptions need pointing out. First, contrary to the idea that what characterizes synthetic biology is a higher level of creativity than previous forms, we must acknowledge that there may be highly creative (innovative, original) processes without control. Second, there may be processes with a high degree of control (e.g., the industrial production of cars or refrigerators) involving a very low creative range, as identical lineages are reproduced (Lewens 2013). In short, a greater degree of control – which synthetic biology advertises – does not in itself guarantee greater creativity.

Moreover, one of synthetic biology's features that supposedly makes it better than other modalities of intervention is the superiority of its forms of selection. Craig Venter has announced, for example, that his goal is to short-circuit evolution and create his own version of a "second genesis". Synthetic biology thus presupposes a rational view of creativity governed not by a random system of generation of variants (as nature does), but by a system of planning and foresight built on the knowledge of natural laws.

Nevertheless, as Lewens (2013) rightly asserts, the particular conditions under which natural selection unfolds (not constrained by a prior commitment to rationality) tend to produce traits that remain robust against a wide range of environmental perturbations. This implies that "forms of redundancy and complexity that follow from blind design may often be superior in performance to the products of rational methods" (Lewens 2013, p. 647). This feature recalls Orgel's famous second law, summarized in the thesis "Evolution is cleverer than you are", according to which trial-and-error strategies are often more effective than intelligent, centralized, human planning. Thus, it seems that the degree of creativity of the process is not undermined by a lack of control over the availability of the raw materials (Lewens 2013, p. 645).

In fact, this phenomenon has been implicitly acknowledged by those involved in synthetic biology projects, as they frequently experiment with the design of evolving systems in order to observe their behaviour. For instance, "directed evolution" has been used to "convert a non-functional two-stage transcriptional cascade (i.e. inverter module) into a working one by mutating the repressor protein and its corresponding ribosome-binding site" (Andrianantoandro 2006, p. 5). In several of these cases, directed evolution has resulted in mutations that would have been difficult to design rationally.

Two ideas emerge from this section. First, if we consider the robustness of its products in relation to environmental demands, it is not clear that synthetic biology implies an authentic surpassing of previous forms of natural selection. Second, neither is it clear that it involves a completely radical break with previous modalities. This becomes explicit when we see that a bioengineer's imagination must inevitably return to models of blind evolution and simulate them in the laboratory to contribute new knowledge that helps optimize synthetic design.

3.4 The Idea of a 'Design *Continuum*'

We have argued, on the one hand, that synthetic biology produces some philosophically relevant 'scale differences', including a greater degree of control over created products, a new speed of change (compared to the way natural evolution works), and the disconnection of synthesized organisms from a historical evolutionary process. These three elements do represent a significant difference with respect to the Neolithic's modes of selection, or even with respect to genetic engineering.

On the other hand, this paper attempts to show that synthetic biology does not have the ontological impact intended by its supporters (i.e., it does not tear down the natural/artificial divide) and, at the same time, it does not represent a radical transformation in the sense that it is not able to create a type of 'living machine' whose laws of deployment are qualitatively different from those of a traditional organism.

If we admit this argument, where should we place this synthetic approach in the history of our interventions in the sphere of the living? One possibility is to place these changes introduced by synthetic biology in what Lewens (2013) has called a "design continuum". Rather than a break, there is, in fact, a continuity of design that goes from the ‘blind’ processes of organic modification at one extreme to the rationally designed processes located at the other.

As I have explained elsewhere regarding the nature of bioartifacts (Parente 2014), this design continuum reveals a progressive increase both in the capacity to control the production process and in the conditions, capacities, and deployment of (bio)products. At one of the poles of this continuum, we find a set of stages characterized by unintended selection: non-conscious natural selection, followed by ‘unconscious selection’ made by humans, and, finally, early naive efforts at domestication. This last stage gradually led to a subsequent stage in which control over production increased. In this stage we first find virtuous domestication (which operates with phenotypes with some systematization), then artificial selection with intervention on the genome and, nowadays, the current practices of synthetic biology.

That said, in which sense are these different phases that, nevertheless, represent a design continuum? First, certain unconscious selection phenomena – such as those that anticipated processes of domestication – are qualitatively similar to natural selection, except that they occur in a ‘human-made’ environment. In this case, humans cannot be said to completely control the scenario as they are not aware of the changes they actually introduce. This phase of domestication should be understood as a co-evolutionary phenomenon within which humans and other species rearrange selective forces, acting on them in unintentional and unanticipated ways (Boivin 2008, pp. 201–202). Subsequently, refinement at the level of selection led to an increasingly conscious and deliberate mode of domestication of certain animals and plants – as illustrated in the Neolithic – using a craft-based trial-and-error technology of breeding. A substantial change in bioartifactual product control occurred with the emergence of Mendelian hybridization technology, still operating “through the whole organism in reproduction, but nevertheless, at the same time, focusing on the gene-chromosome (cellular level) as the unit of genetic transmission” (Lee 2009, p. 101). Finally, the arrival of DNA genetics produced a change in scale because it operates at the molecular level, thereby granting an even greater level of artifacticity in its results.

The previous characterization reveals that this historical journey through the design continuum implies at least three essential features: an increase in the predictability of the created systems, a greater degree of control over both the production and the intended capacities of organisms, and a greater speed of change (compared to “low” methods used by natural selection with no human intervention). The different moments included in this gradual process of nature domestication must be located transversely between two poles, the ‘blind’ processes of organic modification at one extreme, and the completely rationally designed processes at the other.

3.5 Final Remarks

This paper has attempted to situate the true innovations of synthetic biology products within the broader and more comprehensive framework of a design continuum. This does not imply the rejection of the genuine differences between synthetic biology, traditional domestication practices, and modern genetic engineering, however.

One way of assessing the strength of this argument is to take the opposite stance and focus on the following question: What exactly would a radical discontinuous argument need to guarantee powerful evidence? It would require the creation of a fully designed kind of life, i.e., the creation of an 'organism' whose architecture challenges certain basic conditions of our conception of the organic (e.g., the plasticity of organisms in relation to environmental perturbations, their intrinsic normativity, etc.). The problem is that, if this were possible, we would not recognize such a product in terms of a genuine 'living being', nor would we find a way to coherently relate it to our minimalist sense of life.

However, given the laws of the materials with which synthetic biology works, the ideal of a completely designed modification as represented in a 'living machine' constitutes a conceptual impossibility. If we manufactured genuine living beings, regardless of their conditions of production and internal constitution, there is an underlying intrinsic normativity in organisms that prevents complete control and predictability. This in turn makes it difficult to reach the completely designed modification pole that appears as an intended effect.

If we follow this argument, breaking the design continuum would hence imply creating a system that behaves as a living being in several ways (self-generation, self-organization, self-repairing), except for intrinsic normativity, which would eliminate the system's capacity to be out of control. It would be an entity made of biological material which, paradoxically, would function exclusively as an 'instrument', not as a Simondonian 'machine'. That is to say, it would be an entity whose mode of existence would be reduced to a sphere of pure heteronomy, whose behaviour would be thoroughly determined by an extrinsic normativity assigned by its creators. In short, in order to break the design continuum, it would not suffice to instrumentalize an organism's capacity, i.e., to co-opt certain aspects of its intrinsic purposiveness to introduce it into plans that are suitable for human agents. If this is what synthetic biology considers 'revolutionary', strictly speaking, there would be no qualitative differences, because we have already co-opted various functions of organs of animals and plants since the Neolithic. We have instrumentalized the capacities of certain organisms by introducing them into action plans which are particularly beneficial to us.

In the absence of this kind of entity which could actually represent a gap in the design continuum, synthetic biology has no choice but to operate with greater sophistication and control over the immanent framework of life, which comprises a series of processes, laws, and objective constraints whose structure does not depend on our practices in any relevant sense. In brief, synthetic biology, both in its current state and in its various unfulfilled desiderata, should not be understood as representing a radical change of category, but rather as a gradual change within a design continuum.

References

Andrianantoandro, E. et al. (2006). Synthetic biology: new engineering rules for an emerging discipline. *Molecular Systems Biology, 2*. doi:https://doi.org/10.1038/msb4100073.

Baker, L. R. (2008). The shrinking difference between artifacts and natural objects. *Newsletter on Philosophy and Computers. American Philosophical Association, 07*(2), 2–5.

Boivin, N. (2008). *Material cultures, material minds. The Impact of things on human thought, society, and evolution*. New York: Cambridge University Press.

Carlson, R. (2010). *Biology is technology: The promise, peril, and new business of engineering life*. Cambridge: Harvard University Press.

Church, G., & Regis, E. (2012). *Regenesis. How synthetic biology will reinvent nature and ourselves*. New York: Basic Books.

Hagen, K., et al. (2016). *Ambivalences of creating life. Societal and philosophical dimensions of synthetic biology*. Heidelberg: Springer.

Kaebnick, G. (2014). *Humans in nature. The world as we find it and the world as we create it*. New York: Oxford University Press.

Lee, K. (2003). *Philosophy and revolutions in genetics deep science and deep technology*. New York: Palgrave Macmillan.

Lee, K. (2005). *Zoos. A philosophical tour*. New York: Palgrave Macmillan.

Lee, K. (2009). Biology and technology. In J. K. Berg Olsen et al. (Eds.), *A companion to the philosophy of technology*. Oxford: Wiley-Blackwell.

Lewens, T. (2013). From bricolage to BioBricksTM: Synthetic biology and rational design. *Studies in History and Philosophy of Science. Part C: Studies in History and Philosophy of Biological and Biomedical Sciences, 44*(4B), 641–648.

Mackenzie, A. (2010). Design in synthetic biology. *BioSocieties, 5*(2), 180–198.

Nicholson, D. J. (2013). Organisms/machines. *Studies in History and Philosophy of Science. Part C: Studies in History and Philosophy of Biological and Biomedical Sciences, 44*(4B), 669–678.

Parente, D. (2014). El estatuto de los bioartefactos. Intencionalismo, reproductivismo y naturaleza. *Revista de Filosofía. Universidad Complutense de Madrid, 39*(1), 163–185.

Parente, D. (2015). Los bioartefactos en la discusión sobre los límites entre lo natural y lo artificial. In J. E. Linares & E. Arriaga (Eds.), *Aproximaciones interdisciplinarias a la bioartefactualidad*. México: UNAM.

Preston, B. (2008a). The Shrinkage Factor: comment on Lynne Rudder Baker's 'The shrinking difference between artifacts and natural objects'. *American Philosophical Association Newsletter on Philosophy and Computers, 08*(1), 26–28.

Preston, C. (2008b). Synthetic biology: Drawing a line in Darwin's sand. *Environmental Values, 17*(1), 23–39.

Preston, B. (2013). Synthetic biology as red herring. *Studies in History and Philosophy of Science. Part C: Studies in History and Philosophy of Biological and Biomedical Sciences, 44*(4B), 649–659.

Schark, M. (2012). Synthetic biology and the distinction between organisms and machines. *Environmental Values, 21*(1), 19–41.

Schmidt, M., et al. (2009). *Synthetic biology. The technoscience and its societal consequences*. Dordrecht: Springer.

Schyfter, P. (2012). Standing reserves of function: A Heideggerian reading of synthetic biology. *Philosophy & Technology, 25*(2), 199–219.

Part II
Ethical, Political and Regulatory Issues

Chapter 4
Valuing Reprogenetic Technologies: Bringing Insights from the Philosophy of Technology to Bioethics

Inmaculada de Melo-Martín

Abstract Reprogenetic technologies have been enthusiastically embraced by well-known authors who argue that these technologies increase reproductive choice, contribute to a reduction of suffering by eliminating genetic diseases and disabilities, and offer the opportunity to improve the human condition by creating beings who will live much longer and healthier lives, have better intellectual capacities, and enjoy more refined emotional experiences. Indeed, some take reprogenetic technologies to be so valuable to human beings that they believe their use is not only morally permissible but morally required. More often than not, however, proponents of reprogenetic technologies treats these technologies as mere value-neutral tools, limiting their assessments to risk and benefit considerations. In this chapter I bring insights from philosophy of technology regarding the value-laden nature of technologies to bear on bioethical analyses of reprogenetics. I challenge proponents' assumption that an evaluation of risk and benefits is all is needed to determine the moral permissibility or impermissibility of developing and using reprogenetic technologies. I argue that a robust ethical analysis requires attention to the relationship between contextual values and technological development and implementation, as well as to the ways in which technologies reinforce or transform human values by mediating our perceptions of the world and our reasons for action. Ignoring the value-laden nature of reprogenetic technologies results not just in incomplete ethical evaluations but in distorted ones.

I. de Melo-Martín (✉)
Division of Medical Ethics, Weill Cornell Medicine, Cornell University, New York, NY, USA
e-mail: imd2001@med.cornell.edu

B. Laspra, J. A. López Cerezo (eds.), *Spanish Philosophy of Technology*, Philosophy of Engineering and Technology 24,
https://doi.org/10.1007/978-3-319-71958-0_4

4.1 Introduction

Reprogenetic technologies, which combine the power of reproductive technologies with the tools of genetic science and technology, are often sources of newsworthy stories. Discussions about designer babies, cloned humans, and, more recently, three-parent children make for fascinating narratives and interesting movie plots. But they are more than that. These technologies affect some of the most important aspects of human existence: our desire to reproduce, form families and ensure the health and well-being of our children. They give prospective parents an unprecedented and sophisticated level of control not simply over whether and when to have a child, but also over a variety of a future child's characteristics. Prospective parents can now select embryos with or without particular genetically-related diseases and disabilities (Harper and Sengupta 2012) and non-disease related traits such as sex (Sharp et al. 2010). Furthermore, in the future new advances in gene editing techniques might permit prospective parents to alter the genome of gametes or embryos in order to eliminate the risks of some diseases or manipulate certain molecular aspects that affect physical features, cognitive capacities, or character traits (Liang et al. 2015; Cai and Yang 2014).

Well-known authors such as John Harris, Julian Savulescu, and John Robertson have enthusiastically embraced these technologies. According to proponents, these technologies increase reproductive choice, contribute to a reduction in suffering by eliminating genetic diseases and disabilities, and offer the opportunity to improve the human condition by enhancing the human species (Harris 2007, 2016; Robertson 2002, 2003, 2005; Savulescu 2001, 2005; Savulescu and Kahane 2009; Savulescu et al. 2015).[1] In fact, some consider reprogenetic technologies to be so valuable to human beings that they believe their use is not only morally permissible, but morally required (Savulescu 2005; Savulescu et al. 2015; Savulescu and Kahane 2009; Harris 2007).

More often than not, however, proponents of reprogenetics treat these technologies as mere value-neutral tools, limiting their assessments to narrowly understood risk and benefit considerations. In this chapter, I bring insights from the philosophy of technology regarding the value-laden nature of technologies to bear on bioethical analyses of reprogenetics. I present some of the value-laden aspects of reprogenetics and thus challenge proponents' assumption that an evaluation of risk and benefits is all that is needed to determine the moral permissibility or impermissibility of developing and using reprogenetic technologies. I focus here on two such aspects: how these technologies transform concerns about the health and wellbeing of our offspring primarily in individual problems, and the ways in which they challenge current normative conceptions of parenting. These are, of course, not the only ways in which reprogenetic technologies are value-laden (de Melo-Martín 2017), but this limited discussion will suffice to show that analysis of these aspects is relevant to any ethically robust evaluation of reprogenetics. I begin by offering a brief overview of these technologies.

---

[1] These authors' publication record on reprogenetics is extensive; I include here only some of their representative works.

4.2 Reprogenetic Technologies

The term "reprogenetics" generally refers to practices that combine reproductive technologies and genetic tools (Knowles and Kaednick 2007). They allow the creation, storage, and genetic manipulation of gametes and embryos with the aim of reproduction.[2] Some of these technologies, such as in vitro fertilization (IVF), have now been used for almost four decades, although researchers are constantly updating many of the procedures and techniques involved: new fertility drugs, cryopreservation protocols for gametes and embryos, and fertilization methods (Elder and Dale 2011). Use of IVF nowadays accounts for 1.5% of all infants born in the United States (CDC 2014), and over 4% of births in many European countries (Kupka et al. 2014).

Other technologies, such as mitochondrial replacement techniques (MTRs), have only recently being approved for clinical trials (Torjesen 2014). If successful, MRTs will give women at risk of transmitting mitochondrial diseases the opportunity to have unaffected and genetically related children. These techniques involve transferring the nuclear DNA from the egg of a woman who has mutant mitochondrial DNA into a donor egg that has no identified mitochondrial mutations (Craven et al. 2010; Tachibana et al. 2013). Because mitochondria have their own genome, the resulting embryos contain DNA from three different individuals. Significantly, the offspring of women – though not of men[3] – who use these techniques will inherit the mitochondrial DNA from the donor of the eggs, and so will future generations. This makes MRTs germline modifications techniques. In approving their use, the United Kingdom has become the first country to allow these kinds of modifications in human beings (Torjesen 2014).

The majority of the genetic and molecular tools used in reprogenetics are all relatively recent developments. Preimplantation genetic diagnosis (PGD), for example, which involves the removal of one or more cells from an embryo in order to test for chromosomal abnormalities or genetic mutations in the genome (Collins 2013), was introduced only in the early 1990s (Handyside et al. 1990). Until recently, PGD was mainly used to test for disorders caused by chromosomal abnormalities (e.g., Down syndrome), X-linked diseases (e.g., Duchenne muscular dystrophy and hemophilia), and single-gene disorders (e.g., Huntington's disease, cystic fibrosis, bβ-thalassemia, and sickle cell anemia) (Collins 2013). However, the applications for PGD are rapidly expanding and approximately 200 different conditions can now be tested. PGD is presently also being used to identify some late-onset, lower-penetrance mutations (such as BRCA mutations associated with hereditary breast and ovarian cancer), to allow sex selection for the purposes of "family balancing," and for human leukocyte antigen (HLA) matching to ensure the birth of a baby who can become a tissue donor for an existing diseased sibling (Brezina and Kutteh 2015).

[2] These technologies can also be used for research purposes. My concern here, however, is limited to their use for reproductive purposes.

[3] Mitochondria are inherited maternally.

Some of the molecular technologies that can nowadays be used to manipulate the genetic make-up of cells are still in the initial stages of research (Cai and Yang 2014; Liang et al. 2015). Current targeted genome editing technologies using restriction endonucleases, for instance, provide the ability to insert, remove, or replace DNA in precise ways. They can be used not only to study gene function, biological mechanisms, and disease pathology, but also to treat or cure particular diseases (Cai and Yang 2014). These systems are significantly more efficient and more accurate than older technologies and have the potential to dramatically increase the power of researchers to manipulate genomes. Indeed, one of these systems, CRISPR/-Cas9, has already been used to create genetically modified macaque monkeys (Niu et al. 2014). Although concerns raised by the use of this technique in human embryos has led the scientific community to call for a moratorium on this type of research (Baltimore et al. 2015), the UK Human Fertilisation and Embryology Authority (HFEA) recently granted permission to edit the genomes of healthy human embryos for research purposes (Callaway 2016).

In general, reprogenetic technologies have been characterized by their rapid transfer from the laboratory to the clinic to routine care; in many cases, with scant evidence of safety and efficacy. Furthermore, although the initial indications for many reprogenetic techniques was limited (i.e., directed toward particular infertility problems or genetic conditions), their application has steadily broadened. These technologies can now be used – and are used – in cases in which neither infertility nor the risk of transmitting some genetic mutation is present. In fact, some companies such as 23andMe and GenePeeks are going further and now offer services aimed at the prediction of disease risks and other phenotypic traits including height, eye color, sex, and a variety of personality characteristics not of *existing* embryos, but of *hypothetical* ones (Couzin-Frankel 2012; DeFrancesco 2014).

4.3 Valuing Reprogenetic Technologies

Discussions in the bioethics literature regarding ethical concerns related to various biomedical technologies tend to focus on the risks and potential benefits of the technologies in question. Those who defend the technologies emphasize the possible benefits, while critics call attention to the risks. The debate over reprogenetic technologies is no different. Advocates of reprogenetics advance their enthusiastic positions about these technologies by presenting us with an array of actual, possible, and imagined benefits that individuals and society can derive from their development and use. Some such benefits include the possibility that our offspring will have longer lives, be free from severe diseases and disabilities, and be able to enjoy life more fully (Harris 2007, 2016; Savulescu 2005, 2006; Savulescu et al. 2015; Silver 1997). Reprogenetic technologies, proponents contend, can also help ensure that our children are more intelligent and better able to deal with their environments. Similarly, some advocates hold that these technologies could be directed at enhancing human traits that are constitutive of autonomy, such as our concept of self, the capacity to form and act on conceptions of the good life, and the ability to predict and pay attention to the consequences of one's behavior (Savulescu 2006). They could also be aimed at improving our offspring's moral character and thus used to produce people who are more loving, sympathetic, compassionate, or just (Savulescu and Person 2012). Supporters argue that all of these improvements will provide benefits, not only to the individuals who have been selected or enhanced, but also to society in general, because, ultimately, these improvements are likely to result in increased productivity and in more just societies (Buchanan 2008). In fact, the presumed benefits of reprogenetic technologies are so expansive that, when reading proponents of reprogenetics' works, it is difficult to avoid the impression that these technologies constitute the solution to all of the problems that humanity faces.

Given the enormous potential benefits that proponents claim these technologies will bring, it is not surprising that they are not particularly impressed by the risks.[4] For them, much of the discussion about risks to individuals and society is grounded on conjectures, religious beliefs, or irrational concerns (Harris 2005, 2007, 2016; Savulescu 2005; Savulescu and Dahl 2000; Savulescu et al. 2015). Undoubtedly, proponents of reprogenetics do contemplate some risks, but such considerations tend to be quite limited. Although advocates of reprogenetics generally do not address them in any significant way, they consider risks to human health to be relevant. For the most part, however, proponents simply acknowledge that the use of these technologies is predicated upon their being safe – or safe enough – and effective. Undeniably, during the research stage there will be uncertainty about the safety of these procedures, but advocates contend that this is the case for all interventions. As far as they are concerned, if appropriate preclinical studies exist to show that the technologies are reasonably safe and effective and that women and their partners are appropriately informed about risks and uncertainties, then proceeding with clinical research in reprogenetics does not raise unusual moral concerns (Savulescu et al. 2015; Harris 2016).

The attention proponents give to other types of risks usually depends on the kinds of objections they are attempting to address. Their discussions tend to be directed towards either dismissing critics' concerns about most risks or offering assurances that such risks can be appropriately managed. For instance, critics have argued that some parents, having spent a considerable amount of time and presumably money in selecting and enhancing their offspring, might come to see their children as mere products of their wills (Kass 2002). They might also become overbearing, disillusioned, and unloving if their children do not live up to their expectations. Advocates often dismiss these worries by indicating that the existence of overbearing parents does not depend on the use of reprogenetics (Robertson 2003; Green 2007; Harris 2007). Furthermore, they claim that rather than creating overbearing and disappointed parents, the use of reprogenetic technologies is more likely to result in more loving and dedicated ones, partly because some parents

[4] Advocates of reprogenetics problematically reduce uncertainties, ambiguities, and ignorance regarding many of the effects of new reprogenetic technologies to issues of risk. I put this concern aside here.

might find it easier to care for and love a child who is brighter, more beautiful, and healthier than she would have been without the use of reprogenetics (Robertson 2003; Green 2007). Advocates also tend to be dismissive of concerns by critics that these technologies might contribute to discrimination against people with disabilities (Asch 2003; Parents and Asch 1999; Scully 2008), contending that such fears are overstated (Harris 2001). For advocates, the fact that prospective parents use reprogenetic technologies to have children free of disabilities is in no way an instance of discrimination against disabled people or an indication that people with disabilities are not equal and should not be treated as such (Harris 2001; Savulescu and Kahane 2009; DeGracia 2012). Some proponents of reprogenetics also argue that because these technologies will likely result in a decreased incidence of certain disabilities, they could in fact lead to more assistance for those still affected, so that they would be more likely to live full lives with the help of a variety of technological and social support (Bostrom 2003).

Reprogenetics advocates' claim that assessments of risks and possible benefits are sufficient to determine the moral permissibility of developing and using these technologies betrays a conception of science and technology as value-neutral (Parens 2015; de Melo-Martín 2017). For them, reprogenetic technologies are mere means to what they consider to be ethically appropriate ends: increasing reproductive choice, reducing diseases and disabilities, and ultimately improving human wellbeing (Harris 2007; Savulescu 2005). Whatever problems these technologies raise, they argue, are the result of faulty use or problematic public policies. Similarly, the best way to deal with potential harms is to implement regulations that attempt to prevent such harms without excessively infringing on people's freedoms, rather than developing different types of technologies (Harris 2007; Savulescu et al. 2015; Robertson 2003).

However, their value-neutral conception of science and technology in general and reprogenetics in particular is implausible. Science and technology are value-laden activities. They embody particular values, shape and transform our reality, influence our practical options and our reasons for action, and affect what we take to be morally permissible or obligatory (see, eg., Swierstra and Waelbers 2012; Latour 2005; Verbeek 2005; Waelbers 2011; Borgmann 1984; Ihde 1990; Winner 1986; Habermas 1971; Feenbergn 1999). Values affect the decisions that various stakeholders – including engineers – make when developing particular technologies (van de Poel 2001). Technologies can exhibit political qualities (Winner 1980, 1986; Feenberg 1999; Habermas 1971). They influence our perceptions of the world and our behavior and thus affect both the ways we conceptualize problems and the solutions we might think are possible; they mediate our practical options and, with them, the reasons we have to act; they alter and shape our moral beliefs, affect our ideas of what is right or wrong, virtuous or vicious, good or bad, morally obligatory or forbidden (Swierstra and Waelbers 2012; Verbeek 2005; Ihde 1990).

But if technologies are not value-free, then considerations other than risks and potential benefits are germane to any ethically robust assessment of these technologies. Indeed, insofar as technologies embody particular values, shape and transform others, change the ways we perceive and interact with reality, and affect the kinds of answers we give to certain problems, a sound evaluation of scientific and technological developments requires attention to these technological effects.

One of the ways in which reprogenetic technologies are value-laden is that they transform concerns about health, disease, and disability primarily into *individual* problems. These technologies charge prospective parents with managing risks to the health and quality of life of their offspring: from diseases such as Tay Sachs, Huntington's, heart disorders, or cancer, to mental health problems such as anxiety, neuroticism, or schizophrenia, to impairments such as deafness or blindness, through to conditions such as alcoholism, criminal behavior, and substance addiction. If proponents are to be believed, reprogenetics will also allow prospective parents to manage various traits that affect human character and temperament, such as aggression, optimism, sense of humor, shyness, sympathy, novelty seeking, or altruism, and, of course, assorted cognitive capacities such as intelligence and memory. Even sexual preference is included as a trait that prospective parents can choose to "treat" (Savulescu 2001, p. 417). The task of fulfilling the promises of reprogenetic technologies – minimizing the existence of diseases, disorders, and disabilities, increasing the welfare of their children, reducing social problems, and outperforming evolution by choosing specific traits – thus becomes the exclusive responsibility of prospective parents.

Of course, the emphasis on individualizing health and wellbeing is hardly new (Clarke 2010; Rose 2006; Foucault 2008). Nevertheless, this is precisely the consequence of the value-laden nature of science and technology. Reprogenetic technologies are shaped by particular values, such as the emphasis on individual responsibility, and at the same time contribute to a more and more widespread reliance upon individuals to manage risks to their future children's health and wellbeing.

But why is the value-laden nature of reprogenetic technologies important for a sound evaluation of these technologies? It is so for numerous reasons. Consider the individualization of concerns about health, disease, and disability brought about by reprogenetics. First, such individualization has significant implications regarding how to tackle questions about improving health and reducing diseases. Presenting these as individual concerns both removes them from the social context in which they are relevant and overlooks the prospect of collective solutions. Clearly, however, issues about health, disease, and disability can also be addressed in a collective fashion rather than primarily individually. Arguably, doing so should often be the primary response to such problems. After all, whether many of these traits constitute a disadvantage or negatively affect a person's quality of life depends very much on the social context in which the traits in question are expressed. In a society in which access to gluten-free products is common, for example, having an intolerance to gluten would not constitute much of a problem. Likewise, societies that are completely accepting of homosexuality would undermine any attempt to justify selecting against such a trait on the grounds that it affected people's quality of life.

Even for traits – such as at least some diseases – that would presumably negatively affect people's wellbeing in any human society, minimizing the relevance of collective solutions is problematic. It is well known that conditions of material deprivation, such as inadequate nutrition or lack of access to appropriate housing or clean water and sanitation, constitute health risks and increase mortality (Daniels 2008; Marmot et al. 2012). Similarly, a variety of social determinants such as income, education, occupational rank, social class, and social support greatly affect people's health and quality of life. What societies we have, how they are structured, what resources are available to people, and how they are distributed are all factors that significantly influence people's health and wellbeing. Collective actions, and not simply individual ones mediated by biomedicine, are needed to address concerns related to these social determinants.

Evaluations of reprogenetic technologies that focus exclusively on risks and potential benefits understood in a narrow sense will inevitably fail to attend to the ways in which use of these technologies could inappropriately divert attention and resources away from the social, political, and economic factors that have such a significant impact upon health, disease, and wellbeing. Given that genetic factors are only one of many complex elements affecting health and disease, neglecting the social determinants of health will only serve to worsen people's health and quality of life.

Second, in a context in which responsibility for reproduction falls primarily on women, the individualization of concerns about health and disease is bound to have negative effects on their wellbeing (de Melo-Martín 2017). Women's reproductive decisions are continuously scrutinized, particularly the decisions of women who do not fit the normative understanding of the good mother, such as poor women, women of color, and women considered to have some disability (Roberts 1997; Asch 1999. Pregnant women must control what they put into their bodies, sacrifice their pleasures and desires in order to limit even the slightest of risks to their fetuses, and submit to expert medical knowledge (Kukla 2005). Even choices that might appear insignificant at any other time take on exaggerated importance when women are pregnant or are considering becoming pregnant. Pregnant women are strongly – and anxiously – advised to take folic acid, exercise regularly but not excessively, and avoid alcohol, tobacco, caffeine, unpasteurized milk, and X-rays. They are encouraged – sometimes strongly so when they are thought to belong to an "at risk" category (e.g., those over 35 years of age) or to be at risk of transmitting diseases or disabilities – to have prenatal screening to check for fetal abnormalities. They are instructed to breastfeed their newborn so as to both create appropriate bonds with the child and ensure adequate nutrition. Studies purportedly showing the significant effects of women's decisions during pregnancy on the wellbeing of their children are commonplace. What women do or fail to do in their youth, while pregnant, and after a baby is born are said to have serious – sometimes devastating – consequences on children's nutritional status, risks of obesity, allergies, IQs, learning and developmental delays, or behavioral problems (Grzeskoviak et al. 2015; Cao et al. 2014; Tryggvadottir et al. 2016). As has been widely discussed in the feminist literature, these norms, which women are expected to internalize, can be oppressive for any woman, but they can be experienced as particularly punishing by less privileged women, who often lack the resources and the life circumstances that would allow them to have and exert this degree of control over their pregnancies (Roberts 1997; Richardson et al. 2014; Kukla 2005). Clearly, in this context, an emphasis on the individual responsibility of prospective parents to manage the health and wellbeing of their offspring can only serve to increase the scrutiny over women's lives.

As stated previously, evaluations – such as the ones proponents of reprogenetics provide – which are limited to assessing risks and benefits will inevitably ignore the fact that, whatever degree of control these technologies afford prospective parents in their attempts to improve the health and quality of life of their future children, the focus on individual solutions unjustifiably overburdens them and will be particularly oppressive to women. Moreover, proponents' assessments implicitly and uncritically sanction the status quo. In ignoring the ways in which women's bodies and decisions are already subjected to intense public scrutiny and censure and how reprogenetics contribute to and intensifies such scrutiny, advocates of these technologies have failed not only to draw attention to the social conditions and institutions that systematically disadvantage women, but also to challenge such conditions and offer constructive solutions to them.

Another value-laden aspect of reprogenetic technologies involves the way in which they challenge normative notions of parenting. Normative notions of parenting involve claims about the kind of love and care that we believe parents ought to have for their children, simply by virtue of being parents; that is to say, the kind of love that we believe *good* parents have. Of course, many real-life parents can fall short of these normative ideals, though this does not invalidate the ethical value the normative notions in question involve. One such notion takes proper parental love to require parents to love their children simply because they are their children. Accordingly, because being someone's child is a sufficient condition to be loved by one's parents, the sort of love that good parents have for their children does not require their children to have any particular personal characteristic or traits in order to merit being their child and thus being loved (Herissone-Kelly 2007a, b). This normative notion is actually consistent with the widely held belief that choosing trivial characteristics of a child – such as sex for social reasons – simply to fulfill ones' preferences is incompatible with being a good parent (Scully et al. 2006a, b).

Reprogenetic technologies now present prospective parents with the option of having a *chosen* child. They can do so in two ways: by selecting particular embryos by virtue of having or lacking certain desirable or undesirable traits, and, possibly in the future, by requesting modifications to alter the embryos genome so as to have a child with particular characteristics. Moreover, not only are prospective parents able to use these technologies to choose particular embryos, they are morally required to do so (Harris 2007; Savulescu 2001). Granted, common normative notions of parenting are not inconsistent with prospective parents making certain choices. Clearly, our ordinary understanding of what makes a good parent is compatible with choosing *when* to have a child or *how many* children to have. In fact, a variety of technological innovations, from contraceptive methods to reproductive technologies, now allow prospective parents to make just such choices and to do so in more efficient ways. Similarly, one might argue that choosing some embryos rather than others when some of them are destined to live lives beset with excruciating pain is also consistent with being a good parent and hence that reprogenetic technologies are also consistent with this normative understanding of parenting.

However, these technologies can go significantly further by allowing prospective parents to select or modify what are arguably trivial characteristics, such as the child's sex. Advocates actually believe that, in the future, similarly trivial traits such as hair and eye color, height, or perfect pitch will also be available for selection or modification. By introducing the possibility of choosing among embryos based on what constitute trivial characteristics that have nothing to do with the wellbeing of the future child, these technologies challenge a widely accepted – and arguably appropriate – normative notion of parenting: one that calls for parents to love whatever child comes along (Herissone-Kelly 2007a, b; McDougall 2005). Notice that this is not to say that reprogenetic technologies will make parents more critical of their children if they ultimately do not express the traits their parents were attempting to select, or that parents will love their children less. It might well be the case that prospective parents end up loving their children even if they are not what they wished for (Green 2007). Rather, the effects that reprogenetic technologies have on normative conceptions of parenting illustrate their value-laden nature. The possibility that parents will still love their children is thus beside the point and in no way makes the effects of these technologies on parenting notions any less real. After all, it is one thing to be loved *in spite* of one's characteristics, and quite another to be loved *irrespective* of such characteristics. Reprogenetic technologies thus challenge widely accepted normative conceptions of parenting.

Whether or not one takes these concerns about the individualization of health and disease and the challenging of normative conceptions of parenting to be sufficient to oppose the development and use of reprogenetic technologies, there is little doubt that they are relevant to any ethically sound evaluation of reprogenetics. An appropriate assessment of these technologies must thus attend to them. Insofar as proponents of reprogenetics fail to do so, their evaluations are lacking.

4.4 Conclusion

In this chapter, I have challenged the business-as-usual way in which proponents of reprogenetics assess these technological innovations. Their evaluations consist in presenting a list of potential benefits and some risks that are promptly minimized as irrelevant or, when considered serious, they are thought manageable by appropriate regulations. Such assessment leads proponents to embrace reprogenetic technologies with enthusiasm and to characterize their critics as callous, irrational, obstructionists, Luddites, or religious zealots.

Of course, we have good reasons to reject as faulty at least some of the claims about the potential benefits and harms proponents advance (de Melo-Martin 2017). Moreover, the belief that an analysis of risks and potential benefits is all that is needed for a sound evaluation of these technologies is grounded on an inadequate understanding of reprogenetics as value-neutral tools. Yet, reprogenetic technologies are not value-neutral. They embody particular values, shape our worlds, affect our practical options and our reasons for action, and alter our moral norms and beliefs. However, if the assumption made by proponents about the value-neutrality of science and technology is incorrect, then their analyses – even if they were sound – of the risks and benefits of using these technologies not only fail to provide a compelling ethical assessment of these techniques, but rather offer a distorted one. Moreover, ignoring the values that shape and are shaped by technological developments does not make these values disappear. What it actually does is to conceal their presence and relevance, thereby hindering critical scrutiny. On the other hand, recognizing that science and technology in general and reprogenetic technologies in particular are not value-neutral allows us to unveil the values that influence and are influenced by reprogenetics and open up such values to critical examination.

References

Asch, A. (1999). Prenatal diagnosis and selective abortion: A challenge to practice and policy. *American Journal of Public Health, 89*(11), 1649–1657.

Asch, A. (2003). Disability equality and prenatal testing: Contradictory or compatible?, Florida State University law review. *Florida State University. College of Law, 30*(2), 315–342.

Baltimore, D., Berg, P., Botchan, M., Carroll, D., Charo, R. A., Church, G., Corn, J. E., Daley, G. Q., Doudna, J. A., Fenner, M., Greely, H. T., Jinek, M., Martin, G. S., Penhoet, E., Puck, J., Sternberg, S. H., Weissman, J. S., & Yamamoto, K. R. (2015). Biotechnology. A prudent path forward for genomic engineering and germline gene modification. *Science, 348*(6230), 36–38.

Borgmann, A. (1984). *Technology and the character of contemporary life: A philosophical inquiry*. Chicago: University of Chicago Press.

Bostrom, N. (2003). Human genetic enhancements: A transhumanist perspective. *The Journal of Value Inquiry, 37*(4), 493–506.

Brezina, P. R., & Kutteh, W. H. (2015). Clinical applications of preimplantation genetic testing. *Bmj-British Medical Journal, 350*.

Buchanan, A. (2008). Enhancement and the ethics of development. *Kennedy Institute of Ethics Journal, 18*(1), 1–34.

Cai, M., & Yang, Y. (2014). Targeted genome editing tools for disease modeling and gene therapy. *Current Gene Therapy, 14*(1), 2–9. https://doi.org/10.2174/1566523214021403181654 50.

Callaway, E. (2016). UK scientists gain licence to edit genes in human embryos. *Nature, 530*(7588), 18.

Cao, B., Stout, M. J., Lee, I., & Mysorekar, I. U. (2014). Placental microbiome and its role in preterm birth. *NeoReviews, 15*(12), e537–e545. https://doi.org/10.1542/neo.15-12-e537.

CDC, Centers for Disease Control and Prevention, American Society for Reproductive Medicine, and Society for Assisted Reproductive Technology. (2014). *2012 assisted reproductive technology national summary report*. Atlanta: US Dept of Health and Human Services.

Clarke, A. (2010). *Biomedicalization: Technoscience, health, and illness in the U.S.* Durham: Duke University Press.

Collins, S. C. (2013). Preimplantation genetic diagnosis: Technical advances and expanding applications. *Current Opinion in Obstetrics & Gynecology, 25*(3), 201–206.

Couzin-Frankel, J. (2012). Genetics. New company pushes the envelope on pre-conception testing. *Science, 338*(6105), 315–316.

Craven, L., Tuppen, H. A., Greggains, G. D., Harbottle, S. J., Murphy, J. L., Cree, L. M., Murdoch, A. P., Chinnery, P. F., Taylor, R. W., Lightowlers, R. N., Herbert, M., & Turnbull, D. M. (2010). Pronuclear transfer in human embryos to prevent transmission of mitochondrial DNA disease. *Nature, 465*(7294), 82–U89.

Daniels, N. (2008). *Just health: Meeting health needs fairly*. New York: Cambridge University Press.
de Melo-Martín, I. (2017). *Rethinking reprogenetics: Enhancing ethical analyses of reprogenetic technologies*. New York: Oxford University Press.
DeFrancesco, L. (2014). 23andMe's designer baby patent. *Nature Biotechnology, 32*(1), 8–8.
DeGrazia, D. (2012). *Creation ethics: Reproduction, genetics, and quality of life*. New York: Oxford University Press.
Elder, K., & Dale, B. (2011). *In-vitro fertilization* (3rd ed.). New York: Cambridge University Press.
Feenberg, A. (1999). *Questioning technology*. New York: Routledge.
Foucault, M. (2008). *The birth of biopolitics: Lectures at the Collège de France, 1978–79*. New York: Palgrave Macmillan.
Green, R. M. (2007). *Babies by design: The ethics of genetic choice*. New Haven: Yale University Press.
Grzeskowiak, L. E., Hodyl, N. A., Stark, M. J., Morrison, J. L., & Clifton, V. L. (2015). Association of early and late maternal smoking during pregnancy with offspring body mass index at 4 to 5 years of age. *Journal of Developmental Origins of Health and Disease*, 1–8.
Habermas, J. (1971). Technology and science as "ideology". In *Toward a rational society* (pp. 81–122). Boston: Beacon Press.
Handyside, A. H., Kontogianni, E. H., Hardy, K., & Winston, R. M. (1990). Pregnancies from biopsied human preimplantation embryos sexed by Y-specific DNA amplification. *Nature, 344*(6268), 768–770.
Harper, J. C., & Sengupta, S. B. (2012). Preimplantation genetic diagnosis: State of the art 2011. *Human Genetics, 131*(2), 175–186.
Harris, J. (2001). One principle and three fallacies of disability studies. *Journal of Medical Ethics, 27*(6), 383–387.
Harris, J. (2005). No sex selection please, we're British. *Journal of Medical Ethics, 31*(5), 286–288.
Harris, J. (2007). *Enhancing evolution: The ethical case for making better people*. Princeton: Princeton University Press.
Harris, J. (2016). Germline modification and the burden of human existence. *Cambridge Quarterly of Healthcare Ethics, 25*(1), 6–18.
Herissone-Kelly, P. (2007a). Parental love and the ethics of sex selection. *Cambridge Quarterly of Healthcare Ethics, 16*(3), 326–335.
Herissone-Kelly, P. (2007b). The "parental love" objection to nonmedical sex selection: Deepening the argument. *Cambridge Quarterly of Healthcare Ethics, 16*(4), 446–455.
Ihde, D. (1990). *Technology and the lifeworld: From garden to earth*. Bloomington: Indiana University Press.
Kass, L. (2002). *Life, liberty, and the defense of dignity: The challenge for bioethics* (1st ed.). San Francisco: Encounter Books.
Knowles, L. P., & Kaebnick, G. E. (2007). *Reprogenetics: Law, policy, and ethical issues*. Baltimore: Johns Hopkins University Press.
Kukla, R. (2005). *Mass hysteria: Medicine, culture, and mothers' bodies, Explorations in bioethics and the medical humanities*. Lanham: Rowman & Littlefield.
Kupka, M. S., Ferraretti, A. P., de Mouzon, J., Erb, K., D'Hooghe, T., Castilla, J. A., Calhaz-Jorge, C., de Geyter, C., Goossens, V., & European IVF Monitoring Consortium for the European Society of Human Reproduction and Embriology. (2014). Assisted reproductive technology in Europe, 2010: Results generated from European registers by ESHRE. *Human Reproduction, 29*(10), 2099–2113.
Latour, B. (2005). *Reassembling the social: An introduction to actor-network-theory*. Oxford: Oxford University Press.
Liang, P., Xu, Y., Zhang, X., Ding, C., Huang, R., Zhang, Z., Lv, J., Xie, X., Chen, Y., Li, Y., Sun, Y., Bai, Y., Songyang, Z., Ma, W., Zhou, C., & Huang, J. (2015). CRISPR/Cas9-mediated gene editing in human tripronuclear zygotes. *Protein & Cell, 6*(5), 363–372.

Marmot, M., Allen, J., Bell, R., Bloomer, E., Goldblatt, P., & Consortium for the European Review of Social Determinants of Health and the Health Divide. (2012). WHO European review of social determinants of health and the health divide. *Lancet, 380*(9846), 1011–1029.

McDougall, R. (2005). Acting parentally: An argument against sex selection. *Journal of Medical Ethics, 31*(10), 601–605.

Niu, J., Zhang, B., & Chen, H. (2014). Applications of TALENs and CRISPR/Cas9 in human cells and their potentials for gene therapy. *Molecular Biotechnology, 56*(8), 681–688.

Parens, E. (2015). *Shaping our selves: On technology, flourishing, and a habit of thinking*. New York: Oxford University Press.

Parens, E., & Asch, A. (1999). The disability rights critique of prenatal genetic testing. Reflections and Recommendations, *The Hastings Center report, 29*(5).

Richardson, S. S., Daniels, C. R., Gillman, M. W., Golden, J., Kukla, R., Kuzawa, C., & Rich-Edwards, J. (2014). Society: Don't blame the mothers. *Nature, 512*(7513), 131–132.

Roberts, D. E. (1997). *Killing the black body: Race, reproduction, and the meaning of liberty* (1st ed.). New York: Pantheon Books.

Robertson, J. A. (2002). Sex selection for gender variety by preimplantation genetic diagnosis. *Fertility and Sterility, 78*(3), 463.

Robertson, J. A. (2003). Procreative liberty in the era of genomics. *American Journal of Law & Medicine, 29*(4), 439–487.

Robertson, J. A. (2005). Ethics and the future of preimplantation genetic diagnosis. *Reproductive Biomedicine Online, 10*(Suppl 1), 97–101.

Rose, N. S. (2006). *Politics of life itself: Biomedicine, power, and subjectivity in the twenty-first century*. Princeton: Princeton University Press.

Savulescu, J. (2001). Procreative beneficence: Why we should select the best children. *Bioethics, 15*(5–6), 413–426.

Savulescu, J. (2005). New breeds of humans: The moral obligation to enhance. *Reproductive Biomedicine Online, 10*, 36–39.

Savulescu, J. (2006). Justice, fairness, and enhancement. *Annals of the New York Academy of Sciences, 1093*, 321–338.

Savulescu, J., & Dahl, E. (2000). Sex selection and preimplantation diagnosis: A response to the Ethics Committee of the American Society of Reproductive Medicine. *Human Reproduction, 15*(9), 1879–1880.

Savulescu, J., & Kahane, G. (2009). The moral obligation to create children with the best chance of the best life. *Bioethics, 23*(5), 274–290.

Savulescu, J., & Persson, I. (2012). Moral enhancement, freedom and the god machine. *The Monist, 95*(3), 399–421.

Savulescu, J., Pugh, J., Douglas, T., & Gyngell, C. (2015). The moral imperative to continue gene editing research on human embryos. *Protein & Cell, 6*(7), 476–479.

Scully, J. L. (2008). *Disability bioethics: Moral bodies, moral difference*. Lanham: Rowman & Littlefield.

Scully, J. L., Banks, S., & Shakespeare, T. W. (2006a). Chance, choice and control: Lay debate on prenatal social sex selection. *Social Science & Medicine, 63*(1), 21–31.

Scully, J. L., Shakespeare, T. W., & Banks, S. (2006b). Gift not commodity? Lay people deliberating social sex selection. *Sociology of Health & Illness, 28*(6), 749–767.

Sharp, R. R., McGowan, M. L., Verma, J. A., Landy, D. C., McAdoo, S., Carson, S. A., Simpson, J. L., & McCullough, L. B. (2010). Moral attitudes and beliefs among couples pursuing PGD for sex selection. *Reproductive Biomedicine Online, 21*(7), 838–847.

Silver, L. M. (1997). *Remaking Eden: Cloning and beyond in a brave new world*. New York: Avon Books.

Swierstra, T., & Waelbers, K. (2012). Designing a good life: A matrix for the technological mediation of morality. *Science and Engineering Ethics, 18*(1), 157–172.

Tachibana, M., Amato, P., Sparman, M., Woodward, J., Melguizo Sanchis, D., Ma, H., Marti Gutierrez, N., Tippner-Hedges, R., Kang, E., Lee, H.-S., Ramsey, C., Masterson, K., Battaglia,

D., Lee, D., Wu, D., Jensen, J., Patton, P., Gokhale, S., Stouffer, R., & Mitalipov, S. (2013). Towards germline gene therapy of inherited mitochondrial diseases. *Nature, 493*(7434), 627–631.
Torjesen, I. (2014). Government gives the go ahead for mitochondrial donation during IVF. *Bmj-British Medical Journal, 349*.
Tryggvadottir, E. A., Medek, H., Birgisdottir, B. E., Geirsson, R. T., & Gunnarsdottir, I. (2016). Association between healthy maternal dietary pattern and risk for gestational diabetes mellitus. *European Journal of Clinical Nutrition, 70*(2), 237–242.
van de Poel, I. (2001). Investigating ethical issues in engineering design. *Science and Engineering Ethics, 7*(3), 429–446.
Verbeek, P.-P. (2005). *What things do: Philosophical reflections on technology, agency, and design*. University Park: Pennsylvania State University Press.
Waelbers, K. (2011). *Doing good with technologies: Taking responsibility for the social role of emerging technologies*. Dordrecht: Springer.
Winner, L. (1986). *The whale and the reactor: A search for limits in an age of high technology*. Chicago: University of Chicago Press.
Winner, L. (1980). Do artifacts have politics? *Daedalus, 109*(1), 121–136.

Chapter 5
Regulatory Science: Between Technology and Society

José Luis Luján and Oliver Todt

Abstract The rapidly growing importance and impact of technology have made it necessary to regulate technological development in order to maximize its advantages while controlling its possible negative effects on human health and the natural environment. Scientific research has become a fundamental tool for the governance of technology. It provides knowledge about (a) the positive and negative impacts of technology and (b) the best means (public policy, regulation, etc.) for maximizing the desired effects of technology while minimizing the undesired consequences, as well as (c) monitoring regulated technologies (in order to establish the effects of regulation itself and evaluate its efficacy). Since the inception of technology regulation, its development and application have been accompanied by debate and even controversy about the importance and role of scientific knowledge in regulatory decision making. Two of the most relevant questions are: (1) what function, if any, does scientific knowledge have in regulatory data generation and decisions, and (2) whether and under what circumstances could regulatory science be considered a new type of science, distinct from (traditional) academic science. In this chapter, we will analyze these questions via two case studies, one on risk assessment (toxic substances) and the other on benefit assessment (functional foods). As our analysis shows, it is justified (while at the same time, inevitable) to take into account the ultimate goal of regulatory science, namely that of facilitating regulatory decision making, as well as the social and environmental consequences of uncertainty and inductive errors, in the resolution of methodological controversies in this sphere.

J. Luis Luján (✉) · O. Todt
Department of Philosophy, University of the Balearic Islands, Palma, Spain
e-mail: jl.lujan@uib.es; oliver.todt@uib.es

B. Laspra, J. A. López Cerezo (eds.), *Spanish Philosophy of Technology*,
Philosophy of Engineering and Technology 24,
https://doi.org/10.1007/978-3-319-71958-0_5

5.1 Introduction: The Function of Scientific Knowledge in the Regulation of Technology

The ubiquitous presence of science and technology in everyday life has turned technology into an important subject of and objective for regulation. There are several reasons for subjecting technological processes and products to control and regulation. One, of course, is the need to provide a regulated space in which technology can develop and be applied. Without proper regulation of radio frequencies, air traffic or the rules of the road, none of the related technologies (wireless communication, air or road travel) would be able to function properly. Barriers to the introduction of new technology are often related to the absence of adequate standardization and regulation, from video cassette formats, to color television systems and cell phone frequencies. Development of a technology implies development of a regulatory framework that facilitates its application. A second type of regulation is that related to the undesired or unforeseeable negative effects of technological development. These negative impacts are mostly related to human health and the environment, although, in certain cases, they may also affect social or cultural spheres (for instance, regulation or restriction of media content). A third kind of regulation, the most recent of the three, concerns the desired or positive impacts of technology. Here, the question is how regulation can help in obtaining or boosting the benefits derived from certain technological applications. In this last category can be found pharmaceuticals and medical procedures, functional foods, as well as, for instance, environmental legislation. Our interest in this chapter concerns the second and third kinds of regulation.

Science and technology are thus important objectives of what we may call the Regulatory State. Regulating technology requires scientific knowledge for decision making. In this chapter, we will present an analysis of the controversies that arise in technology regulation in relation to the generation and application of scientific knowledge. We will make use of two case studies: (a) risk assessment in the regulation of toxic or otherwise harmful substances, and (b) benefit assessment in the regulation of functional foods. The analysis of the controversies in these two cases will focus on the scientific methodologies on which data generation is based, particularly their role and legitimacy. We conclude that non-epistemic values exert an influence in both risk and benefit assessment and that their influence can have positive effects, as they may enable the generation of knowledge that is more directly relevant for regulatory decision making.

5.2 Regulatory Science and the Governance of Technology

Growing preoccupation with the negative impact of technology on health and the natural environment has led to a process of intense regulatory activity starting in the 1960s and 1970s (Hackett et al. 2008; Sarewitz 1996; Dickson 1988). An entire new field of science developed due to the need for scientific data as an important input for decision making in the regulatory context, namely regulatory science (Jasanoff 1990; Bipartisan Policy Center 2009). This field includes research into risks and impacts of specific technological applications (risk assessment), from nuclear power generation to pesticides, but also other, related studies, such as environmental impact assessment and technology assessment. Regulatory science thus constitutes scientific research whose goal is to generate knowledge that can be used in the definition of public policy and regulation. While public policies related to scientific-technological development had already existed before the 1960s, these had been concerned almost exclusively with the promotion of research by way of direct financial support (*policy for science*). Public concerns about health and environmental impacts gave rise to what is called *science for policy*: using scientific knowledge in decision and policy making (Luján and Moreno 1996).

By becoming more directly involved in the processes of decision and law making, science has entered the public sphere (Todt and Luján 2008; Luján and Todt 2000), with all the concomitant controversies and public debate concerning its objectives and results, as well as its function in regulatory decisions. This public dimension of science, and even more so regulatory science, has become increasingly relevant (Luján and López Cerezo 2004; López Cerezo et al. 1998).

The diverse tiers of government concerned with regulating science and technology have generally responded to citizen concerns about the negative impacts of technological development by implementing "technocratic" policies and regulations, in the sense of a rapidly growing relevance of expert knowledge in policy (Luján and Briggle 2005; Levidow and Carr 2007). On the operational level, they have introduced the concept of the (regulatory) agency. These agencies, such as the U.S. Food and Drug Administration (FDA), the U.S. Environmental Protection Agency (EPA), the European Food Safety Authority (EFSA) and the European Medicines Agency (EMA), are organizations established by a government or an intergovernmental body. They form part of the executive branch, but tend to have a semi-autonomous administrative status. Agencies are created to answer to specific policy and regulatory needs (Jordana and Levi-Faur 2004).

Basing public policy, regulatory decision making and advice from governmental or semi-autonomous agencies on scientific knowledge is in fact a product of the stated aim of depoliticizing debate about technology and its impacts: science as a (supposedly) neutral arbiter will allow for independent decisions that all stakeholders can accept, precisely because the advice is independent of government and day-to-day politics. In practice, however, using science to depoliticize decisions about technology has led to the politicization of scientific knowledge itself. In other words, the controversies about technical change have shifted from the technology under scrutiny to its regulation and, more specifically, to the scientific knowledge on which regulation is based (Todt 2011). Such debate covers all aspects of science and scientific practices, including their accountability, legitimacy, neutrality and credibility (Jasanoff 1990; Cranor 1993; Michaels 2008).

Another important consequence of the involvement of science in the public sphere is related to the methodological characteristics of regulatory science. The question is whether regulatory science possesses (or could possess) methodological characteristics that distinguish it from academic science (Todt et al. 2010). A number of authors have given an affirmative answer to this question (Elliott 2011; Cranor 1993; Shrader-Frechette 2004; Wandall 2004; Douglas 2009; Todt and Luján 2014a). Most of their proposals are based on the notion of taking into account the social, health and environmental consequences of scientific uncertainty and inductive errors. The latter are directly related to the possible errors that scientists may commit when generating decision-relevant knowledge (Mayo 1991). An example of such an error could be the (regulatory) decision of not restricting the placing on the market of a particular chemical compound because the scientific data has (erroneously) indicated that the substance is harmless. Another example would be recommending the consumption of a functional food because of its (alleged) positive health effects, even though the data on which this recommendation is based are incorrect. The authors we have cited above argue that taking into account the consequences of these kinds of errors in decisions implies that certain scientific methodologies are more (or less) suitable than others in the context of regulatory decision making (Todt and Luján 2011).

5.3 Risk Assessment: Toxicity and Regulation

We will now analyze the principal controversies in risk assessment with respect to four aspects: the conceptualization and identification of risk, the selection of alternative guidelines for data extrapolation, the establishment of standards of evidence and, finally, the determination of the burden of proof.

5.3.1 Conceptualizing Risk

"Risk" does not exist in the world. It depends on our conceptualizing events or phenomena of the world as such, because they affect our health, natural environment, economy, religious or ideological beliefs, or institutions (López Cerezo and Luján 2013). In different contexts and at different times, particular events or phenomena have been characterized as risks in different (or even vastly different) ways. In one place and time, mad cow disease has been considered a risk (Europe, in the 1980s and 1990s), while in other places and times, dying due to the lack of sufficient or adequate food or water would be understood to be a risk (Sahel zone famines in the 1970s). In this sense, the "risk society" (Beck 1986) could be interpreted as a concept mainly applicable to affluent, technologically advanced societies.

This introduction already points to the social nature of the concept of risk. In fact, conceptualizing something as risk – and evaluating and managing it as such – is subject to all kinds of social dynamics and is highly influenced by norms and values. Rescher (1983) argues that risk assessment is not – and cannot be –

value-free. Simply determining the "magnitude" or "relevance" of a specific risk is impossible without some kind of normative valuation, as it implies comparatively assessing the negative impact or harm of an event. Depending on the social actors involved, this evaluation may lead to very different outcomes.

In the particular context of industrialized (and even some industrializing) societies, whatever is considered a risk is usually subject to regulation. Regulating or controlling risks, however, does not resolve the social controversies related to risk. Debate typically shifts from the risks in question to the regulatory processes. Different social actors commonly have different interpretations: for some, a proposed regulation could constitute a case of "overregulation", while for others, the very same regulation may be concomitant to "underregulation" (Luján and Todt 2015).

Underregulation leads, for instance, to substances that entail risks being authorized for market release, thereby imposing (social, environmental or health) costs on consumers or society at large. Overregulation, to the contrary, imposes costs on innovators and producers and leads to products that are beneficial and without risk not being commercialized. Over- and underregulation are directly related to the frequency of type I and II errors. Debate in the field of risk assessment is commonly about different methodological strategies for data generation and the implications for the frequency of occurrence of these two types of statistical errors (Steel 2010; Douglas 2000).

We will now analyze three of the most important debates in risk assessment, all of which constitute different forms of a single concern: how to distribute the generation of type I and type II errors in regulation-relevant data acquisition and analysis. The debate has to do with inference guidelines, standards of evidence, and the burden of proof (see also Todt and Luján 2015a).

5.3.2 Inference Guidelines

A typical (though inevitable) problem in risk assessment is the need to extrapolate data (Mayo 1991; Shrader-Frechette 1994). On the basis of data obtained through laboratory research under controlled conditions, it is usually necessary to hypothesize the effects a substance will have under real-life exposure conditions, i.e., when used by its intended consumers in a foreseen context. In many cases, it is simply not possible to generate the regulation-relevant data under real-life exposure scenarios, which usually imply long-term and very low dose exposures.[1] Regulatory agencies therefore have to adopt models explicitly for extrapolating data, which they codify as inference guidelines (Shrader-Frechette 1994). The decision to adopt one or

[1] This is due to the sources of information. These are mostly bioassays (research on animal models) and epidemiological data (real exposure situations, but with different levels of exposure as compared to the desired data, including the effects of industrial accidents, exposure to other, similar substances, etc.).

another guideline for extrapolating data necessarily implies different frequencies of false positives or false negatives, leading in the one case to overregulation, and in the other to underregulation (Steel 2008, 2011).

5.3.3 Standards of Evidence

To classify something as a risk will depend on the level of evidence we demand in order to consider that there is sufficient proof that, for instance, a substance has negative health effects (López Cerezo and Luján 2000). A simple example is saccharine, which produces cancer in laboratory animals. It is currently banned from the market in Canada yet authorized in the United States (Vallverdú 2005), showing that different regulators may interpret the level of evidence necessary for taking a particular decision differently.

Standards of evidence differ not only from one country to the next, but also vary over time. The standards of evidence required by U.S. regulatory authorities to establish the carcinogenic potential of chemical substances are a case in point. In the 1970s, human evidence from epidemiological studies was considered necessary to establish that a substance was carcinogenic. In the 1980s, data from bioassays were also admitted as sufficient evidence. In the 1990s, the recommendations changed once again, admitting all types of evidence in humans and animals, including mechanistic data, finally basing decisions on the aggregation of all the available data (weight-of-evidence approach, see next paragraph) (Gibb 1997; IARC 2015).

A number of researchers and regulators whose concerns focus on underregulation, i.e., false negatives, have proposed methodological changes in risk assessment that imply a certain relaxation of the standards of evidence. These new regulation-relevant methodologies do not have the same level of demand for accuracy that is typical of academic laboratory research. Rather, they are geared towards facilitating the production of relevant data for regulation, as well as the use in regulation of already existing data. These methodologies include: (a) short-term tests (designed for rapidly testing large numbers of substances that may be harmful for health or the environment (Cranor 1995), (b) the weight-of-evidence approach (taking decisions on the basis of large numbers of different studies produced under differing circumstances and based on different methods (Haack 2014; Luján and Todt 2008), and (c) structure-activity relationships (extrapolating the effects of a particular substance on the basis of its structural or molecular similarity to other substances known to possess harmful effects (Cranor 2008).

5.3.4 *Burden of Proof*

Among those concerned with underregulation, one important methodological proposal is the shifting of the burden of proof from regulators to producers (Tickner 1999; Wandall 2004). Normally, it is the regulators of a new technological product or process who have to demonstrate its harmfulness in order to justify subjecting it to regulation. Regulatory agencies, however, have limited resources at their disposal and are usually unable to conduct sufficiently conclusive scientific tests (among other reasons, because some of the effects may only show up after years or decades). Shifting the burden of proof to the producer means that it will be the developers or advocates of a new product or technological process who have to demonstrate that it does not entail any risks. One current example of the adoption of this methodology in regulatory decision making is the European Union (EU) REACH regulation (Registration, Evaluation, and Authorization of Chemical Substances) (European Parliament and Council 2006a). REACH adopts a precautionary stance that in practice involves shifting the burden of proof: all chemical substances, including those that are already authorized, are to be subjected to evaluation for their possible negative health and environmental effects (Luján and Todt 2012).

5.3.5 *Non-epistemic Values and Methodological Controversy*

These examples show that the concern with overregulation and underregulation, as well as their respective social consequences, leads to methodological controversies in risk assessment. As we have seen, a number of authors argue that overregulation can impose additional costs on technology innovators. Thus, overregulation may lead to a reduction in a society's capacity for generating wealth, which ultimately means fewer resources for social services, public health and education (Sunstein 2005, 2014). Underregulation, on the other hand, leads to the authorization of substances that may eventually turn out to be harmful for health and the environment. In short, controversies about over- and underregulation are directly related to non-epistemic values such as the protection of human health and the environment. This means that methodological debate in this context is never value-free.

5.4 Benefit Assessment: Health Claims Regulation and Functional Foods

Our case study on health claims shows that the controversies concerning the function of scientific knowledge in benefit assessment are similar to those in risk assessment, as they are related to methodological choice and the social consequences of inductive errors.

5.4.1 Benefits and Functional Foods

The interrelation between methodological controversy and the social consequences of regulation can be seen not only in risk assessment, but also in benefit assessment. Benefit assessment consists in acquiring and evaluating data about the possible benefits of the consumption or use of particular products. Besides drug trials, the most important current example for benefit assessment is the regulation of functional foods (Hobbs et al. 2014; van der Meulen 2014). These are food products which, in addition to their nutritional value as foods, confer upon the consumer specified additional health benefits, such as the reduction of cholesterol levels via the ingestion of plant sterols or beta-glucans.

Food products that are designed to have these additional properties are subject to regulation in order to be labeled as beneficial. The statements about their beneficial properties are called health claims. Such health claims can influence consumption patterns for functional foods. While they are aimed at the individual benefits obtained by the consumer of a particular food, they also imply potentially important effects for overall public health, due to the aggregate benefits of widespread consumption of such food products (Gregori and Gafare 2012, Todt and Luján 2016).

5.4.2 European Health Claims Regulation and Controversy

As in the case of risk, whatever is considered a benefit depends on the demanded level of required evidence (Tijhuis et al. 2012). Since their introduction in Japan in the 1980s, health claims have been subject to regulation in most industrialized countries (Bagchi 2014). In the EU, they are subject to a European Commission regulation (NHCR, Nutrition and Health Claims Regulation: European Parliament and Council 2006b) that is applicable in all EU Member States. Health claims are authorized based on prior scientific substantiation (benefit assessment).

The European regulatory framework implicitly demands the establishment of a causal relationship between intake and outcome in order to be able to authorize a claim on a food or ingredient (Aggett 2012). The regulatory operationalization of this demand for causality is embodied in a hierarchy of evidence, organized according to which data most convincingly show the existence of a causal link. This hierarchy of evidence is part of the European regulatory process. The evaluation of the data, however, depends on the level of confidence with respect to the particular methodology used for generating such data. In other words, the hierarchy of evidence turns into a hierarchy of methodologies for data generation that can legitimately be used for the purpose of substantiation. At the top of this hierarchy, we find Randomized Controlled Trials (RCTs),[2] followed by other types of human

[2] RCTs are human intervention studies based on subdividing the sample by way of a random process into an experimental group (members receive the substance under test) and a control group (members receive a placebo).

intervention studies. The next level down comprises observational studies. Finally, there are mechanistic analyses and other types of methodologies, such as computer simulations.

The demand by regulators for causal data has contributed to a controversy between many nutrition scientists and the European regulators. Nutrition scientists criticize the EU regulation for conferring – as they understand it – special privilege on one specific methodology (RCTs), which they do not consider the most suitable procedure for producing data in nutrition research (Blumberg et al. 2010; Biesalski et al. 2011). Specifically, nutrition scientists argue that RCTs in food research are limited by: (a) severe difficulty in designing control groups, as there is no "nutrient-free" state similar to a drug-free state in pharmaceuticals trials, (b) the functional complexity and multifunctionality of food ingredients, as well as the interaction of the ingredient in question with the entire food matrix, (c) not being able to provide answers to "how" and "why" questions (they only show that an association exists, but do not explain why it exists, i.e., they do not provide data on the underlying mechanisms), and (d) operational complexity, due to the enormous resources needed to conduct RCTs and problems such as the non-compliance of participants (Heaney 2008; Richardson 2012).

5.4.3 Issues in the Controversy

There are three main issues in this controversy (Todt and Luján 2015b):

1. Standards of evidence. European benefit assessors demand evidence that allows the minimizing of false positives, in other words, RCTs. The critics consider the RCT standard very (or even too) demanding in the context of nutrition science. They argue that the RCT methodology was adopted in the field of nutrition uncritically, even though it was developed in – and for – a very different context (drug trials).
2. Methodologies. The EU regulatory framework ranks methodologies according to type, with RCTs at the top due to their (supposed) inherent capacity of producing causal data. However, the critics of EU regulatory practice consider that RCTs are not a suitable methodology for generating data in nutrition science because this methodology is very difficult to apply to nutrition studies and is incapable of correctly capturing aspects that are crucial in nutrition (for instance, the interaction between different ingredients).
3. Extrapolation. To obtain EU authorization, it is necessary to present data that allow the unequivocal determination of an improvement in some concrete and measurable parameter as a result of the consumption of the ingredient in question. In many cases, it is not possible to measure the desired final outcome directly; for instance, when this outcome lies decades in the future (prevention of

the risk of contracting cancer). In such cases, biomarkers serve as a substitute for the final outcome. This is comparable to extrapolation in risk assessment and entails similar limitations. Another related problem is that ingredients often produce multifactorial effects, which cannot be measured by one single outcome or marker.

4. Weight of Evidence. Counter to EU regulators, the critics argue that the assessments of all available studies should be taken into account, even those based on different methodologies. Individual study quality and the relevance of the data will be decided by applying a weight-of-evidence approach in data analysis.

5.4.4 Analysis of the Controversy

From the EU regulatory standpoint, all the effects for which a causal relationship between consumption of a particular, defined ingredient and a single outcome measure (for instance, a biomarker), cannot be established unequivocally are nonexistent. This leads to an increase in false negatives, because some ingredients that are positive for health will not be considered as such in assessments.

The methodological controversy is directly related to different regulatory strategies. One such strategy is to protect consumers from false or erroneous information (minimizing false positives, i.e., not authorizing products without benefits). The alternative strategy consists in giving consumers access to a large number of products that carry health claims (minimizing false negatives, i.e., avoiding not conferring claims on products that are beneficial).

In the former case, consumers can be sure that the products carrying claims really have the advertised health properties. This comes at a price, however: some of the products that have beneficial effects and whose consumption would have positive effects on individual and public health do not carry claims. This means that the consumption patterns related to these products may not be changed and the associated (individual and general) benefits may not be generated.

In the latter case, many more food products will carry health claims, even though some of them are not actually beneficial for health. However, the overall change in consumption patterns related to the labeled products is ultimately hoped to have more beneficial effects than that of only labeling those products whose beneficial properties we can be absolutely sure of. In other words, the price that consumers pay in this case is that some of the products they buy – obviously at a premium – will not have any additional benefits.

5.5 Conclusions: Is It Possible to Improve on Current Regulatory Science?

As we have seen, the relationship between decisions regarding scientific methodologies for data generation and regulatory decisions is not straightforward. In regulation, whether of risks or benefits, there always exists the possibility of error, either of underregulation (too permissive) or overregulation (overly restrictive). The social and environmental consequences (which flow from the concomitant regulatory decisions) of these errors are not the same (Todt and Luján 2014b). In other words, decisions regarding whether to apply one scientific methodology or another may lead to the generation of different data sets, which may subsequently justify one regulatory decision or another (producing different non-epistemic outcomes).

As we have already seen, it is always necessary in the process of scientific research to take decisions on inference guides, biomarkers, standards of evidence, and other elements that invariably exert an influence on the data generated. The process of research is not epistemically determined (López Cerezo and Luján 2012). Appealing to non-epistemic values in methodological decisions is thus legitimate. In regulatory science, this means that non-epistemic concerns about the ultimate social and environmental effects of regulatory decisions can legitimately influence methodological choice. Furthermore, this influence is precisely what makes regulatory science useful for decision making, i.e., it facilitates the generation of the type of knowledge necessary for regulatory decisions.

Acknowledgments This study has received support from the Spanish Government's State Secretariat for Research, Development and Innovation (research projects: *La evaluación de beneficios como ciencia reguladora* [FFI2013-42154-P], *Estandáres de prueba y elecciones metodológicas en la fundamentación científica de las declaraciones de salud* [FFI2017-83543-P]) and European Commission ERDF funds.

References

Aggett, P. (2012). Dose-response relationships in multifunctional food design: Assembling the evidence. *International Journal of Food Science and Nutrition, 63*(S1), 37–42.

Bagchi, D. (Ed.). (2014). *Neutraceuticals and functional food regulations in the United States and around the world.* Amsterdam: Elsevier.

Beck, U. (1986). *Risikogesellschaft.* Frankfurt A.M.: Suhrkamp.

Biesalski, H. K., et al. (2011). 26th Hohenheim consensus conference, September 11, 2010 Scientific substantiation of health claims: Evidence-based nutrition. *Nutrition, 27*(10 Suppl), S1–20.

Bipartisan Policy Center. (2009). *Improving the use of science in regulatory policy.* http://bipartisanpolicy.org/wp-content/uploads/sites/default/files/BPC%20Science%20Report%20fnl.pdf

Blumberg, J., Heaney, R., Huncharek, M., Scholl, T., Stampfer, M., Vieth, R., et al. (2010). Evidence-based criteria in the nutritional context. *Nutrition Review, 68*, 478–484.

Cranor, C. (1993). *Regulating toxic substances.* New York: Island Press.

Cranor, C. (1995). The social benefits of expedited risk assessment. *Risk Analysis, 15*, 353–358.

Cranor, C. (2008). *Toxic torts: Science, law, and the possibility of justice*. Cambridge: Cambridge University Press.
Dickson, D. (1988). *The new politics of science*. Chicago: University of Chicago Press.
Douglas, H. (2000). Inductive risk and values in science. *Philosophy of Science, 67*, 559–579.
Douglas, H. (2009). *Science, policy, and the value-free ideal*. Pittsburgh: University of Pittsburgh Press.
Elliott, K. (2011). *Is a little pollution good for you? Incorporating societal values in environmental research*. New York: Oxford University Press.
European Parliament and Council. (2006a). Regulation (EC) No. 1907/2006 of 18/12/2006, concerning the registration, evaluation, authorisation and restriction of chemicals. *Official Journal of the European Communities, L.136*, 3–280.
European Parliament and Council. (2006b). Regulation (EC) 1924/2006 of the European Parliament and of the Council of 20 December 2006 on nutrition and health claims made on foods. *Official Journal of the European Union, L.404*, 9–25.
Gibb, J. (1997). Epidemiology and cancer risk assessment. In V. Molak (Ed.), *Fundamentals of risk analysis and risk management* (pp. 23–31). Boca Raton: Lewis Publishers.
Gregori, D., & Gafare, C. (2012). Multifunctional food: Medical evidence and methodological notes on substantiating health claims? *International Journal of Food Sciences and Nutrition, 63*(S1), 29–36.
Haack, S. (2014). *Evidence matters. Science, proof, and truth in the law*. Cambridge: Cambridge University Press.
Hackett, E. J., Amsterdamska, O., Lynch, M., & Wajcman, J. (2008). *The handbook of science and technology studies* (3rd ed.). Cambridge, MA: MIT Press.
Heaney, R. (2008). Nutrients, endpoints, and the problem of proof. *Journal of Nutrition, 138*, 1591–1595.
Hobbs, J., Stavroula, M., Sogah, E., & Yeung, M. (2014). *Regulating health foods. Policy challenges and consumer conundrums*. Cheltenham: Edgar Elgar.
IARC, International Agency for Research on Cancer. (2015). *IARC monographs on the evaluation of carcinogenic risk to humans*. Lyon: World Health Organization.
Jasanoff, S. (1990). *The fifth branch. Science advisers as policy makers*. Cambridge, MA: Harvard University Press.
Jordana, J., & Levi-Faur, D. (Eds.). (2004). *The politics of regulation. Institutions and regulatory reforms for the age of governance*. Cheltenham: Edward Elgar.
Levidow, L., & Carr, S. (2007). Europeanising advisory expertise: The role of "independent, objective and transparent" scientific advice in agri-biotech regulation. *Environment and Planning C, 26*, 880–895.
López Cerezo, J. A., & Luján, J. L. (2000). *Ciencia y política del riesgo*. Madrid: Alianza.
López Cerezo, J. A., & Luján, J. L. (2012). Ciencia y valores en la regulación del cambio tecnológico. In E. Aibar, M. A. Quintanilla, & M.A. (Eds.), *Ciencia, tecnología y sociedad. Enciclopedia Iberoamericana de Filosofía* (pp. 277–302). Madrid: Trotta.
López Cerezo, J. A., & Luján, J. L. (2013). A philosophical approach to the nature of risk. In F. Lan, F. G. Wallner, & A. Schulz (Eds.), *Concepts of a culturally guided philosophy of science: Contributions from philosophy, medicine and science of psychotherapy* (pp. 161–179). Frankfurt am Main: Peter Lang Verlag.
López Cerezo, J. A., Méndez, J. A., & Todt, O. (1998). Participación pública en política tecnológica: problemas y perspectivas. *Arbor, 159*, 279–308.
Luján, J. L., & Briggle, A. (2005). Environmental regulation. In C. Mitcham (Ed.), *Encyclopedia on science, technology and ethics* (pp. 670–676). New York: Macmillan Reference USA.
Luján, J. L., & López Cerezo, J. A. (2004). De la promoción a la regulación. El conocimiento científico en las políticas públicas de ciencia y tecnología. In J. L. Luján & J. Echeverría (Eds.), *Gobernar los riesgos. Ciencia y valores en la sociedad del riesgo* (pp. 75–98). Madrid: Biblioteca Nueva.

Luján, J. L., & Moreno, L. (1996). El cambio tecnológico en las ciencias sociales: el estado de la cuestión. *Revista Española de Investigaciones Sociológicas, 74*, 127–161.

Luján, J. L., & Todt, O. (2000). Ciencia, tecnología y nuevos movimientos sociales. In M. Medina & T. Kwiatowska (Eds.), *Ciencia, tecnología/naturaleza, cultura en el siglo XXI* (pp. 95–109). Anthropos: Barcelona.

Luján, J. L., & Todt, O. (2008). Ciencia precautoria y la "fabricación de incertidumbre". *Theoria, 23*, 307–317.

Luján, J. L., & Todt, O. (2012). Precaution: A taxonomy. *Social Studies of Science, 42*(1), 143–157.

Luján, J. L., & Todt, O. (2015). The role of values in methodological controversies: The case of risk assessment. *Philosophia Scientiae, 19*, 45–56.

Mayo, D. G. (1991). Sociological versus metascientific views of risk assessment. In D. G. Mayo & R. D. Hollander (Eds.), *Acceptable evidence: Science and values in risk management* (pp. 249–279). Oxford: Oxford University Press.

Michaels, D. (2008). *Doubt is our product*. Oxford: Oxford University Press.

Rescher, N. (1983). *Risk: A philosophical introduction to the theory of risk evaluation and management*. Langham: University Press of America.

Richardson, D. (2012). Preparing dossiers: Strength of the evidence and problems of proof. *Proceedings of the Nutrition Society, 71*, 127–140.

Sarewitz, D. (1996). *Frontiers of illusion: Science, technology, and the politics of progress*. Philadelphia: Temple University Press.

Shrader-Frechette, K. (1994). *Ethics of scientific research*. Lanham: Rowman & Littlefield.

Shrader-Frechette, K. (2004). Using metascience to improve dose-response curves in biology: Better policy through better science. *Philosophy of Science, 71*, 1026–1037.

Steel, D. (2008). *Across the boundaries. Extrapolation in biology and social science*. Oxford: Oxford University Press.

Steel, D. (2010). Epistemic values and the argument from inductive risk. *Philosophy of Science, 77*, 14–34.

Steel, D. (2011). Extrapolation, uncertainty factors, and the precautionary principle. *Studies in History and Philosophy of Biological and Biomedical Sciences, 42*, 356–364.

Sunstein, C. (2005). *Laws of fear: Beyond the precautionary principle*. Cambridge: Cambridge University Press.

Sunstein, C. (2014). *Valuing life. Humanizing the regulatory state*. Chicago: University of Chicago Press.

Tickner, J. (1999). A map toward precautionary decision making. In C. Raffensberger & J. Tickner (Eds.), *Protecting public health and the environment. Implementing the precautionary principle* (pp. 162–186). Washington, DC: Island Press.

Tijhuis, M. J., de Jong, N., Pohjola, M., Gunnlaugsdóttir, H., et al. (2012). State of the art in benefit-risk analysis: Food and nutrition. *Food and Chemical Toxicology, 50*, 5–25.

Todt, O. (2011). The limits of policy: Public acceptance and the reform of science and technology governance. *Technological Forecasting and Social Change, 79*, 902–909.

Todt, O., & Luján, J. L. (2008). A new social contract for technology? – On the policy dynamics of uncertainty. *Journal of Risk Research, 11*, 509–523.

Todt, O., & Luján, J. L. (2011). Dinámica de los conceptos reguladores. Factores cognitivos y no cognitivos en el contexto de la toma de decisiones sobre riesgos tecnológicos. *Éndoxa, 27*, 317–336.

Todt, O., & Luján, J. L. (2014a). Analyzing precautionary regulation: Do precaution, science and innovation go together? *Risk Analysis: An International Journal, 34*, 2163–2173.

Todt, O., & Luján, J. L. (2014b). Values and decisions: Cognitive and non-cognitive values in knowledge generation and decision making. *Science, Technology, & Human Values, 39*, 720–743.

Todt, O., & Luján, J. L. (2015a). Non-cognitive values and methodological learning in the decision-oriented sciences. *Foundations of Science*. https://doi.org/10.1007/s10699-015-9482-3.

Todt, O., & Luján, J. L. (2015b). The role of epistemic policies in regulatory science: Scientific substantiation of health claims in the European Union. *Journal of Risk Research*. https://doi.org/10.1080/13669877.2015.1100661.

Todt, O., & Luján, J. L. (2016). ¿Bueno para la salud? Un análisis de los requisitos de sustanciación científica en la regulación europea de las declaraciones de salud. *Salud Pública de México, 58* (in press).

Todt, O., Rodríguez Alcázar, J., & Luján, J. L. (2010). Practical values and uncertainty in regulatory decision making. *Social Epistemology, 24*, 349–362.

Vallverdú, J. (2005). La difícil consecución de la evidencia científica: La evaluación de riesgos de la sacarina. *Redes, 11*(21), 77–118.

van der Meulen, B. (Ed.). (2014). *EU Food Law handbook*. Wageningen: Wageningen Academic Publishers.

Wandall, B. (2004). Values in science and risk assessment. *Toxicology Letters, 152*, 265–272.

Chapter 6
Practices and Knowledge: Philosophy of Biomedicine, Governance and Citizen Participation

Esther Ortega Arjonilla, Silvia García Dauder, Nuria Gregori Flor, and Eulalia Pérez Sedeño

Abstract The present chapter addresses the emergence of new forms of governance and citizen participation in the context of certain biotechnologies. On the basis of three case studies, we have mapped different models of public participation in health issues (especially health activism and Internet users, though not exclusively): identifying and analyzing actors, their mutual relationships (mainly those between "patients"/activists groups and the biomedical community), strategies and forms of participation, the exchange and circulation of "expert/lay" knowledge (and the role of the Internet in these processes), and the different activities and forms of knowledge production by "lay" citizen groups (or individuals). We draw special attention to the epistemic challenge resulting from these mixed forms of knowledge production derived from "experiential expertise", "epistemic communities" and "evidence-based activism". We detect "epistemic correctives", demands regarding "undone science", and hidden innovations in these types of citizen participation.

E. Ortega Arjonilla (✉)
Tufts-Skidmore Spain, Madrid, Spain
e-mail: esther.ortega@tufts-skidmore.es

S. García Dauder
Social Psychology, Rey Juan Carlos University of Madrid, Madrid, Spain
e-mail: dau.dauder@urjc.es

N. Gregori Flor
Health Department Elche-Hospital General, University of Valencia, València, Spain
e-mail: interemergen@gmail.com

E. Pérez Sedeño
CCHS-CSIC, Madrid, Spain
e-mail: eulalia.psedeno@cchs.csic.es

B. Laspra, J. A. López Cerezo (eds.), *Spanish Philosophy of Technology*,
Philosophy of Engineering and Technology 24,
https://doi.org/10.1007/978-3-319-71958-0_6

6.1 Introduction

Science and technology play a crucial role in our lives and in both social and economic development. This has led Western countries – in our case, the EU – to pose a need for citizen participation in the creation and implementation of public actions as well as in the construction of a *European research area* derived from new forms of cooperation at different government levels: European, national, regional and local. This has been called *governance*.

According to the *European Commission White Paper* (COM 2001), "good governance" is a set of rules and procedures that influence the exercise of power within the European area from a top-down approach. However, biomedicine has taken on a new dimension due to the multiplicity of actors and levels involved. Consequently, 'science and technology regulation systems are being reformulated: they no longer operate under "government" and "public policies", but start to assume new characteristics encompassed within the concept of "governance"' (Todt 2006), also in a down-top fashion.

The new rationales underlying governance and citizen participation have produced several transformations and hence new forms of knowledge creation and disclosure. There is an increasing number of examples in which the democratization of "expertise", due to the inclusion of "lay" knowledge (Epstein 1996), is contributing to a "better science" (as far as objectivity, responsibility and social inclusion are concerned), thereby shifting from "deficit to dialogue and from dialogue to participation" (Bucchi 2015). Over the past few decades, efforts by patient groups and citizen movements have taken on increasing importance by participating in and sometimes even challenging current health care practices and biomedical knowledge. At the same time, the information and communication technologies (ICTs) have contributed to wider-ranging access to and distribution of knowledge, as well as to the participation of these new actors within virtual communities. This collection of multiple voices (active patients or caregivers, support groups, patient associations, activist groups and health social movements) calls for new rationales of democratic participation, healthcare system organization and both the regulation and distribution of medical knowledge.

Different classifications have been established concerning health movements and patient associations and their interaction with medical research and health care. Callon and Rabeharisoa (2003) differentiate between "auxiliary, partner and opponent organizations". This classification enables us to identify the extent to which such scientific knowledge production is complementary, instrumental or opposing (productive and/or conflictive interactions). Further analyses refer to either *biosocialities* (Rabinow 1996), social communities based on genetic or biological conditions (Rose 2007), or patient communities in their role as potential users of new biomedical technologies. On the other hand, Brown (2008) identifies three types of health social movements: movements which seek equitable *access* to health care; *embodied health movements*, which challenge mainstream science regarding the etiology, diagnosis, treatment and prevention of certain diseases; and health movements based on *differences,* who report health inequalities based on categories such as "race"/ethnicity, gender, class or sexual orientation.

However, these new participation procedures also reveal an epistemic dimension, questioning the unique legitimacy of biomedical sciences and knowledge to describe, know and prescribe everything related to health. New knowledge arises from this multiplicity of voices and the use of new technologies: the recognition of those affected as experts on themselves and their "illnesses" has meant the claim of experience and its associated knowledge as evidence. We likewise find the joint participation of affected people, relatives and researchers in a new hybrid scientific research approach. Wehling et al. (2015) list a variety of science activities connected to patient associations and health social movements: fostering research demands, providing scientific resources, activating research lines and identifying blind spots, mediating and contacting researchers from other countries or fields of knowledge, and producing changes in knowledge and paradigm shifts.

In the light of this citizen participation context, it is interesting to analyze the production, circulation and exchange of knowledge between "experts" and "laypeople", and hence the questioning and breaking down of this rigid distinction through patient "expertise" processes (Epstein 1996). Likewise, new forms of hybrid knowledge emerge, such as "evidence-based activism" (Akrich et al. 2015) and "epistemic communities" (Akrich 2010), in which "lay'" expertise and biomedical knowledge merge. These lay experts 'are not anti-science but pro-knowledge: they seek to re-value forms of knowledge that professional science has excluded' (Epstein 1996, p. 13).

6.2 Intersex Groups, Citizen Participation and Health Paradigm Shifts

We are currently facing a paradigm shift regarding the managing of intersexual variations, from a biomedical paradigm to one that is both psycho-social and human rights-oriented, varying according to different cultural, political and health settings. Social participation processes in health matters have been crucial in this shift, together with the circulation and exchange of "expert/lay" knowledge. Health and social activists, in particular, have played a key role, not only in the production of new knowledge, but also in the transformation of practices.

Traditionally, people with intersex variations used to appear in scientific literature as "exotic pathological cases", subject to medical interventions without a voice in the matter. These practices were supported by a biomedical handling protocol, "Money's Protocol", in which prescription secrecy and stigmatization ruled over the medical process (lack of information, iatrogenic doctor-patient communication and "parades" of doctors and residents that observe and photograph the wrongly called "ambiguous" genitalia) (Kessler 1998).

In her article "Hermaphrodites with Attitude", ISNA (Intersex Society of North America) promoter Cheryl Chase (1998) described her process of biographic transformation from silence and loneliness to action and collective mobilization via meetings over the Internet with other people with similar experiences. Thus, in the US context, the concept *intersex*, which initially arose within clinical practice, ended up supporting a collective and political identity (García-Dauder and Romero 2012). From ISNA, different groups have organized intersex activism. OII (Organization Intersex International) is worthy of special mention in this respect. This organization fundamentally claims for: a transformation of medical protocol aimed at avoiding irreversible and unnecessary surgeries, especially in the case of infants; the destigmatization of medicalization processes; and self-determination, i.e., the right of intersexual infants to determine their own sexual identity at a suitable time, without governmental or medical interference.

The Internet and other forms of participation have enabled "medical knowledge objects" to *object* (Latour 2000) and resist medical protocols and discourses, seeking to exchange knowledge and promote a collectivized and politicized identity (Still 2008). This has triggered changes in our knowledge of subject-object relationships. In some cases, the "objects of knowledge" (now activists or intersex groups) have become real "experts" on the "subjects of knowledge" (the doctors), i.e., "experts on experts", forming groups that keep up-to-date with respect to the scientific literature and who engage in critical discussion (García-Dauder and Romero 2012). In many individual countries as well as on a global scale, what is known as "evidence-based activism" (Akrich et al. 2015) has been paramount in changing protocols and the centrality of the normative body that govern them, under the premise of "no harm", thereby also transforming science and society (Kessler 1998).

If we focus on the Spanish context, however, the scenario is quite different. Despite attempts by the LGBT movement to include the letter I (*intersex*) in its acronym, the voices and experiences of people living with an intersex condition or with differences of sex development (DSD) have rarely been heard. They have only occasionally been asked about their willingness to include their claims under the shelter of the rainbow umbrella (in most cases, the letter I is included with ignorant good will) (Gregori 2016). The changes that international human rights legislation is bringing about in other countries (like Argentina or Malta) encourage initiatives in the Spanish context, though yet again, without listening to the actual protagonists. Meanwhile, the needs of "affected" people (which is how they call themselves) and their relatives concerning particular intersex conditions/DSD are addressed in small support groups (e.g., the androgen insensitivity syndrome association *Grapsia*, with whom our research team has established a close bond of cooperation). These groups supply medical information (about conditions, treatments, assistance networks, rights, etc.) and offer adapted knowledge (expressed in accessible language) to contribute to a better understanding and hence better choices. Furthermore, they provide invaluable emotional support, which breaks down isolation and stigma, offering referents and a "second family", a secure place where they can openly talk and hear about their conditions and where they can build up their self-esteem and empower themselves (García-Dauder et al. 2015; Grapsia 2015).

Besides, these support groups, as in the case of Grapsia, provide information to health professionals and the rest of society, eventually breaking down the distinction between expert and lay knowledge. For example, they produce detailed psychoeducational material, both accessible and visible (mainly via the Internet) under the basic principle that learning and information mean the power to change. They have made professionals shift from talking *about* them to talking *with* them, as well as listening *to* them (for example in their own meetings) (Grapsia 2015). Through their claims and thanks to their research collaborations, they play a key role in correcting practices and discourses.

Likewise, associations related to specific intersex/DSD conditions as well as different intersex activist organizations have participated in a wide variety of activities related to the production of scientific knowledge (Wehling et al. 2015). They have started or sponsored research projects, established multidisciplinary cooperation networks (from social sciences, the field of bioethics, law or health sciences), and promoted psychosocial research lines, identifying "undone science" (Hess 2009), etc. Moreover, they have built their own forms of research and knowledge production forms ("lay epidemiology", collecting statistics never compiled before, creating clinical practice guidelines or designing health care procedures, etc.). Finally, especially those groups of a clearly "opponent" nature (Callon and Rabeharisoa 2003), in terms of the demedicalization of intersex variations, have introduced epistemic correctives within mainstream research. They have challenged medical definitions of health and illness, certainties about behaviors, identities, "normal" bodies and the need to undergo medical intervention. They have exerted great pressure on medicine and science to act in a socially responsible manner, thereby expanding biomedical practices and knowledge via a social, ethical and legal approach; for instance, in relation to informed consent or medical interventions in infants (Gregori 2015).

We are thus facing *hybrid* forms of knowledge: scientific/biomedical knowledge enriched with experiences and health care practices shared and exchanged collectively. Focusing on the practices of some groups and intersex associations (whether virtual or face-to-face), we can observe several changes: spaces where they can share and exchange advices and experiences; disclosure of accessible, translated biomedical knowledge imbued with experiential expert knowledge; and, finally, the production of their own knowledge as a resource to vindicate the positions these groups defend. This is what Akrich et al. (2015) have called evidence-based activism: knowledge not as something external used for the purpose of defending their claims, but as something *produced* by the groups themselves as an argumentative resource at the time of defending their claims and positions.

The discussion here invites us to reflect not only on the role of groups and social movements in the construction of "better medical science", but also as to how science itself affects these groups (e.g., co-optation risks during their negotiations to obtain improvements) (Epstein 1996). However, we cannot hide the complexity and heterogeneity of the "subjects" within these movements and the different strategies and alliances undertaken by different actors in their attempts to claim their rights and make requests. In some contexts, such as the US, we find groups like *ISNA* and

Accord Alliance which have created alliances (of either an "auxiliary" or a "partner" type) with the medical community in order to change *how* intersex is medicalized. Others, like *OII*, call for the demedicalization of intersex as well as the acknowledgement of the diversity and variability of sexed bodies (bereft of medical terminology implying sexual development "disorders"). In the Spanish context, we have followed Grapsia's progress for more than a decade. This association started out as an "auxiliary" organization to later adopt a "partner" stance. Nowadays, it may be said to have a triple role, as it also carries out "opponent" functions (even though the association itself does not claim to be activist, it has brought about changes in terms of depathologization thanks to its mediations and activities). All that has been outlined so far shows that the position of a group is not static; it depends on its own dynamics and history. We can even find a great deal of diversity among its members in terms of their own experiences as part of a group and as regards their specific intersex "condition".

The discussion here likewise leads us to face the dilemma of both participatory *representation* and *legitimacy* in the light of such a multiplicity of voices and perspectives, not only among the different groups, but also within the groups themselves. How can we represent "genuine" needs and interests in the context of such a diverse scenario without lapsing into "unique" stories? The unavoidable epistemic question emerges: who speaks on behalf of the affected people? Especially in those situations when they are not heard (because they prefer to safeguard their identity) or when they are under guardianship, "unable to talk" (as in the case of medical interventions in infants). The issue is all about including the greatest number of socially relevant perspectives and developing new research tools that allow the participation of those traditionally excluded from knowledge production. We can thus conclude that we are experiencing a transformation of the mechanisms via which someone is recognized as an "intersex expert", broadening the boundaries of legitimacy and credibility beyond the medical sector.

6.3 Trans Collectives, Citizen Participation and the Paradigm Shift in Governance Models

At least since the late 1980s, Spain has witnessed a change in the public perception of identity and differences in sexual orientation. Trans people are no longer considered "socially dangerous" either by Spanish society or by the legal system (Pineda 2008; Platero 2009; Platero and Ortega-Arjonilla 2016). However, this shift in social and legal recognition came hand in hand with the inclusion of trans people within a new biopolitical framework, in which transsexuality is considered a mental problem – transsexuality or gender dysphoria – which can only be solved via medical intervention upon the physical body.

This governance model concerning gender transition started in the 1990s and ended in 2007, with the coming into force of *Law 3/2007, dated March 15, regulating the rectification of the registered entry concerning peoples' sex,* better known as the "gender identity law". This governance framework rests on both legal and biomedical aspects, interconnected in Law 3/2007. The law regulates the change of the person's first name and the mention of gender in all official documents in compliance with two pre-requisites and two requisites per se.

The two pre-requisites are related to the people subject to this new regulatory framework, those with Spanish nationality and of legal age. At the same time, the legal requisites link the regulatory framework with its biopolitical counterpart, establishing the need for a medical diagnosis of "gender dysphoria" and 2 years of medical treatment to have access to this legal change (Law 3/2007).

From the very moment Law 3/2007 was passed, trans and queer groups claimed it was pathologizing, as it ascribed the right to identity – and thus its change – within a biopolitical framework of sexual difference, in which gender flexibility was a problem to be treated by medical intervention upon the physical body. The sexual difference paradigm states that there are only two possible sexes and genders; so anything that falls outside these categories must be pathologized and subject to medical intervention.

Biomedical practices for the regulation of gender transition have been implemented by the Spanish National Healthcare System since 1999. "*Gender Identity Disorder* Units" (UTIGs, for its Spanish acronym) have been responsible for supervising these transitions in certain autonomous communities. As it is not a medical benefit included in the general range of services provided by the National Healthcare System, these units have been created according to the criteria of each autonomous community. Currently, there are nine units in Spain (Esteva de Antonio et al. 2012), offering "medical treatment" to trans people. The UTIGs are multidisciplinary units formed by specialists belonging to the three disciplines that intervene in gender transition: psychiatry and psychology, responsible for the psychological evaluation and the diagnosis of "Gender Dysphoria"; endocrinology, which deals with hormone treatments; and plastic surgery, responsible for all the interventions included within the concept of sex reassignment surgeries (SRS). However, not all units offer the three aspects of the so-called "triadic treatment".

The role of trans activism in Spain has undergone variations from the first organized groups in the late 1980s to the broad variety of groups and associations we encounter nowadays. Likewise, positions concerning the trans issue are also diverse in the light of the biomedical paradigm. Following the classification by Callon and Rabeharisoa (2003) regarding interaction between patient associations, research and biomedical practice, we could say that the different trans groups and associations have veered from an auxiliary organization role, in both clinical and medical practice, to assume a more active stance.

However, in this chapter we will focus on the influence of those activists and groups, closer to a *queer* standpoint, who have been very critical towards Law 3/2007 since it was passed. Some of these activists created the *Spanish Network for Depathologization of Trans Identities* that subsequently had a major international impact, becoming the well-known *International Campaign Stop Trans Pathologization-2012*, currently called *International Campaign STP*. From the very beginning, the aforementioned network adopted the stance of an opponent group – according to Callon and Rabeharisoa's classification–, confronting not only clinical practice related to trans people, but also, in a broader spectrum, biomedical and legal governance towards trans issues within the State (Ortega Arjonilla et al. 2014). The main characteristic of these opponent associations, according to Callon and Rabeharisoa (2003), is that they reject any definition of their condition emanating either from the scientific or the medical community, placing their identity in the centre of the discourse and never in terms of a "disease" or "condition".

The major claims to the Spanish State pursued by the STP Campaign focused on the modification of Law 3/2007 to withdraw the requisites of diagnosis and medical treatment for the change in the registered entry concerning the person's name and gender, as well as the inclusion of reassignment processes within the benefits of the National Healthcare System. Above all, in line with our approach, they seek a change in the medical practice governing reassignment processes, from a 'medical authorization model to a more autonomous one where trans people can exert their right to decide upon their own body modification process' (Spanish Network for Depathologization of Trans Identities 2010a).

In fact, both the Network for Depathologization and the subsequent International Campaign STP can be classified under the label of opponent organizations. However, as opposed to Callon and Rabeharisoa (2003), we have argued the case that these opponent organizations have a further goal beyond their detracting nature (Ortega Arjonilla 2014). In the case of the *Spanish Network for Depathologization of Trans Identities,* this organization developed an alternative healthcare model based on the experiences of its members within the model of the Spanish Public Healthcare System called *Best Practices Guide to Trans Health Care in the National Healthcare System* (2010).

This guide was a document that intended to offer an alternative proposal to the *Standards of Care* internationally in force at that time since 2001: the *Standards of Care for Gender Identity Disorders* (6th Version). The *Guide* represents a milestone for the inclusion of those who lack previous scientific expertise, so-called *laypeople*, in the discussion and creation of a health care framework from the concerned group perspective. Akrich et al. (2015) call it evidence-based activism. Furthermore, it also offers an alternative standpoint towards trans issues, different from that of the healthcare sphere and especially from that of the State. The LGBTQ movement had already raised a type of discourse based on the human rights paradigm of those with a different sexual orientation or gender identity, the so-called *Yogyakarta Principles* (2007), concerning the enforcement of international legislation on human rights regarding sexual orientation and gender identity. However, as far as trans health care is concerned, the discourse was simply subject to a *Gender Identity Disorder* (GID) diagnosis according to DSM-IV (American Psychiatric Association 2000) or to one of *Gender Dysphoria* according to DSM-5 (APA 2013) updated version.

In the *Guide*, variability in gender identity is understood neither as a mental disorder nor as a physical or organic disease. Trans identities are acknowledged as part of human diversity. This document also gave rise to fresh contributions. Firstly, the diagnosis process has been unlocked and is to be replaced by a *counseling* approach from professionals in psychology to trans people during their own process. Secondly, there has been *a reformulation of the demand for intervention* as opposed to what SOC 6 prescribes (Spanish Network for Depathologization of Trans Identities 2010b).

Thus, via the *Spanish Network for Depathologization of Trans Identities*, trans activism has led to what we understand as epistemic corrections. The replacement of diagnosis by a *counseling follow-up* is currently a growing practice among health professionals in Spain. Moreover, new intervention practices, ones that are more flexible and respectful of trans people as real experts on their own processes, are being carried out outside the UTIGs, though still within the National Healthcare System; for instance, TRANSIT in Barcelona and other "alternative networks" in Madrid (Ortega Arjonilla 2014).

6.4 Citizen Participation and Knowledge Production in Virtual Communities: The *Personas Que* (People Who) Case

Virtual communities as well as social networks are built by groups of people pursuing common goals; however, while a "*virtual* community" is strictly linked to the Internet, a "*social* network" can exist either on the Internet or elsewhere. As Vázquez Atochero suggests (2008), 'A social network is a wider concept that would include a virtual community, i.e., a social network based on telematic media.' The main goals of any virtual community are the exchange of information, support and conversation.

Easy access to information is transforming not only social behavior, but also knowledge production. The endless array of possibilities that the Internet offers to create a whole new world make it an unparalleled cultural invention developing in cyberspace, though constrained by a given socio-cultural context. This has contributed to the creation of cyberspace communities, who can be simply defined as social groups that get together in pursuance of a common goal, beyond space and time, where they not only communicate and exchange information, but also produce it. The subject of our study deals with this cyberspace where communication is mediated by electronic devices, representing a different case of individual and collective knowledge production, governance and participation by community members.

We have chosen the virtual community *Personas Que* (*People Who (live with an illness)*), which arose from a digital platform: 'We are all people. And over the course of our lives, we all live with illnesses. We know how hard it is, but we truly believe that if we get involved and participate, we can ease the burden.'

The platform is addressed to those people who live with an illness in one way or another, either as patients, caregivers or simply as people who show a special interest in a particular disease. The ultimate goal of the community is to reach all people, regardless of the illness they live with, which is why new illnesses are being gradually incorporated in the platform. In fact, since the moment we commenced the study and follow-up of this community in June 2014, the number has increased as of June 2016 from six to twenty-one illnesses, including: allergies, Alzheimer's disease, arthritis, ovarian cancer, skin cancer, celiac disease, depression, diabetes, epilepsy, schizophrenia, atrial fibrillation, hepatitis, infertility, chronic myeloid leukemia, obesity, psoriasis, transplants, bipolar disorder, HIV and HPV. We have chosen the arthritis community for several reasons: arthritis is a chronic inflammatory autoimmune disease with different unspecified general symptoms, a high prevalence in women and no known cure.

The terms of use of this platform are open and free access (though registration is required to access certain areas), being of legal age, maintaining a respectful attitude towards the community, and the use of adequate and appropriate language. In terms of intellectual propriety, *People Who Global, S.L.* is the owner of all the intellectual and industrial property rights of the *People Who* platform and of all the elements contained in it, though the platform's users are the owners of the content they decide to publish in any of the sections (called "services" on the platform) and are responsible for exercising the rights of use of any content that they decide to include in these sections.

In the section called "Community", you can learn and share experience knowledge and also knowledge emanating from people who have lived with the illness for a long time. Furthermore, it provides the possibility of searching for and offering information by writing and publishing articles. It is worth mentioning that the contents are always reviewed and "adapted" to the platform; therefore, the community is always under the surveillance of experts. In the section called "Control", you can keep track of an illness, create a record of relevant data and treatments on a day-to-day basis, etc. There is also an application that allows you to monitor an illness on a daily basis wherever you are. The third and last major section is the "Magazine", which allows users to keep up-to-date with respect to any given illness, offering information in a lightweight format, such as short articles, presentations, computer graphics and videos.

In the "Help" section, they offer a guide that helps people to use *Personas Que (People Who)*; it also enables registration as a member of a Patients' Association. "Adverse Effects" allows users to report any side effect of a drug, either through the Spanish Agency for Medicines and Medical Devices (where you are required to fill out an electronic application form) or through the surveillance departments of the pharmaceutical companies that collaborate with the association directly (namely Lilly, AstraZeneca and Daiichi-Sankyo when we last accessed the website).

The ethical code covers a set of values, rules and moral principles that may be summarized as follows: it is all about people, not patients; the right to have access to information and knowledge; equality and non-discrimination; promotion of dignity and empathy, confidentiality, privacy and freedom of expression; withdrawal of brand names from conversations; cooperation with minorities; a sustainable and free project that is both transparent and independent, but which always seeks new sponsors who can ensure the provision of quality contents while placing the interests of *People Who* users first and foremost.

The content analysis allows users to check the type of community produced by the platform, whether new knowledge is built or not and the relation between experts and laypeople. As far as scientific content is concerned, the most interesting sections are "Community", "Control" and "Magazine". In the first one, we find what are known as "conversations", the Q&A section and finally the groups and associations that have been created. Conversations are interactions among several users, but they are also questions that can be answered either by an expert or by another user. Advertisements for any activity offered by the associations involved can also be found here. In the second section, questions are usually more precise and therefore answered by experts, although some users respond and share their experiences. The most frequent questions are related to medicines and alternative treatments.

The "Experiences" section in the "Magazine" includes articles written by people who either live with the illness or take care of those who suffer it. In general, they talk about the moment of diagnosis, treatment, and attitude towards the illness. Some people also provide information about alternative or complementary treatments. Finally, the "Magazine" includes articles written by experts or science journalists: essential articles (including fundamental aspects of the illness and the treatment); expert opinion articles; articles about treatment, medicines, food and exercise; advice to feel better; and the culture and leisure section (for example, ideas about possible presents for people suffering from arthritis). Essential articles have scientific content, although they may also include advice or general information (most of the articles are written by experts).

The relationships that emerge among the different actors (patients, caregivers, doctors, the health industry and communication professionals) are of an informative, cooperative and unifying nature. Challenging positions from patients towards the scientific community and pharmaceutical companies can also be seen; even challenges within the scientific community itself. For example, a very illustrative case found under "Experiences" relates the case of a person who started with outbreaks of arthritis at the age of 16, suffered a severe crisis when he was nineteen when many parts of his body were affected. The prediction was that he would soon end up in a wheel chair. 'I started to research on my own. I thought that if the origin of the illness is autoimmune, in which our own defenses attack us, we should find a way to modify them…' Following his research, he started treatment based on different products and 'supervised by a doctor'. This was not a conventional doctor; 'he went *against the system and the stated protocols*'.[1] He resorted to an innovative treatment, but always under the supervision of an expert.

[1] http://www.personasque.es/experiencias/articulos/alternativas-o-complementarias-dentro-del-sistema. Last accessed June 5, 2015. Emphasis added by the authors.

6.5 Conclusions

Through the three case studies presented here, we have analyzed different forms of participation concerning health and hence the production, disclosure and exchange of "expert" and "lay" knowledge. First, "opponent" trans activism; second, the mainly "partner" role of intersex condition/DSD support groups; and, lastly, a virtual platform that brings together patients, relatives and caregivers providing a predominantly "auxiliary" function. In the third case, the platform allows the creation of a biosociality. In the first two cases, there is an already existing organized group that uses technology to achieve its aims and goals.

Personas Que (People Who) moves within the dominant paradigm, producing auxiliary experiential knowledge. Even in those cases in which innovative knowledge is generated, it is always under the supervision of an expert (who relocates it within the paradigm). The knowledge produced by the platform is circumscribed to its internal use within the virtual community. In the other two cases (trans and intersex), however, the knowledge produced goes beyond the community itself; its ultimate goal being social transformation. In the trans case, "evidence-based activism" has led to a shift from a biomedical paradigm to one of human rights. In the intersex case, the shift has been from a biomedical to a psychosocial paradigm. The *Network for Depathologization of Trans Identities*, understood as an opponent organization, uses biomedical knowledge, though reformulated within a new paradigm. The result is "hybrid knowledge" (experiential, political and biomedical), transforming activists into lay experts. In the case of intersex, we have studied a support group which, although it holds the status of a partner organization, sharing cooperative research activities with health professionals, it also adopts a critical and transforming stance. However, this stance is not sufficient for it to be understood as an opponent social movement (unlike other intersex activisms in contexts such as the US).

Furthermore, we have highlighted how trans and intersex groups produce their own knowledge, in many cases identifying and developing "undone science", acting as epistemic correctives and producing hidden innovations. This has led to important bioethical and philosophical debates regarding the definition of health and illness, the socio-medical norm, patient autonomy and integrity and issues related to informed consent or medical intervention in infants. It has also opened up research to a greater number of voices and perspectives, especially those traditionally excluded, thereby avoiding unique narratives and experiences about conditions. All this draws attention to the power relationships among the multiplicity of actors that take part in the processes of biotechnological governance, in which economic (pharmaceutical companies), legal (legislation and rules) and political (institutions at different levels) actors play a key – and sometimes even a co-opting – regulatory role.

Overall, these different forms of social and health care participation have produced a democratizing redefinition of socio-scientific knowledge and governance. New forms of organization on the part of the different actors involved generate knowledge and bring about alliances that produce minor innovations or epistemic breaks. Consequently, a diversity of voices and perspectives has emerged concerning social, healthcare and legal issues that spark off controversies and debates, now opened to a multiplicity of actors, not just health care experts. In conclusion, there has been a reformulation of what is understood as expert knowledge, as well as a challenge to the expert/lay dichotomy.

References

Akrich, M. (2010). From communities of practice to epistemic communities: Health mobilizations on the internet. *Sociological Research Online, 15*(2), 10.

Akrich, M., O'Donovan, O., & Rabeharisoa, V. (2015). The entanglement of scientific and political claims: Towards a new form of patients' activism. In P. Wehling, W. Viehöver, & S. Koenen (Eds.), *The public shaping of medical research* (pp. 72–88). London: Routledge.

APA, American Psychiatric Association. (2000). *Diagnostic and statistical manual of mental disorders, DSM-IV-TR*. American Psychiatric Publications.

APA, American Psychiatric Association. (2013). *Diagnostic and statistical manual of mental sisorders: DSM-5*. American Psychiatric Publications.

Brown, P. (Ed.). (2008). *Perspectives in medical sociology*. Long Grove: Waveland Press.

Bucchi, M. (2015). Changing contexts for science and society interaction: From deficit to dialogue, form dialogue to participation -and beyond? In P. Wehling, W. Viehöver, & S. Koenen (Eds.), *The public shaping of medical research* (pp. 211–225). London: Routledge.

Callon, M., & Rabeharisoa, V. (2003). Research "in the wild" and the shaping of new social identities. *Technology in Society, 25*, 193–204.

Chase, C. (1998). Hermaphrodites with attitude. *Journal of Lesbian and Gay Studies, 4*(2), 189–211.

COM. (2001). *European governance: A white paper* (p. 428). http://europa.eu/rapid/press-release_DOC-01-10_en.htm. Accessed 13 Jul 2016.

Epstein, S. (1996). *Impure science. AIDS, activism, and the politics of knowledge*. Berkeley: University of California Press.

Esteva de Antonio, I., Gómez-Gil, E., Almaraz, M. C., Martínez-Tudela, J., Bergero, T., Olveira, G., & Soriguer, F. (2012). Organización de la asistencia a la transexualidad en el sistema sanitario público español. *Gaceta Sanitaria, 26*(3), 203–209.

García-Dauder, S., & Romero, C. (2012). Los desplazamientos políticos de las categorías médicas. In E. Pérez-Sedeño & R. Ibáñez (Eds.), *Cuerpos y diferencias* (pp. 213–240). Madrid: Plaza y Valdés.

García-Dauder, S., Gregori, N., & Hurtado, I. (2015). Usos de lo *psicosocial* en la investigación y tratamiento de las intersexualidades/DSD. *Universitas Psychologica, 14*(5), 1649–1666.

Grapsia. (2015). Tratamiento y afrontamiento del SIA: una tarea compartida. *Revista Española de Endocrinología Pediátrica, 6*(2), 28–34.

Gregori, N. (2015). *Encuentros y des-encuentros en torno a las intersexualidades/DSD: narrativas, procesos y emergencias*. Doctoral Thesis. Universidad de Valencia.

Gregori, N. (2016). *Diagnóstico de los diferentes posicionamientos, realidades y demandas de la población intersexual/con DSD en la ciudad de Barcelona*. Barcelona: Ayuntamiento de Barcelona.

Hess, D. (2009). The potentials and limitations of civil society research: Getting undone science done. *Sociological Inquiry, 79*(3), 306–327.

Kessler, S. (1998). *Lessons from the intersexed*. London: Rutgers Univ. Press.

Latour, B. (2000). When thinks strike back: A possible contribution of "Science Studies" to the social sciences. *British Journal of Sociology, 51*(1), 107–123.

Ley 3/2007, de 15 de marzo, reguladora de la rectificación registral de la mención relativa al sexo de las personas. (2007). 5585 § Disposiciones generales 11251–11253. http://www.boe.es/diario_boe/txt.php?id=BOE-A-2007-5585

Ortega Arjonilla, E. (2014). *La cuestión de sexo/género en medicina: tecnologías de reasignación de sexo y valores de género en España*. Doctoral Thesis. Universidad de Santiago de Compostela.

Ortega Arjonilla, E., Romero Bachiller, C., & Ibáñez Martín, R. (2014). Discurso activista y estatus médico de lo trans: hacia una reconfiguración de cuidados y diagnósticos. In E. Pérez-Sedeño & E. Ortega Arjonilla (Eds.), *Cartografías del cuerpo: biopolíticas de la ciencia y la tecnología* (pp. 521–572). Madrid: Cátedra.

Personas Que. www.personasque.es

Pineda, E. (2008). Mi pequeña historia del lesbianismo organizado en el movimiento feminista de nuestro país. En *Lesbianas. Discursos y Representaciones* (pp. 31–60). Barcelona: Melusina.

Platero, R. L. (2009). Transexualidad y agenda política: una historia de (dis)continuidades y patologización. *Política y Sociedad, 46*(1–2), 107–128.

Platero, R. L., & Ortega-Arjonilla, E. (2016). Building coalitions: The interconnections between feminism and trans* activism in Spain. *Journal of Lesbian Studies, 20*(1), 46–64. https://doi.org/10.1080/10894160.2015.1076235.

Principios de Yogyakarta. (2007). Principios sobre la aplicación de la legislación internacional de derechos humanos en relación con la orientación sexual y la identidad de género. Access: 17 July 2013, from http://www.yogyakartaprinciples.org/

Rabinow, P. (1996). Artificialidad e ilustración: de la biosociología a la biosocialidad. In J. Crary & S. Kwinter (Eds.), *Incorporaciones* (pp. 201–222). Madrid: Cátedra.

Red por la Despatologización de las Identidades Trans del Estado Español. (2010a). *Best practices guide to trans health care in the national health system*. Access: http://www.stp2012.info/guia/STP_guide_health.pdf

Red por la Despatologización de las Identidades Trans del Estado Español. (2010b). *STP-objetivos*. Access 25 Oct 2013, from http://stp2012.wordpress.com/objetivos/

Rose, N. (2007). *The politics of life itself. Biomedicine, power and subjectivity in the twenty-first century*. Princeton: Princeton UP.

Still, B. (2008). *Online intersex communities*. Amherst: Cambria Press.

Todt, O. (2006). La gobernanza tecnocientífica en la Unión Europea. *Revista CTS, 7*(3), 21–42.

Vázquez Atochero, A. (2008). *Ciberantropologia: Cultura 2.0*. Barcelona: UOC.

Wehling, P., Viehöver, W., & Koenen, S. (Eds.). (2015). *The public shaping of medical research: Patient associations, health movements and biomedicine*. London: Routledge.

Chapter 7
The Culture of Risk: STS Citizens Facing the Challenge of Engagement

José Antonio López Cerezo and Belén Laspra

Abstract Plurality and divergence of opinion, based on freedom of thought and information, are nowadays broadly recognized as requirements for the healthy performance of a democratic society. The aim of this chapter is to take this idea somewhat further by characterizing the agents of such performance. We do so by highlighting a culture of risk with respect to the Science and Technology System. In our view, risk culture implies having a skeptical awareness grounded on sound information towards science and technology, combining an overall positive attitude and awareness of their limitations and threats, and adapting one's behavior accordingly. We argue that risk culture is a key element for democratic governance in contemporary risk societies with increasingly pressing technical issues open to social debate. Accordingly, we first review the concept of risk culture within the framework of scientific and technological culture, and then proceed to examine the role of this risk culture in social engagement. Finally, we reflect on some challenges of the implications of risk culture for the relation between society, science, and technology.

7.1 Risk Culture

In recent decades, research on scientific and technological (S&T) culture has generated a considerable amount of literature in the field of Science, Technology and Society (STS), as well as a diversity of related areas like Science Communication, Science Policy, Science Education and Public Understanding of Science (PUS). Despite the many attempts to reach an operational concept of S&T culture, or at least to develop a commonly shared framework, the question of what scientific

J. A. López Cerezo (✉)
Department of Philosophy, University of Oviedo, Oviedo, Asturias, Spain
e-mail: cerezo@uniovi.es

B. Laspra
Center for Political Studies, University of Michigan, Ann Arbor, MI, USA
e-mail: laspra@umich.edu

B. Laspra, J. A. López Cerezo (eds.), *Spanish Philosophy of Technology*, Philosophy of Engineering and Technology 24,
https://doi.org/10.1007/978-3-319-71958-0_7

culture is still remains open. For example, the absence of a theory that enables a suitable interpretation of the results of public understanding surveys has been pointed out as one of the major weaknesses of the PUS field (Pardo and Calvo 2002). In this chapter, based on previous research (Cámara Hurtado and López Cerezo 2012), S&T culture is understood as a process that involves the consumption of S&T information and entails the cognitive enrichment of the individual, the readjustment of his or her beliefs and attitudes according to the information acquired, and the resulting generation of behavioral dispositions and changes in habits.

Research on S&T culture usually refers to three types of elements: epistemological, axiological, and praxeological.[1] The epistemological elements include knowledge of science and technology (S&T), both concerning scholar science[2] and cutting-edge science,[3] as well as knowledge about political, economic, cultural, or ethical issues in S&T (i.e., meta-scientific knowledge). The axiological elements refer to interests, values, and attitudes associated with S&T. The praxeological elements are those related to behavioral dispositions and human action. Here, we shall focus on a cross-cutting axis of S&T culture, present in each of these dimensions: the culture of risk. We shall use the term 'risk culture' to refer to the epistemological, axiological, or praxeological elements of S&T culture, thus related to knowledge, valuation, and action or disposition to action, in relation to negative effects and risks produced by S&T change.

We understand the culture of risk as framed within S&T culture. Although there is some overlap between the two concepts, the culture of risk has certain features that make it different. Specifically, as far as the epistemological elements are concerned, in the case of risk culture, these logically comprise more relevant knowledge concerning current S&T issues, such as knowledge related to scientific controversies or the existence of potential risks and benefits of specific fields of technological change. This does not mean that a basic level of scientific literacy is not required to make sense of current debates on issues involving negative or potentially negative aspects related to S&T (Miller 2012). As to the axiological elements, from the perspective of the culture of risk, the focus lies on the perception and valuation of risks versus the benefits of S&T in general and on the perception of the risks or negative impacts associated with specific applications of S&T. Also within the axiological dimension, although the focus is on interactions between interests, values, and attitudes, trust and distrust are particularly important in risk culture, as both play a key role in the praxeological dimension, i.e., in decision-making and the

[1] These terms are understood in their Greek sense. Episteme: concerning justified beliefs. Axiology: concerning values. Praxis: concerning actions.

[2] 'Scholar science' refers to that basic body of knowledge originally set by the American Association for the Advancement of Science in 1989, through the document entitled *Science for All Americans*. This body of knowledge is measured in PUS surveys by quiz tests on basic scientific literacy.

[3] 'Cutting-edge science' refers to S&T results that appear in the newspapers and tend to raise social controversy. There are many examples in the fields of fracking, vaccines, car engines, nuclear energy, cybersecurity, human reproduction, etc.

generation of behavioral dispositions based on our beliefs and valuations concerning the threats or negative effects of S&T.

Although common surveys on the public perception of S&T do not usually pay much attention to the negative aspects of S&T (Cámara Hurtado and López Cerezo 2014), according to the available data-survey for Spain, people have a low perception of risks, at least when a balance between positive and negative effects is asked for. Results of Special Eurobarometer 340 (EC 2010) show that 68% of Spanish respondents totally agree or tend to agree with the statement "the benefits of science are greater than any harmful effects it may have". In contrast, 9% totally disagree or tend to disagree with the same sentence. These results are slightly higher than those from Special Eurobarometer 224 (EC 2005), where 57% totally agree or tend to agree, while 8% totally disagree or tend to disagree with the above statement. In both surveys, the level of "neither agree nor disagree" was significantly high (19% in EB 2010; 28% in EB 2005). The high perception of the benefits of science on the part of Spanish society is also shown in national surveys (see the FECYT 2002–2014 series). Results differ interestingly if risk and benefits are measured as two separate dimensions. According to the 2007 Ibero-American survey (FECYT-RICY-OEI 2009), a good percentage of people tend to perceive both many risks and many benefits when asked about the effects of S&T. As regards the case of Madrid (Spain), 17.1% perceive many risks, 42.2% some risks, 32.2% few risks, and 8.5% none; while 36.2% perceive many benefits, 47.7% some benefits, 13.7% few benefits, and 2.4% none.[4]

The logical context for reflecting on the political significance of the culture of risk is that of the risk society. Since it was proposed by the German sociologist Ulrick Beck in 1986, the concept of the risk society has spread widely and has served to highlight the existence of the hazards that accompany the benefits derived from S&T development. One salient feature characterizing the risk society is the recognition of global risks that transcend spatial and temporal boundaries. An accident at a nuclear power plant such as the one that occurred in Fukushima, a health alarm such as the one triggered by the Zika virus, or legislation on the commercialization of transgenic foods all have repercussions that go beyond national, gender, or generational borders. Besides, at the individual level, risk currently lies at the core of everyday life. Given the wide diversity of courses of action opened up by current S&T change, the binding traditions of the past have now lost their strength to regulate individual behavior. We thus have to constantly face taking risky decisions in our lives. For example, when deciding in the supermarket whether or not to

[4]This survey was promoted by the Spanish Foundation for Science and Technology (FECYT), the Ibero-American Organization of States (OEI) and the Network of Indicators for Science and Technology (RICYT). The survey was conducted in autumn/winter 2007 in seven major cities in Ibero-America: Bogota, Buenos Aires, Caracas, Madrid, Panama City, Santiago, and Sao Paulo. Here the traditional question regarding the balance between benefits and risks was split into two individual questions addressing the individual issues of benefits and risks. The salient fact was that, in all cases, the corresponding results comprised percentage sums above 100% in the perception of very many plus many risks *and* very many plus many benefits.

buy genetically modified corn, submit to a medical technique, or consume an artificial sweetener.

However, society is not merely a recipient of the risks and benefits of S&T. Its role is not limited to making one decision or another in contexts of risk. Current threats are no longer conceptualized as hazards, i.e., as unavoidable harm. Virtually all the evils that threaten us today are understood as risks, i.e., as harm resulting from the action or omission to act of any human being. In the past, and possibly still in some cultures strongly anchored in tradition or in remote corners of the globe, evils were attributed to fate, nature, or some supernatural will. Today, they are a common cause of attribution of responsibility to some social agent. Consequently, risk assessment and risk management have become high-visibility issues in the political arena, in response to this 'politicization' of threats conveyed by the notion of risk and an increase in social awareness of the impacts and potential risks of S&T change. Accompanying this phenomenon and fueled by the growing political leadership of a great diversity of social movements, society nowadays demands a more active role in decision-making processes whenever risks and benefits are at stake. How to build a scenario that enables social engagement in S&T is currently one of the challenges of contemporary S&T policies.

7.2 The Challenge of Engagement

'Engagement' refers to an aspiration, to a concrete way of understanding how the relationship between science and society should be modulated, and to the strategies developed to foster this relationship. It is, in a way, a talisman term, as is 'innovation', whose inclusion in certain documents often responds to the need to attract funding. To engage society more broadly in research and innovation activities is one of the goals of the strategy on Responsible Research Innovation (RRI) promoted by the European Union through the Horizon 2020 program. From a linguistic perspective, 'engage' has a French origin and meant 'to pawn or pledge something'. It entered English in the fifteenth century via French influence, adding the suffix '-ment' in the seventeenth century and meaning, in the general sense, 'a legal or moral obligation'. It is currently commonly used in English and means to be involved, interested, or engaged, in the sense of acquiring a formal agreement, e.g., to get married. It is a term frequently used in the field of labor relations and organizational culture, but has been gaining ground in the fields of STS and PUS. From a PUS perspective, the lineage of engagement could be said to be rooted in the enlightened aspiration of a literate society. When exploring its genealogy, we find names like John Dewey, Ronald Davis, Benjamin Shen, Jon Miller, and Walter Bodmer; institutions such as the National Science Foundation, the Royal Society, the House of Lords, and the European Commission; and documents such as *The Public Impact of Science in the Mass Media* (Davis 1958), *Public Understanding of Science* (Royal Society 1985), *Science and Society* (House of Lords 2000), and *Public Engagement in Science* (European Commission 2008).

The biography of engagement has been addressed by many authors. One of the most influential versions is that of Bauer et al. (2007). In this paper, the authors describe three paradigms in PUS that divide the evolution of the field into three main periods: scientific literacy (1960s–1980s), public understanding (1985–1990), and science-in-society (1990s–Present). Each of the paradigms includes a particular conception of the relationship between science and society. The paradigms point out the limitations that hinder the mutual approach between science and society, and the strategies to overcome these constraints. In short, the first paradigm emphasizes the low level of scientific knowledge on the part of society (cognitive deficit); the second stresses the lack of social interest in S&T issues (attitudinal deficit); while the third highlights the mistrust between political leaders, the scientific community, and society (trust deficit). The respective strategies to address each deficit are the promotion of scientific literacy, the reinforcement of science communication mechanisms, and the promotion of public participation.

The social engagement paradigm can thus be understood as an evolution of the third paradigm, or even as a fourth paradigm in the PUS field. In general terms, engagement entails the need for a new social agreement for science.

Some steps have already been taken in this direction. For instance, the *Public Engagement in Science* report (EC 2008) seems to constitute a step forward. This document acknowledges the shift from Public Understanding of Science (PUS) to Public Engagement with Science (PES) and the need to renew the social contract for science. Echoing earlier milestones – such as the Budapest World Congress for Science 1999, promoted by UNESCO and ICSU–, the report states, "There is an increasing body of evidence showing that interactions between science, civil society and the wider public can generate new forms of social intelligence and create mutual benefits by stimulating new directions for innovation" (EC 2008, p. 10). It also warns about the need for a more sophisticated view of the relationship between governance, ethics, and competitiveness in global innovation networks. Innovation and science cannot be considered a quantitative issue, as a simple race between competitors. The most important question is not 'how much?' or 'how fast?', which seems to prevail in political discourse, but 'where?', contemplating this process via a variety of possible directions that should be the subject of debate by civil society (EC 2008: 11).

There seems to be a common feeling supporting the demand that science and society tighten their ties for the benefit of both, and yet, despite the efforts, on looking back a question arises: Are we moving forwards or in circles? This is the question posed by Alan Irwin (2008, 2014), who, via his query, channels the distress of other authors such as Sheila Jasanoff (2014), Brian Wynne (2014), and Stilgoe et al. (2014) concerning what appears to be yet another unsuccessful attempt to bridge the gap between society and science. Criticisms of the linear model of communication, the sacralization of expert knowledge, or the view of the public as a homogeneous natural kind are still in force today, more than 30 years after their formulation, as well as the constant claims on the value of non-expert knowledge, the overcoming of the dichotomy between experts and laymen, the promotion of dialogue, and so on. As Irwin puts it, "I can on a good day claim partial progress" (Irwin 2014, p. 73).

Why does engagement seem to be having such limited success? From our point of view, the key problem resides in how 'the public' is represented. There is a certain misunderstanding of the public by the scientific community and political managers that has made the goal of engagement only a partial success. We shall defend a more complex view of the relationship between science and society that is based on dialogue, but on a dialogue that includes many voices. Science does not speak with a single voice, nor does the public, and dissent, criticism, plurality of voices, and divergence of opinion are precisely the main indicators reflecting engagement as well as the main vectors promoting it among citizens.

A germane document to understand the fundamental issues of this debate is provided by an influential booklet entitled *The Public Value of Science*, authored by Wilsdon, Wynne and Stilgoe and published by the British think-tank Demos in 2005. These authors ask about the contribution of S&T to more general social ends and point out that this debate always runs the risk of ending up in one of two dead ends: determinism or reductionism. The former refers to the error of thinking that what is politically correct is to always define oneself as being pro-science and pro-innovation, without asking what science or what innovation, thus favoring by inaction the fostering of certain technological trajectories as if they were inevitable, and feeding a polarized pro-innovation versus anti-science discourse. The latter constitutes the error of believing that the purposes and orientation of science and innovation must be set by experts through tools such as the economic calculation or technical calculation of risks, thus excluding social debate about the ultimate purposes of S&T change.

The concept of 'public value', expressed through public preferences and materialized through the services implemented by public policies, defines a space for participatory deliberation which, according to these authors, helps to avoid the risks of determinism and reductionism. This concept also stresses the importance of promoting a model of participation in which this is not seen as a brake on progress, but as a way of maintaining and reinforcing the social contract that supports science, with the possibility to modify techno-scientific trajectories under the light of the discussion on public values that should guide S&T change (Wilsdon et al. 2005). For these authors, "disagreement and protest, as well as participation, are signs of a healthy democracy" (Wilsdon et al. 2005, p. 29).

In a chapter of the aforementioned report, *Public Engagement in Science*, James Wilsdon (2008) argues that 15 years of social debates about mad cows, transgenic foods, or nuclear waste do not seem to have been of much use in dealing with current nanotechnologies, fission nuclear energy, neuroscience, or synthetic biology. More that understanding or dialogue, what we really need, Wilsdon argues, is 'upstream engagement', with new forms of accountability and public involvement bringing to the forefront values, purposes, and underlying interests, and taking place in the initial stages of the research-development-exploitation cycle. The author thus aligns with the stated objectives of the Horizon 2020 program, in which the RRI strategy is understood as "an inclusive approach to research and innovation, to ensure that societal actors work together during the whole research and innovation process. It aims to better align both the process and outcomes of research and innovation, with the values, needs and expectations of European society" (EC, Online). According to Wilsdon and Willis (2004), we need to go beyond the risks versus benefits debate, beyond the question of whether a technology is safe or not, towards issues such as: What we want, what alternatives there are, what interests the technology promotes, and what long-term consequences it may have for my family and society.

Early involvement renders it possible to make these basic issues visible and to reveal the values and visions that guide science to public scrutiny, creating a valuable opportunity for the orientation of S&T change in line with social sensitivities and concerns. Nonetheless, it is important not to lose sight of the ultimate goal of engagement, namely to avoid the dead ends of determinism and reductionism in the negotiation of a new social contract for science. Emerging concepts such as RRI can lead to new forms of the deficit model, silencing social voices and closing the dialogue through a supposed business or scientist's assimilation of (their own representation of) public values. Similarly, it is important not to lose sight of the public of engagement, recognizing its diversity and heterogeneity: the public is not a homogenous, pre-existing entity. Gluten allergic individuals or astronomy lovers are not fixed natural kinds, but spaces in constant transformation, publics that are condensed around themes and technoscientific objects that arouse interest or concern, selectively intensifying the consumption of scientific information, and who demand opportunities in the political arena to participate in the construction of technoscientific scenarios of the future (Jasanoff 2014).

7.3 The Role of Critique

As basic ground for the construction of engagement, the dialogue format entails the assumption of responsibilities on the part of both parties. The fact that dialogue stresses the need for greater sensitivity on the part of the scientific community towards social concerns and demands does not eclipse the part concerning society. Although greater social involvement in decision-making are targets of the political agenda, scientific literacy and public understanding in S&T continue to be agenda items. While the achievement of these objectives depends to a large extent on the political fabric making them possible, society must also have an interest in achieving them. Enabling mechanisms for people to participate in decision-making is of little use if the individuals involved are not sufficiently concerned parties. The challenge of dialogue between science and society thus seems to have a number of checkpoints. On the one hand, the promotion of scientists who are more sensitive to social demands and more committed to the dissemination of their research, i.e., the promotion of civic scientists (Lane 1997). On the other, the promotion of citizens who are in the best conditions to participate in decision-making processes, i.e., citizens with a critical and informed view of S&T, ready to contribute to the creation of technological scenarios of future by means of their oriented support of the S&T system.

The evolution of the PUS field makes it possible to detect a trend highlighting the value of criticism, scrutiny, and social implication in relation to the general objective of promoting the rapprochement between science and society. The vindication of disagreement, social protest, and informed criticism as signs of a healthy democracy is present in recent contributions such as those of Wilsdon, Jasanoff and Wynne, and is useful in dealing with powerful, ingrained preconceptions such as the dangers of determinism and reductionism mentioned above. However, long-standing inertias are difficult to overcome. In fact, the evolution of science policies since the 1960s has followed the patterns of social exclusion and rejection of external criticism that tended to adopt general technocratic policies, thus protecting themselves from social protest movements that demanded openness and accountability. This is a situation that has produced dysfunctional effects such as the so-called 'science wars' and fueled the post-positivistic contempt between the natural sciences and the social sciences (Snow's famous two cultures), in addition to stimulating a reductionist and inadequate view within the technical study of transdisciplinary problems such as those related to risk (e.g., Althaus 2005; Greenberg and Lowrie 2013). This situation also negatively affects the relationship between science and society, fueling a denatured image in the media both of science (mythical science) and of society itself (according to the deficit model): you are either pro-science or you are an ignorant devotee of anti-science. The results that have been feeding on public opinion is not surprising, namely alienation and distrust.

Although still partially veiled, especially from the public opinion studies on the public perception of science, society shows a type of critical awareness concerning science, of risk culture arising from familiarity and interest in science. It is the skepticism of the well-informed citizen that leads him or her to be aware of uncertainty, to compare and contrast sources, to seek a second medical opinion, etc. It is the kind of criticism expressed from the point of view of public opinion studies by the loyal skeptics of Martin Bauer et al. (2012) or the distrustful engagers of the 2014 PAS survey in the UK. For example, in the aforementioned survey, the profile of "distrustful engagers" (17% of the total population) corresponds to citizens who are very interested in science and who feel informed about it; think that science is beneficial to society, although they are cautious about scientists and regulation; consider that the public should play a role in decisions on science-related issues; and have a high educational level and generally a good social position. They are the citizens that we have called the 'many-many population' because they are distinguished in surveys (which are able to detect them)[5] as people who appreciate many benefits in S&T but also many risks (López Cerezo and Cámara Hurtado 2014).[6] It is a population with an appreciable level of schooling, which keeps itself informed and has qualified and differentiated opinions regarding different areas and applications of

[5] See the aforecited 2007 FECYT-OEI-RICYT Ibero-American Survey (FECYT-OEI-RICYT 2009).

[6] With a clear-cut trend of a parallel percentage increase with increasing schooling level, this population segment comprises 50% of the population with university level of schooling in the aforementioned survey.

S&T. These are also citizens usually inclined to give their opinion on controversial social issues related to the application of science or technological development. It is the population located to the far right of Bauer's so-called inverted U, in the association between attitude and knowledge characteristic of post-industrial societies.[7]

7.4 The Challenge of Fostering STS Citizens

As we pointed out at the beginning of the chapter, the culture of risk focuses specifically on the perception and acceptability of risks, based on beliefs, valuations, and decisions/actions. A clear manifestation of the presence of a culture of risk is the trend towards high discrimination in the assessment of benefits and risks in diverse fields of S&T, i.e., in a disaggregated assessment by specific areas of application. It is thus a population that does not manifest itself uniformly or as many-benefits/few-risks (enthusiastic pro-science), nor as few-benefits/many-risks (pessimistic anti-science). These individuals believe in science, but not in the myths of science; they do not place religious faith in science, as in the unconditional support of enthusiastic pro-scientists (Evans 2014). For example, according to data from the International Social Survey Program (ISSP) and the US General Survey (GSS) between 1993 and 2010 for 12 Western countries (including Spain, the UK, and the USA), although there is a positive significant association between educational level and general confidence in science (in the sense that it provides values and goals and can solve human problems in general), this association is still significant but reverses its sign concerning the relationship between educational level and faith in the ability of science to solve particular problems related to technological applications in the physical world (e.g., those related to health or the environment). If it were not somewhat irreverent, one would be tempted to call them 'STS citizens'.

We use the adjectival term STS to refer to individuals, materials, events, contents, etc. that fulfil certain features very familiar for the STS readership (Spiegel-Rösing and Solla Price 1977; Jasanoff et al. 1995; Mitcham 2005; Felt et al. 2017): they acknowledge the deep-rooted interdependence between the S&T system and the social context in which it is framed, an interdependence that means that the former cannot be understood outside the latter and affects the understanding of both; they break with the expert/lay dichotomy, moving away from a scientistic view and the risk of reductionism (Wynne 1992); they acknowledge that S&T development is not determined by some kind of inner logic, but rather is open to a diversity of influences, including those arising from non-expert knowledge, and, accordingly, they

[7] A typical public opinion survey result is the significant positive association between the level of scientific knowledge and a favorable attitude towards science. In the more industrialized countries, however, this association usually reverses its sign from a certain level of knowledge onward, revealing the existence of a cultured, but cautious population, aware of the great potential of S&T, familiar with both, but also aware of the risks and negative effects of technology-based industrial development, as well as skeptical of the mythical images of science (see, e.g., Bauer et al. 2012).

are in favor of the incorporation of the public in all stages of the R&D process; finally, they are aware that S&T development has a strong impact on society, an impact that implies both benefits and harms (Funtowicz and Ravetz 1993).

In post-industrial democratic societies, the quality of society's voice and the mechanisms available to include this voice in decision-making processes are indicators of the health of the democratic system. In this respect, dissent and freedom of opinion, accompanied by upstream participation, are basic elements for the proper functioning of a democratic society. A society with an excessively positive attitude towards science will show an uncritical attitude towards it. Better information and a less acritical image of science are conditions for greater personal leadership and greater social participation. The challenge is to stimulate personal empowerment built on a critical and informed attitude, in line with the ethics of non-power espoused by Jacques Ellul (1954), the rebellion against the tyranny of things in Herbert Marcuse (1964), the active implication in the S&T process described by Callon and Rabeharisoa (2008), and the like.

Nonetheless, a number of difficulties should be pointed out. Incorporating society's voice in decision-making may trigger answers which run countercurrent to what would be considered a pro-science attitude. Attitudes of distrust and suspicion may respond to ignorance, but they may also be the product of the acquisition of scientific information and be present in informed individuals who hold critical attitudes. A potential secondary effect of fostering a critical attitude can sometimes be the generation of truly unscientific behavior. This point deserves some consideration.

The anti-vaccines movement is a good example in this regard. This movement is made up of generally well-informed people who, on the basis of the available information and their own judgment, freely decide not to vaccinate their children. The presumed origin of the anti-vaccination movement can be found in an article published by Andrew Wakerfield in the medical journal *The Lancet* in 1998, where it is argued that the famous triple-virus (measles-rubella-mumps) causes autism in children. Faced with the avalanche of criticism of negligence and concealment of data that he received from the scientific community, the author retracted and the magazine eventually removed the article. However, the fuse of distrust and suspicion of institutional messages on vaccines was lit in the population. Side effects, pharmaceutical interests, and disagreements between pediatric associations, the World Health Organization, drug agencies and political leaders contributed to fueling skepticism. A key element in this process, in our view, was the use of the traditional diffusionist conception of public communication of science (Bucchi 2008). The anti-vaccination movement is nothing more than the fruit of the tree that has nurtured institutions over recent decades via the diffusion of a distorted image of science and the assumption of a poor image of the public: the tree of suspicion and distrust. However, these same suspicions of institutions are what have made nuclear power plants safer, that succeeded in suppressing DDT, that made genetic engineering a safe working field, and which are behind the many agencies and instruments which contribute to making today's technological products and installations much safer.

As Ragnar Löfstedt (2009) points out, the lack of trust in society is not necessarily bad. Between visceral rejection and emotional acceptance lies a wide-ranging and fertile territory of what we may call "critical trust". Trust really does not disappear; it is only qualified through criticism of traditional actors (government, industry) and redirected as an asset to new actors (NGOs, universities, judiciary). It is a redirecting in which, moreover, trust often changes its nature: from an understanding of trust as technical competence, it becomes understood as independence or integrity (if not as empathy), or vice versa.[8] No longer trusting the government due to its opaque and controversial management of an environmental threat echoed in the media may lead us to start trusting ecological groups regarding the matter in hand because we consider the latter to be more independent of industry and therefore more credible (trust transference: same type of trust, different target group). Or no longer trusting physicians because of the bad news on one's terminal illness and their lack of sensibility to their patients may redirect a new type of trust towards doubtful healers in order to be able to preserve some hope (trust transformation: different type of trust, different target group).

7.5 Final Remarks

From an etymological point of view and in line with Aristotle, the human being is a political animal, as it only develops its capacities in association with other human beings, in the community of the *polis*. The contemporary city, that of modern societies in industrialized countries, defines an environment of potentialities and threats that requires scientific culture and a culture of risk, as well as the political coordination of individuals for the full exercise of citizenship. However, the new social realities defined by a vertiginous technological transformation of our living conditions, in addition to the obsession for security and aversion to risk generated by the evolution of the recent global geopolitical board, have created new challenges that must be faced by our post-industrial society.

The *Science and Society* report opened with the following diagnosis:

> Society's relationship with science is in a critical phase. [...] On the one hand, there has never been a time when the issues involving science were more exciting, the public more interested, or the opportunities more apparent. On the other hand, public confidence in scientific advice to Government has been rocked by a series of events [...]; and many people are deeply uneasy about the huge opportunities presented by areas of science including biotechnology and information technology, which seem to be advancing far ahead of their awareness and assent. In turn, public unease, mistrust and occasional outright hostility are breeding a climate of deep anxiety among scientists themselves. (House of Lords 2000, art. 1.1)

[8] Something similar happens with risk: it is not destroyed; it is only transformed. When we try to minimize it, we often only manage to redistribute it, transforming the original risk into a new risk that may affect the original population or a new population (Graham and Wiener 1995).

Social mistrust has usually been understood as problematic and undesirable. In the literature of political theory and risk management (e.g., Slovic 1999), we often find the idea that trust is a capital asset for the proper functioning of a society. Trust is certainly a key element of social capital in contemporary society. However, besides being misunderstood, it is possibly overestimated to the detriment of public scrutiny and accountability. Trust is not an on/off attribute: it may have different modalities and easily migrates from one actor to another. Mad cows, dioxins in food, and other recent crises that have particularly affected European countries, coupled with the intensification of citizen activism that demands participation, have created a new scenario in which it no longer seems possible to maintain a traditional view of institutional trust as the main support for efficient social functioning.

Thus, in a world in continuous and accelerated transformation due to S&T change, with a diversity of actors struggling for limited resources in the public arena, a certain amount of skepticism and caution is fundamental today to generate transparency, accountability, and spaces for citizen participation. It also seems to be a good indicator of a mature attitude that contributes to democratic governance (Poortinga and Pigdeon 2003). The new trends in the evolution of the PUS field, in particular the research on the culture of risk, allow us to espouse the value of social criticism for the advancement and good health of science itself. Just as literary or cinematographic criticism lend a good service to the novel or the cinema, critical and well-informed attitudes towards science (in the sense of critical awareness, not anti-scientific rejection) contribute to rapprochement between science and society, to the good health of public policies in the matter, and to strengthening science in the service of society.

Acknowledgments The writing of this chapter was supported by the research projects: "Concept and Dimensions of Risk Culture" (MINECO-15-FFI2014-58269-P) from the Spanish Ministry of Economy and Competitiveness, and "Grupo de Estudios CTS" (FC-15-GRUPIN14-128) from the FICYT Agency of the Regional Government of Asturias, as well as ERDF funds from the European Union.

References

AAAS, Association for the Advancement Of Science. (1989). *Science for all Americans*. New York: Oxford University Press. http://www.aaas.org/report/science-all-americans. Accessed 12 Aug 2016.

Althaus, C. E. (2005). A disciplinary perspective on the epistemological status of risk. *Risk Analysis, 25*(3), 567–588.

Bauer, M., Allum, N., & Miller, S. (2007). What can we learn from 25 years of PUS survey research? Liberating and expanding the agenda. *Public Understanding of Science, 16*(1), 79–95.

Bauer, M. W., Shukla, R., & Allum, N. (2012). Towards cultural indicators of science with global validity. In M. W. Bauer, R. Shukla, & N. Allum (Eds.), *The culture of science. How public relates to science across the globe* (pp. 39–54). New York: Routledge.

Beck, U. (1986). *Risk society: Towards a new modernity*. London: Sage.

Bucchi, M. (2008). Of deficits, deviations and dialogues. Theories of public communication of science. In M. Bucchi & B. Trench (Eds.), *Handbook of public communication of science and technology* (pp. 56–76). New York/London: Routledge.

Callon, M., & Rabeharisoa, V. (2008). The growing engagement of emergent concerned groups in political and economic life: Lessons from the French Association of Neuromuscular Disease Patients. *Science, Technology & Human Values, 33*, 230–261.

Cámara Hurtado, M., & López Cerezo, J. A. (2012). Political dimensions of scientific culture: Highlights from the Ibero-American Survey on the Social Perception on Science and Scientific Culture. *Public Understanding of Science, 21*(3), 369–384.

Cámara Hurtado, M., & López Cerezo, J. A. (2014). Cultura científica y percepción del riesgo. In B. Laspra & E. Muñoz (Eds.), *Culturas científicas e innovadoras. Progreso social* (pp. 159–178). Buenos Aires: EUDEBA.

EC, European Commission. (2005). *Special Eurobarometer 224: Europeans, science and technology* (EB-63.1). European Commission. http://ec.europa.eu/commfrontoffice/publicopinion/index.cfm.

EC, European Commission. (2008). *Public engagement in science. Report of the science in society session.* Portuguese Presidency Conference. "The future of science and technology in Europe", Lisbon, October 8–10, 2007.

EC, European Commission. (2010). *Special eurobarometer 340: Science and technology* (EB-73.1, 340). European Commission. http://ec.europa.eu/commfrontoffice/publicopinion/index.cfm.

EC, European Commission (Online). (n.d.). *Horizon 2020. The EU framework programme for research and innovation.* https://ec.europa.eu/programmes/horizon2020/en/h2020-section/science-and-society. Acessed 10 Dec 2016.

Ellul, J. (1954). *The technological society*. New York: Vintage. 1964.

Evans, J. H. (2014). Faith in science in global perspective: implications for transhumanism. *Public Understanding of Science, 23*(7), 814–832.

FECYT, Fundación Española para la Ciencia y la Tecnología. (2003, 2005, 2007, 2009, 2011, 2013, 2015). *Percepción social de la ciencia y la tecnología en España*. Madrid: FECYT.

FECYT-OEI-RICYT. (2009). *Cultura científica en Iberoamérica. Encuesta en grandes núcleos urbanos*. Madrid: FECYT.

Felt, U., Fouché, R., Miller, C. A., & Smith-Doerr, L. (Eds.). (2017). *The handbook of science and technology studies* (4th ed.). Cambridge, MA: MIT Press.

Funtowicz, S. O., & Ravetz, J. R. (1993). Science for the post-normal age. *Futures, 25*, 735–755.

Graham, J. D., & Wiener, J. B. (Eds.). (1995). *Risk versus risk: Tradeoffs in protecting health and the environment*. Cambridge, MA: Harvard University Press.

Greenberg, M., & Lowrie, K. (2013). Sheila Jasanoff: Culture and diversity in risk management. *Risk Analysis, 33*(7), 1179–1181.

House of Lords. (2000). *Science and society. Third report*. London: House of Lords Select Committee on Science and Technology/HMSO. http://www.publications.parliament.uk/pa/ld199900/ldselect/ldsctech/38/3801.htm. Accessed 14 Feb 2015.

Irwin, A. (2008). Moving forwards or in circles? Science communication and scientific governance in an age of innovation. In R. Holliman, E. Whitelegg, E. Scanlon, S. Smidt, & J. Thomas (Eds.), *Investigating science communication in the information age: Implications for public engagement and popular media* (pp. 3–17). Oxford: Oxford University Press.

Irwin, A. (2014). From deficit to democracy (re-visited). *Public Understanding of Science, 23*(1), 71–76.

Jasanoff, S. (2014). A mirror for science. *Public Understanding of Science, 23*(1), 21–26.

Jasanoff, S., Markle, G. E., Petersen, J. C., & Pinch, T. (Eds.). (1995). *Handbook of science and technology studies*. Thousand Oaks: Sage.

Lane, N. F. (1997). *An open letter to scientists and engineers: Let's get the world out together about why science matters*. National Science Foundation. http://www.upenn.edu/almanac/v44/n03/benchmar.html. Accessed 9 Sept 2014.

Löfstedt, R. E. (2009). *Risk management in post-trust societies*. London: Earthscan.
López Cerezo, J. A., & Cámara Hurtado, M. (2014). Cultura científica y percepción del riesgo. In B. Laspra & E. Muñoz (Eds.), *Culturas científicas e innovadoras. Progreso social* (pp. 159–177). Buenos Aires: EUDEBA.
Marcuse, H. (1964). *One-dimensional man: Studies in the ideology of advanced industrial society*. London: Routledge.
Miller, J. D. (2012). The sources and impact of civic scientific literacy. In M. W. Bauer, R. Shukla, & N. Allum (Eds.), *The culture of science. How public relates to science across the globe* (pp. 218–240). New York: Routledge.
Mitcham, C. (Ed.). (2005). *Encyclopedia of science, technology and ethics*. Farmington Hills: Macmillan Reference USA.
Pardo, R., & Calvo, F. (2002). Attitudes toward science among the European public: A methodological analysis. *Public Understanding of Science, 11*(2), 155–195.
Poortinga, W., & Pigdeon, N. (2003). Exploring the dimensionality of trust in risk regulation. *Risk Analysis, 23*(5), 961–972.
Slovic, P. (1999). Trust, emotion, sex, politics, and science: Surveying the risk-assessment battlefield. *Risk Analysis, 19*(4), 689–701.
Spiegel-Rösing, I., & de Solla Price, D. (Eds.). (1977). *Science, technology and society: A cross-disciplinary perspective*. London: Sage.
Stilgoe, J., Simon, J., & Wilsdon, J. (2014). Why should we promote public engagement with science? *Public Understanding of Science, 23*(1), 4–14.
Wilsdon, J. (2008). Public engagement in science across the European Research Area. In EC. *Public engagement in science. Report of the science in society session*. Portuguese Presidency Conference. "The future of science and technology in Europe", Lisbon, October 8–10, 2007.
Wilsdon, J., & Willis, R. (2004). *See-through science. Why public engagement needs to move upstream*. London: Demos.
Wilsdon, J., Wynne, B., & Stilgoe, J. (2005). *The public value of science. Or how to ensure that science really matters*. London: Demos.
Wynne, B. (1992). Public understanding of science research: New horizons or hall of mirrors? *Public Understanding of Science, 1*(1), 37–43.
Wynne, B. (2014). Further disorientation in the hall of mirrors. *Public Understanding of Science, 23*(1), 60–70.

Part III
Development and Innovation

Chapter 8
Engaging Technologies: Criteria for an Alternative Model of Technological Development

Miguel Ángel Quintanilla Fisac

Abstract At each stage of technological development, decisions are continuously taken that help shape the final result. However, the prevailing model of technological development in our society aspires only to maximise economic benefit, a state of affairs that determines the type of decisions taken. The result is technological alienation, i.e., the situation in which the users or operators of a technical system must abandon all hope of controlling the technology they are using if they wish to take full advantage of it. In contrast to the prevailing model of technological development leading to alienating technologies, a set of criteria is put forward here to assess technology projects that define an alternative model for development of engaging, non-alienating technologies.

8.1 Engaging Technologies

One of the characteristics of the predominant discourse regarding the philosophy of technology is its **pessimistic** tone (Ellul 1954, 1964). The same pessimistic outlook can be found in the majority of representations of technology in film and literature throughout the twentieth century. To a great extent, this pessimistic view of technology and science is also seen in surveys on public perception of science and technology (Bauer 2009; López Cerezo and Cámara Hurtado 2014): the majority of people accept the contributions of science and technology to welfare, health and wealth, but are wary of the dangers of a technologically advanced society and the undesirable consequences of technological development, particularly in terms of people's safety and environmental damage as a result of industrial applications of science and technology (FECYT 2011).

M. Á. Quintanilla Fisac (✉)
ECyT Institute, University of Salamanca, Salamanca, Spain
e-mail: maquinta@usal.es

B. Laspra, J. A. López Cerezo (eds.), *Spanish Philosophy of Technology*, Philosophy of Engineering and Technology 24,
https://doi.org/10.1007/978-3-319-71958-0_8

It is not easy to analyse the social mechanisms or logical structure accounting for the prevalence of this ambiguous and pessimistic discourse regarding technology. It would appear that, underlying these pessimistic views, there always exists a perception of technology and its development characterised by technological **determinism** and the idea of the **autonomy** of technique. According to this conception, technology evolves blindly, leaving no room for conscious human intervention.

In previous publications (Quintanilla 1997, 2005), I have discussed the non-deterministic nature of technological development and have always presented a somewhat more optimistic view of technology, as an arena in which we can explore the possibilities of what humanity can achieve, rather than as a source of danger and threats. In defending this view, I take up the baton from Ortega y Gasset, who was surely the only philosopher of the first half of the twentieth century to develop an optimistic and humanist philosophy regarding technology (Ortega y Gasset 1939) and the scientific tradition of epistemology and philosophy of technology, represented, by Mario Bunge (1966, 1985, 2014), among others.

Unlike what might have occurred with the most significant industrial and military technologies of the twentieth century, today's technology offers striking examples of highly sophisticated technical devices which are fully integrated in our day-to-day lives. This is an important point to consider, as one of the reasons for the technological pessimism of twentieth century philosophers was the **strange and alienating** nature of the most significant industrial technologies of the time. These were technologies which had an impact on people's daily lives, although their origins and structures were entirely alien from the lives they enriched. The railways of the nineteenth century constituted an incredible advance in transport and communication, but a rail transport system was something as utterly beyond the grasp of its users as a pyramid would have been to those building such structures in the Mayan or Egyptian civilizations. The great technological achievements, the "Machine" of authoritarian technology outlined by Mumford (Mumford 1967, 1970), exist right there in front of us, conditioning or determining our existence, but completely alien to our desires or our means to influence them.

Nowadays, however, the problem has become far more complicated. We now have technologies that are even more ubiquitous than traditional industrial technologies and – in a way – more impenetrable. Nonetheless, we have fully welcomed them into our daily lives, as extensions of ourselves, and it seems we have given ourselves over to their control. Think of the advanced, modern-day technology of the mobile phone. Thirty years ago, such a thing would have been barely conceivable. Nowadays, the mobile telephone network has broken down all imaginable technological, social, political and cultural barriers. Never has a technology developed and spread throughout the world at such a pace.

The idea discussed in this chapter pertains to all of the above. Our starting point is the assumption that technological pessimism arises from what could be called the 'alienating' nature that predominates most technologies and the development of contemporary technology in general. We shall analyse the structure of **technological alienation** and demonstrate that determinism and the apparent autonomy of technology are effects of this alienation, rather than the causes. At the same time, we shall put forward a model of technological development to prevent estrangement and alienation towards technology, a model which, by **contrast**, we shall call **engaging technologies,** i.e., **non-alienating** technologies. This article is not the first to air this model of non-alienating technological development (see also Quintanilla 2002, 2009a), but it is the first time that it has been so systematically set out. The idea of engaging technologies has already borne fruit to some degree, even before the formal birth of the concept. It has been noted that the concept of technological alienation here is not akin to that of worker alienation in the Marxist sense (Correa Lucero and González 2014) and that different implications of technological alienation must be understood depending on whether the technology is viewed from the perspective of the designer or the user (Parselis 2016a, b).

8.2 How Does Technological Alienation Occur?

Rooted in Marxism, the idea of alienation is very apt to describe a significant part of the relations that occur among users of modern technological systems. Marx's concept of alienation essentially refers to the experience of the worker whose action creates an object (a commodity) in which he himself is reflected and embodied, though at the same time the commodity is both alien and an independent entity, disconnected from its creator (Marx, *Economic and Philosophic Manuscripts of* 1844). This same experience, with the necessary nuances, is that which generally occurs in the relationship between the users of many technological devices and the devices themselves. The most complex technical systems are undoubtedly systems of actions, organised and intentionally designed to achieve the goals that the user is assumed to value (Quintanilla 2005). Thus, the user (not just the manufacturer or designer) is actually an essential part of any technological system, as a deliberate participant who determines the objectives and actions of manipulation and control that define the system itself. Yet at the same time, the user is only constituted as such to the extent that his or her objectives and actions are adapted and included in the structure of the system as a whole. It is this inter-dependency that makes technological alienation possible. The operator of the technical system is actively included in a network of interactions that exist only because that operator has activated them; yet, once produced, they can neither be substantially altered nor avoided. This is the essence of what we understand by technological alienation, which only partially coincides with the Marxist concept. The fundamental difference is that technological alienation does not occur solely through the use of machinery (under the control of the capital holder) by the worker in the production process, but also through the use of technology in general, if it has not been designed in such a way that the user retains control: a circumstance which, in principle, is independent of the ownership regime within which the technology operates.

The two main sources of technological alienation are firstly, the **lack of transparency** in the design of technological systems and secondly, simplification of the **assessment criteria** used to assess the processes of design and dissemination of technologies.

It could be said that the **lack of transparency** in a technological design is the concealment of its function and structure, its purpose and/or its *affordances* ("An affordance is a relationship between the properties of an object and the capabilities of the agent that determine just how the object could possibly be used." (Norman 2013, p. 11)) Generally, technological devices are the deliberate creations of human agents that are characterised by being designed to achieve certain objectives in a supposedly efficient way. A simple device usually indicates its purpose and how it should be used by way of its structure and shape. A typical example is a latch or handle: we hardly need to ask what it does or how it works. If the door is closed, it is obvious that it is kept closed because the latch is set in its holder, and if we want to open it, we must obviously move the latch, free it from its holder in the doorframe and pull or push the door. A handle is obviously a device for opening and closing doors: its benefits are obvious. In one way or another, all technical systems convey information about their roles and structure (Lawler 2007), but this information ceases to be so obvious and easily accessible as systems become more complex. Most likely, the function of an electronic lock on a safe will still be obvious, but instructions will be required to understand how it works. A bicycle is undoubtedly one of the few transportation devices (others include the scooter and the African cargo vehicle, the *chukudu*[1]) which is entirely accessible in terms of its structure, and which can easily be fully understood in terms of its operations (at least until the recent introduction of new gears, typical of today's mountain bikes, and electric scooters). A motorcycle is a far more complex, opaque system. In the case of the car (even including the traditional motor car, which could be disassembled in a garage), its complexity is closer to that of an organism than to that of its immediate functional predecessor, the horse-drawn carriage.

Lack of transparency is an obvious source of frustration – and, by extension, alienation – for the user, as revealed by scholars in both industrial and commercial design (Norman 2013). This kind of alienation is easily overcome, however. The operating instructions for a safe can be limited to a couple of simple instructions. The maintenance manual of a car is somewhat more complex, but, when necessary, we can always go to an expert mechanic whose knowledge would be accessible to us personally if we took the trouble or the time to learn about the workings of a car. This type of opacity is basically surmountable. It is not structural and, although it may be a source of alienation, its remedy can be found in the same ailment from which it arises: more knowledge allows us not only to do more complex things, but also to understand them better.

There is, however, another type of **technological opacity that is both structural and dynamic** and which leads to a strand of technological development that is unavoidably alienating, though not inevitably so. Think of technological interfaces which are 'transparent to the user' (TTTU), specially the GUI (Graphics User Interface), a characteristic of modern information technologies. Naturally, TTTU graphical interfaces are a good technical solution which facilitate the use of technology and thus enable their wider dissemination. However, what I would like to point

[1] https://en.wikipedia.org/wiki/Chukudu. Accessed July 1, 2016.

out here is that they have a paradoxical effect which must not be underestimated: they increase the alienation of technology. The simplicity with which a modern PC user executes the necessary instructions to carry out the simplest of very complicated electronic functions, even going so far as to identify emotionally with a supposed virtual reality (Turkle 2005), is matched only by the ingenuity with which a primitive human relied on the configuration of the constellations as an indication from the gods that the seasons for sowing, harvesting or rain were upon them. Strong convictions, complete ignorance, and an inability to control technology lead to total alienation. Without using the word 'alienation', Ortega y Gasset warned of the danger of this type of unawareness regarding technology almost a century ago:

> [Humans] could lose awareness of technology, and of the conditions (moral conditions, for instance) in which it is produced, coming to view it, as our ancient ancestors did, merely as a natural gift, which we simply have and which therefore does not require active and conscious maintenance. (Ortega y Gasset 1939)

This blindness or irresponsibility toward technology, denounced by Ortega, is an essential component of the kind of technological alienation that the engaging technology model aims to remedy.

A serious aspect of structural opacity as a source of technological alienation is that it is neither a malicious, deliberate implementation nor a design flaw in technical systems. Opacity can end up being more functional than transparency of function.

There are other aspects of technological design that produce similar concealment and inaccessibility – and hence alienation – which also provide the technical needs required for high levels of efficiency and functionality. In this case, we see an example of what might be called the **compositional** nature of technological systems.

One of the most notable features of technical systems is their compositionality (Broncano 1988): the existence of a certain hierarchy of possible levels of composition between technological systems. The simple machines of the Hellenistic tradition form part of any complex machine (any machine consists of wheels, levers, inclined planes, screws and nuts). Furthermore, any complex system can be broken down into several subsystems: a car has a cab, engine, transmission, brakes, steering and so on. An internal combustion engine, in turn, has cylinders, a crankshaft, cooling system, combustion gas exhaust system, ignition system and so on. In principle, what matters for a subsystem to operate as an integral part of another system is simply that it acts as a **black box**: the flow of inputs and outputs. Its internal composition is of interest only insofar as its components may affect that flow. This applies to any level of the composition of a system. What this means is that any technological system can be represented as a hierarchy of black boxes whose internal structures are, in principle, irrelevant for the composition of the whole. Here we see another source of technological alienation which is characteristic of many modern technological systems: the user is not faced with a mechanism whose function can be unravelled and understood by analysing its performance, but instead with an opaque structure of intertwined black boxes. If it works, all is well; but when something goes wrong, there is no other solution than to detect where the anomaly is occurring in the flow of inputs and outputs and then replace the corresponding black box.

We are also presented with another phenomenon that is at once alienating and functional. We have given up on repairing technological products, but instead we are able to update our technology continually in a relatively simple and inexpensive way. It is as though, in medicine, we could remedy any disease by replacing the affected organ with a new one, instead of trying to make it function better.

8.3 Could Things Be Improved?

Pessimists undoubtedly think that these features in some technical systems are the inevitable consequences of the autonomous, uncontrolled development of technology within a social and economic context in which short-term profit-maximisation strategies and private benefits are valued most. This is not true, however. In my opinion, technologies are what they are because there are people who decide that they should be that way. What we can conjecture in light of philosophical and social analysis is that different decisions could be taken without ceasing to drive technological development forward while making the most of its beneficial social and economic effects.[2]

The question is whether or not we are able to conceive of an alternative model of technological development, a model aimed at engaging, rather than alienating technologies. The intuitive idea behind the notion of engaging technology is simple. We have seen that technological alienation manifests in the opacity of technologies and the inability to exercise human control over technological development in terms of moral responsibility. **The notion of engaging technology can be understood as the opposite of alienating technology: it is not friendly, soft or 'intermediate' technologies (although all of these descriptions of alternative technology development should be borne in mind), but rather technologies over which we exercise control and where we take responsibility for their development as enlightened citizens, rather than merely as consumers in the market.**

What would the engaging technology model look like?

It is possible to apply the Darwinian model of natural selection to explain the evolution of technological trajectories (Dosi 1982). A twofold process exists in this model: on the one hand, there is a need for a new idea, goal or technological purpose, to spark the process of change. Starting out from there, the innovation process commences with steps involving research, development, testing, demonstration and evaluation, subsequently progressing to the production of the final prototype, followed by its marketing. Numerous decisions have to be made throughout this pro-

[2] Naturally, once we have accepted the normality of structural opacity, it becomes easier to use it for other purposes which have nothing to do with the functionality of technology, but with its financial profitability. The case of Volkswagen is revealing: the use of technology to mislead users by generating false data for the electronic systems monitoring contaminant gas emissions. It is difficult to imagine how a large-scale fraud such as this could have been perpetrated without the legitimising contribution of the black-box design, where the black boxes become armoured boxes, inaccessible to the user.

cess (which need not be linear, but rather involve a great deal of feedback and openness to influence by external factors: see Kline 1985). Many of these are based on evaluation criteria, some of which are purely technical, others economic, commercial, cultural, and so on. The end result must also undergo the process of 'natural selection', a function which will be fulfilled in this case by the market or any other mechanism of social acceptance. If it succeeds, then new applications of the technology will be established, contributing to its consolidation and wider dissemination, and so on.

Admittedly, there are multiple occasions throughout this process (such as the selection of system objectives, selection and organisation of components, assessment of results, etc.) when evaluation criteria can be introduced. These subsequently contribute to whether the end result approaches the model of engaging technologies, or veers clear of it. On other occasions (Quintanilla 1993, 2005), the outlines of this technology assessment scheme have been proposed based on the combination of two types of criteria: inherently technological; and socio-technical or extrinsically technological criteria. The former refers to technical properties of the elements being assessed, in areas such as feasibility, efficiency and reliability. The latter refers to social and cultural values, such as the suitability or economic viability of a project, its moral, aesthetic and political value, as well as risk assessment, environmental impact and the social implications of a technology. What is presented here is a set of criteria that can help promote technological development based on engaging technologies. Many of them are mixed indicators, affecting strictly technological properties as well as economic, moral, political and social characteristics (Table 8.1).

Table 8.1 Criteria for evaluating technological projects

Intrinsic	Extrinsic	Engaging technology
Feasibility	Economic viability	Open
Effectiveness	Moral suitability	Versatile
Efficiency	Cultural suitability	Controllable
Reliability/likelihood of becoming obsolete	Risk assessment	Limited
	Environmental impact	Reversible
	Social impact	Recoverable
		Understandable
		Participative
		Sustainable
		Socially responsible

8.4 A Set of Criteria for Engaging Technologies

8.4.1 Openness

Open and closed technologies have existed for thousands of years. A technology is open if there are no limits on its ownership, use or development, stemming solely from standards or criteria which are external to the specific functional needs of the technical design. As far back as ancient Egypt, engineers were able to implement operating machines whose mechanisms were kept secret. These are the first examples of closed technologies (Hall et al. 1971). The *s*ame mechanism (e.g., using steam pressure to open the doors of a temple) could remain secret or be explicitly revealed without harming its technical efficiency, although it would certainly lose its mysterious, supernatural air on being revealed. In such a case, the cause of technological alienation is not the size of the machine or the spectacular nature of its operation, but the fact that its internal logic remains concealed, incomprehensible and hence uncontrollable. A Roman aqueduct may be an imposing feat of engineering, but it holds no secrets and therefore should be considered an open technology.

One might ask whether certain types of technologies need to be closed due to their very nature. For example, the central component of a nuclear reactor in which fuel fission occurs is inaccessible while in full operation,[3] and there is reason to believe that the design of this radioactive component would not improve much even if it were made accessible. However, access limits not in the core, but in the vicinity of the nuclear plant may only be justified as safety measures against the danger of attacks or accidents. It is not something inherently required by the technology itself, but by external factors. In principle, it would be conceivable to build a 'theme park' around a nuclear power plant,[4] which would enable civilians and schoolchildren to familiarise themselves with this energy source and the risks it entails.

Today, the term 'open technology' is linked to the free software movement and the GNU/LINUX Project, initiated and promoted by Richard Stallman through the Free Software Foundation (FSF) operating system. Stallman himself (1986) defines the essential feature of free software:

> The word "free" in our name does not refer to price; it sometimes refers to freedom. First, the freedom to copy a program and redistribute it to your neighbours, so that they can use it as well as you. Second, the freedom to change a program, so that you can control it instead of it controlling you; for this, the source code must be made available to you.[5]

In 1998, a split occurred in the free software movement, creating the Open Source Initiative (OSI), which incorporates the essential elements of free software, but emphasises that the source code should be accessible and reproducible by anyone.

---

[3] Of course, the same is true of the boiler of a coal-fired kiln, and therefore it is not considered wrong for the kiln to be kept closed whilst alight.

[4] Whilst not a theme park, as such, it must be acknowledged that tourists can now visit the ruins of the Chernobyl nuclear plant in the Ukraine.

[5] https://en.wikipedia.org/wiki/The_Free_Software_Definition#The_definition_and_the_Four_Freedoms

As Stallman himself puts it, the differences between the two branches of the movement – free software and open-source software – are more nuanced and ideology-based than practical. In both cases, the computer program protected by a free software/open-source licence must be able to be copied, modified and distributed, incorporating any changes and always maintaining the copyright of the original product. Overall, the free software movement, both in its origins and in practice, very well defines the spirit of what we today call an 'open' technology.

The open nature of technology should not be confused with ease of use or private ownership. GNU/Linux technology is complicated and its use requires a certain level of technical expertise. Yet what is definitive and makes it a prototype of engaging technology is that there are no secrets: all of the code is accessible to any user with the appropriate technical skill, and any user who accesses it can use it, change it or improve it and redistribute the result incorporating the changes. There are also rules and institutions concerned with maintaining standards in the GNU/Linux system in order to prevent fraud. That does not make it a closed technology, though: it simply makes its use and development safer and consistent, but none the more restricted.

Along with GNU/Linux, there are many other open source or free software initiatives. These include the widespread applications offered as open technologies (in the sense of 'free'), although they are actually commercial 'claims' to get the user interacting with a program with artificial limitations to try it out with the aim of encouraging the participant to 'upgrade' to a 'premium' or similar version that is no longer 'free'.

Thus, what characterises a truly 'open' application is not so much the cost of accessing it, but rather the fact that the structure of the technology has not been conditioned to limit access and protect intellectual property rights. **The technology is open when it is designed in such a way that anyone can use and develop it**. A fixed bridge over a river, which one must pay a toll to cross, might be an open technology. However, it may become more or less closed by imposing higher or arbitrary tolls, or also by changing its structure to prevent freedom of movement (e.g., by installing artificial mobile barriers or designing the height of the bridge so that large vehicles such as buses cannot pass over it (Winner 1980)).

Open technologies are not limited to software alone. In fact, any piece of technology has design components that can be made available to all. In recent years, free hardware projects such as RepRap 3D printers, RaspberryPi pocket computers and the ARDUINO digital microcontroller have proliferated. In all these cases, the essential feature is that there are no limitations on the usage, modification, copying or distribution of devices based on these technologies: it is not that they are 'free' (clearly they are not), but they are open, insofar as they can be used freely, though there are constraints mandating that all derivatives of the product remain open-source. This openness, in itself, does not fully prevent technological alienation, but it is a basic condition for us to be able to speak of engaging technologies.

We therefore propose a criterion for engaging technology as being open, defined as follows:

> A technology is open when external influences cannot impose their own access restrictions on usage, copyright, modifications or distribution.

8.4.2 Versatility

A technology is versatile if it is designed so that it can be used by different types of agents and for different purposes (i.e., if it is useful in various contexts: Quintanilla 2005).

Of the two dimensions (extension and intension) that we can distinguish between in technological development (Quintanilla 1996), one grows when we design technological systems tailored to specific objectives (intensional growth leads to increased accuracy and control in a given area of reality), and the other, as we deal with broader and more diverse objectives (extensional growth means an increase in the scope of the technology's usage). Take, for example, the tightening of a ¼-inch nut. A spanner calibrated to the size of the nut will be more effective in this case, whilst an adjustable wrench is more versatile because it can handle different size nuts and may even be more practical when the nut has a defect that prevents it from fitting snugly in a spanner of its standard size.

Can we always truly justify prioritizing a versatile technical solution over another which is perfectly fitted to the task at hand? Clearly not. There are situations in which it is necessary to design tailored solutions, and these are all the more viable and efficient the more targeted they are for their purpose. For example, in the pharmaceutical industry, it is almost always preferable to have specific drugs to tackle each variant of a disease, thereby avoiding many of the side effects of 'broad-spectrum' antibiotics. There are many other situations in which the value of a technology can be increased if versatility-oriented design elements are incorporated. There are two basic options: versatility through integration of technologies and versatility through alternative uses.

Consider a spreadsheet. Before spreadsheets were invented, there were specific programmes for each of the tasks now performed by a spreadsheet programme. This is an example of a multipurpose application integrating hundreds of different functions, all related to the handling, processing, management and so on of tabular data. Now, any new data-processing application is considered much more valuable if it can be integrated as an extension of a spreadsheet, thus increasing its versatility.

Another notable case is the proliferation of smartphone applications, all of which are small programmes, usually written in JavaScript, designed to be easily integrated into the operating system of a mobile phone in a way that is practically transparent to the user. Each 'app' usually serves one specific function, and the fact that it is integrated into a single system along with many other 'apps' makes that system – be it a mobile phone or a tablet – a multi-purpose system. Thanks to these 'apps', the smartphone has become the most versatile multi-purpose device in contemporary computer technology.

Another example of versatile technology lies in the possibility of alternative uses for technology (Quintanilla 2009b). Wiki technology, for example, was designed to meet the needs of Wikipedia, but can be applied to many other uses in collaborative content production.

There is undeniably a relationship between the open nature and versatility of technology. In principle, an open technology facilitates the development of alternative uses and therefore makes it potentially more versatile. Conversely, in many cases, the constraints resulting from the application of access control technologies through DRM (Digital Restrictions Management) eliminate certain benefits by restricting a technology's versatility resulting in a gap in the market, to be filled by another targeted solution.

In summary, two types of versatility can be seen in a technical system: versatility via integration of various technological capabilities in a single system, or via the enhancement of a technical system through possible alternative uses arising from the original design. We can thus now include another principle for the development of engaging technologies, namely that of versatility:

> A piece of technology is more versatile when it is able to integrate different objectives into a single technical system or allow operators to make alternative uses of the technology. When all other factors are equal, versatility increases the engaging nature of a technical system.

8.4.3 Controllability

One of the most distinguishing components of technological alienation is that reflected in the myth of the 'Rise of the Machines', also known as the 'sorcerer's apprentice' or the 'Frankenstein syndrome'. This is a very widespread phenomenon. At the most basic level, it holds that additional risks are posed to humanity through the integration of machines into society. That is to say, if there are new machines, then there are also new risks of accidents. The specific idea behind the 'Rise of the Machines' is that the risk arises not so much from the mere existence of machines and their inappropriate use, but from their autonomous operation.

In both literature and film, there are iconic interpretations of the myth of the rise of the machines. We can start by considering the medieval Hebrew myth of the *Golem*: an artificial humanoid made of clay with the ability to act as a kind of soulless robot. Closer to industrial civilisation is the myth of the *Sorcerer's Apprentice*, inspired by a poem by Goethe (1797). In the same breath, we should also mention Mary Shelley's *Frankenstein* (1818) and the cult film *Blade Runner* (1982) based on the novel by Philip K. Dick (1968). The common denominator in all these stories is the idea that machines or devices invented by man to carry out arduous tasks become autonomous and get out of control, with disastrous consequences for their creators. Both the *Golem* and the *Sorcerer's Apprentice* introduce the idea of the machine being out of human control. *Frankenstein* and *Blade Runner* also add the idea of rebellion against the creator and even provide a possible justification as to why machines are rebelling; i.e., the creator is not taking responsibility for his creations (or for their limitations).

These are very good examples of stories illustrating the notion of well-behaved technologies. As in the cases mentioned above, here too we find several complex issues belying the apparent simplicity of the idea. Let us consider Asimov's "Laws of Robotics" (2004). According to Asimov, a robot must strictly comply with the following laws:

1. A robot may not injure a human being or, through inaction, allow a human being to come to harm.
2. A robot must obey orders given it by human beings except where such orders would conflict with the First Law.
3. A robot must protect its own existence as long as such protection does not conflict with the First or Second Law.

The problem is that there are situations where, in attempting to apply these laws to its behaviour, the robot finds itself paralysed; for instance, when it is impossible to avoid harming one human without harming another. In cases such as these, the situation can be clarified somewhat by applying what is known as Law 0 of robotics: *A robot may not harm humanity, or, by inaction, allow humanity to come to harm.*

Actually, what lies behind this discussion about the limitations of robotics is a very powerful idea: *the technology (the robots) cannot be autonomous, but instead must always be under human control.* Having established this principle, the circumstances in which the human operator exerts that control may vary, within the limits imposed by the general principle.

Thus, a technology may be said to be controllable insofar as its structure facilitates human control over its operation. This implies that a human operator should be able to correct possible deviations in the system's operation, including the ability to stop it and restart it efficiently in case of emergency (i.e., the technical system is not entirely autonomous and the final responsibility for a robot always lies with a human being).

In summary, in an engaging technology model, the controllability of technical systems is a parameter to be considered, and is defined as follows:

> A technology is more or less controllable according to the degree to which its operation, functioning and stopping depend directly on a human agent.

8.4.4 Limitation and the Precautionary Principle

There are two types of limitations that are relevant for the design of a technical system: functional or intrinsic limitations, derived from the system's actual structure; and extrinsic limitations, pertaining to the social and environmental consequences of using the system. From the latter is derived a third group of limitations: those that affect the system's design and intrinsic structure, but which are extrinsic in origin. An intrinsic limitation of technology is that the second law of thermodynamics cannot be broken, meaning that we cannot set technological goals which are impossible to achieve, such as creating a perpetual-motion machine. An extrinsic limitation, for example, is the fact that social networks must not be used to harm or defame anyone. Finally, intrinsic limitations caused by extrinsic factors comprise all the limitations imposed on the design of a technology that add constraints on its use, aimed at protecting intellectual property rights (restrictions on the copying of computer files) or preserving a sales quota or allowing the principle of precaution to be applied.

Intrinsic limitations derive from feasibility criteria, which may be material or operational in nature. A technological goal is scientifically feasible if it falls within the limits compatible with the available scientific knowledge that is relevant for the technical system we are attempting to design. It is materially feasible if the material resources needed to build the system (its material components) exist and are available, while it is operationally feasible if operators are available who have the appropriate knowhow to activate and control the system.

In its negative version, we can formulate a principle of intrinsic **non**-limitation of a technology as a variant of the principle of versatility: **technical systems must not be designed in such a way as to prevent their being used for all their potential legitimate functions**.

On the other hand, the extrinsic limitations do need to be established if we wish to preserve the engaging nature of the technological development, especially in terms of the technology's environmental and social impacts (see below). For instance, the production and use of non-recyclable plastic bags as part of the packaging of consumer products should be avoided, because of its environmental consequences, though in technical and economic terms, they may seem a very appropriate solution.

A relevant case is the application of the **principle of precaution**. We are dealing with situations where we do not have sufficient scientific evidence to determine the consequences of the implementation of a new technology or to prevent such consequences. In these cases, the so-called precautionary principle stipulates that such technologies should not be implemented until we have relevant, sufficient and secure information at our disposal. Put differently, the precautionary principle is tantamount to saying that in technology development, our actions should be guided by **prudence**: if we are not sure of what will happen and we have well-founded fears that the consequences may be seriously dangerous, the principle states that we should act with caution. Understood in this way, it can be considered a facet of the principle of limitation of technological development.

As a criterion for the development of engaging technologies, the limitation principle must be understood thus:

> We should act carefully and limit the reach of technology so that its development is compatible with the material and human resources available and hence its social and environmental consequences are predictable.

8.4.5 Reversibility

A process in a system is reversible if, from its end state, it is possible to initiate a sequence of events which will return it to its initial state. The heating and cooling of a liquid in an open system are reversible processes (as can be observed every day in our refrigerators). However, the extinction of a biological species or the appearance of a genetic mutation in a new ecological niche are examples of irreversible processes. The development of a new technology always involves some or other irreversible component, but it is often possible to incorporate characteristics that can help increase or decrease reversibility.

In the process of evaluating technological options, not much attention is generally paid to reversibility. It is assumed that the introduction of a technological innovation is justified by the value of its objectives and the efficiency of the solution delivered. Once the decision to produce the technology has been made and it has been distributed, the world will have changed a little and we suppose that the change will have been for the better. Thus, we do not consider the option of retracting out steps. However, this approach is not always reasonable. In general, human technologies are not very efficient, which means that for each of the goals we pursue by creating them, we also produce a great many unforeseen results. There are two possible solutions to improve the situation thus created: either we adapt our evaluation criteria so the unexpected results fall within the bounds of acceptable change, or we attempt to improve our existing technologies so that, we can improve their efficiency and reduce the undesirable effects of our technical actions via new versions. The strategy usually adopted in these cases involves adding new layers of technological development over those which we have previously built. This strategy is known as **patching**: we improve what we have already made by applying patches to it which remedy some of its limitations. Sometimes, though, it is better to restart the whole process from scratch and take the technological development in another direction. In these cases, it tends to be important to be able to retrace the path we have taken; at least partially. On certain projects, this can prove especially crucial. For example, when evaluating the social dimensions of biofuel production technology, it is easy to see that bioethanol production from urban waste causes few irreversible social changes, as opposed to what occurred with first-generation fuel-production technologies using corn seed, which caused profound changes in the agricultural production of vast swathes of countryside which are difficult to reverse, from both an environmental and social standpoint (Ribeiro and Quintanilla 2015).

Generally speaking, technological development causes many irreversible changes and processes, but as a criterion for engaging technology, we should adopt the following:

> Under equal conditions, preference should be given to those technological options which are partially or totally reversible, meaning that it is possible to restore the natural or social environment in which they are being implemented, and redesign alternative options ab initio, if needed.

8.4.6 Recovery, Maintenance and Obsolescence

For years, the advertising strategies accompanying capitalist economic growth have meant that certain maxims have been adopted as common-sense options. When examined with a critical eye, however, these maxims are evidently cultural and technical aberrations, and even financial absurdities. One of these common-sense maxims of the mass consumption-based economy is the maxim of 'use and throw away': i.e., the proliferation of short-lived consumer goods made of disposable material, which it is not economically profitable to maintain and recover. The developmental model based on engaging technologies is diametrically opposed to this: unlike the throwaway maxim, the model proposes to prolong the use and maintenance of technical systems, minimise non-recoverable technological waste, and avoid planned obsolescence.

A technological item is said to be obsolete when one of the following situations occurs: (a) it has ceased to have the value it originally held for us (we are no longer interested in it because we have ceased to appreciate the function it served); (b) it has stopped working properly and the cost of repairing it is excessive in comparison to other available options; or (c) a new model has come onto the market which fulfils the same specifications and possibly some additional ones, too, making the new model preferable from a technical, aesthetic or simply publicity-based point of view.

To some extent, we are surrounded by gadgets which, by definition, are fated to obsolescence. Everything we have is perishable and this should present no surprise. However, there is a type of obsolescence which should not be accepted as inevitable or harmless: so-called **planned obsolescence**. This is when the items or products sold to us are designed in such a way that, from the very start, they include a planned expiry date, aimed at maximizing the manufacturer's or vendor's profits, rather than the customer's satisfaction. According to innovation historians, the first time this was consciously and systematically done was in the incandescent lightbulb production industry. In 1924, the world's leading manufacturers formed the Phoebus cartel, agreeing to limit the lifespan of their bulbs to 1000 h, to ensure sustained demand, at a pace that was compatible with their interests as manufacturers.[6] The first time that planned obsolescence was publicly advocated as an economically viable strategy to stimulate demand was in 1932 (London 1932). Since then, the practice of planning obsolescence has become widespread and has even become a distinguishing feature of certain industries. Such is the case, for instance, with the fashion industry: the catwalks of the season are merely a ritual of renovation of clothes which, from the moment the new products appear, become aesthetically obsolete. However, this can be deemed a special case, because in fact, what the fashion industry sells only has value because it is short-lived: it falls out of fashion. The problem is that, today, the same publicity rituals found in the fashion industry have become commonplace in the technology industry. For example, for the major leaders in computing, it has become routine practice to schedule launch campaigns for new products which force the obsolescence of earlier models.

The height of refinement comes when technological devices include software instructions which bring about the 'artificial' obsolescence of the whole system. In

[6] As evidence of the untruth, the fire station in Livermore, California has an incandescent bulb (http://www.centennialbulb.org/ctbulb.jpg) which has been shining continuously for over 100 years. The mere existence of that bulb disproves all technical arguments as to the obsolescence of this type of object.

such cases, planned obsolescence may even be an act of fraud. Today, all products that are sold must carry a warranty period during which any defect will be repaired at the manufacturer's expense. There should also be an obligatory period of non-obsolescence, guaranteeing that, for the stipulated period of time, the manufacturer commits to maintaining the technical service and providing spare parts so that the product does not artificially become obsolete. It also ought to be compulsory for each model in a range of products to clearly specify the differences and technical compatibilities with previous models. Ultimately, we must try to prevent 'use and throw away' becoming general practice.[7]

Another strategy with decisive effects in the technological development model pertains to the management of by-products and technological waste. It could be argued that, under certain circumstances, it is easier to replace a broken-down system than attempt to keep it operable by repairing it. Similarly, it is easier to prevent breakdowns by limiting a system's expected lifespan than try to prolong it and, in so doing, incur a higher repair cost. What, though, would happen if any of the following conditions were met? (1) The cost of labour in poor countries where new devices are manufactured were equal to that in the countries where their consumption is widespread. (2) The cost of recovery and management of technological waste were included in the sale price to the end user. (3) The price charged to the consumer for a spare part were equal to the cost of the original part for the manufacturer.[8] If these conditions were realised, it could become financially viable to repair a broken system rather than replace it with a new one, as well as to extend the second-hand market rather than penalising it with aggressive sales strategies. This is built into our new principle for development based on the model of engaging technologies:

> It is essential to facilitate the recovery and maintenance of technological devices, to ban planned obsolescence, and to incorporate the operations of waste management and recycling into the design and marketing of technical systems.[9]

8.4.7 Understandability

A technical system can be designed in a way that facilitates or obfuscates the understanding of the system: its goals, its contributions, its value, etc. The requirement of manifest or understandable design goes one step further than does open technology.

---

[7] Today, there are numerous initiatives to raise awareness of this issue. Of particular note is the Spanish documentary '*Comprar, tirar, comprar*' [Buy, Throw away. Buy], available at: http://www.rtve.es/television/documentales/comprar-tirar-comprar/

[8] Why is it that an original replacement part tends to be sold at a much higher price than that paid when the product is first made on the assembly line? In the automobile market, it is well known that Original Equipment Manufacturer (OEM) spare parts cost, on average, more than those from an After Market Manufacturer (AMM).

[9] These principles already form part of the movement against planned obsolescence, whose organisation issues certification to products and companies which comply with them. (http://feniss.org/sello-issop/). And the European Economic and Social Committee called for a total ban on planned obsolescence in 2013 (http://www.eesc.europa.eu/en/news-media/press-releases/eesc-calls-total-ban-planned-obsolescence)

An open technology makes it **possible** for users to access and understand its operation. An understandable technology, meanwhile, is an open technology which also actively **facilitates** understanding of how it works. Of course, we cannot expect a complex technology which uses principles and processes stemming from the most advanced findings in physics, chemistry or biology to be as equally comprehensible as the operation of a simple machine (e.g., the operation of an old gramophone is more easily understood than is that of a compact disk). However, we must assume that, in general, it is preferable for users to be able to understand how their devices work, and how they are supposed to use them. This is a requirement of 'illustration', which can effectively help prevent alienation in the face of technology.

One way of restricting the understanding of a technology is to use the strategy of black-box design, making each of the black boxes in a complex system an impenetrable armoured box, inaccessible to the user.

One practical consequence of the principle of understandable technologies is that the technical systems should always come with an **instruction manual**. These manuals should be easily accessible and written in an unambiguous, easy-to-understand way.

Furthermore, the very design of the item should take account of what could be called its cultural dimension; i.e., its ability to integrate into broader cultural systems which facilitate appropriate, profitable use. An automatic door, for example, could be designed in such a way that users naturally and effortlessly understand what they need to do to open it, or else it could cause accidents and inconvenience if users have serious difficulties in understanding what they need to do (Norman 2013).

In short:

> An engaging technology should be an understandable technology, which comes with a proper operating manual, designed so that the system's operation and that of its components can be understood by users with a basic cultural background.

8.4.8 Participative Design

Technology can be designed so that a system's potential users either actively participate or so that they may simply react to a previously and independently designed offer. The standard model of technology development is not generally participative. Rather, it corresponds to technocratic and hierarchical behaviour. It is supposed that somebody (no matter who or why) selects an objective for its potential value on the market; technicians design a device which delivers the selected objective; and, finally, the market proves or disproves the value of the end product. The public may partially participate in the selection of the goals on a very broad scale – e.g., through political representatives or civic associations – when decisions are being made about objectives in technology development. The public can also participate in the market's approval of innovations by means of the cumulative effect of individual consumer decisions which determine the success or failure of an innovation. Lastly, it is possible to promote the participation of a broad-ranging audience, involving them in the evaluation and design of a technology throughout its process of development.

In order for technological development to be open to collaborative participation, appropriate institutional agreements are needed, as are design options which facilitate participation. Some experiments have been conducted in this field, which need to be extended further. For example, civic participation in evaluating a technology's environmental impact is a widespread and well-established practice in the judicial norms of the most advanced countries. Other experiments, such as parliamentary science and technology options assessment bureaus (e.g., the STOA[10] Programme run by the European Parliament), also need to be driven forward as part of an engaging technology-based technological development model.

In summary:

> Engaging technologies should encourage civic participation in their development. To this end, institutions and procedures should be set up which enable the potential users of a technology to take part in not only the process of acceptance or rejection of a predefined technological offer, but also in the discussion of the different technological options throughout the development process.

8.4.9 Sustainable Development

According to the definition given by the Brundtland Report (Brundtland et al. 1987), sustainable development is:

> (…) development that meets the needs of the present without compromising the ability of future generations to meet their own needs. It contains within it two key concepts:
>
> The concept of 'needs', in particular the essential needs of the world's poor, to which overriding priority should be given; and the idea of limitations imposed by the state of technology and social organization on the environment's ability to meet present and future needs.[11]

Understood in this way, sustainability must clearly be viewed as an assessment criterion on engaging technology projects. It is merely a question of the process of technology assessment incorporating reference to human needs and to the continuity of technological development in environmental, economic and social terms. If we allow technology to deteriorate the environment until life on Earth becomes unviable, we are reducing the possibilities for future technological development and giving up on the programme of successfully controlling reality. The same occurs if, as a result of our decisions, we destroy our society's ability to survive and develop or our economy's ability to generate wellbeing and wealth. Understood in these terms, the criterion of technological sustainability incorporates into practice the fundamental content of the idea of "appropriate technology" or "intermediate technology" put forward by E. F. Schumacher (1973) in a generic sense (Willoughby 1990, 2004).

---

[10] http://www.europarl.europa.eu/stoa/cms/home

[11] http://www.un-documents.net/ocf-02.htm

8.4.10 Social Responsibility

Of course, technology may develop either in a way which contributes to widening the chasm between peoples and nations, or it may favour the less well off. What is clear is that whether it develops in one direction or the other will depend on the decisions that are made. For example, we could favour drug development projects to combat diseases linked to poverty, or dedicate all our efforts to the fight against disease which are prevalent in the most developed countries. We could promote projects which help bridge the technological divide between rich and poor countries, or focus all our efforts on those innovations which have the greatest commercial potential, though they may be less important in social terms.

> The principle underlying the engaging technology model is the maxim of developing technologies that help combat social inequality and contribute to improving the situation of the less fortunate.

8.5 Looking Forward

The development of a technological path is a complex process with multiple dimensions. One of these – though simply one of many – is the economic dimension: the market value of a new technology. Today, discourses and models of technological development that exclusively focus on the economic dimension prevail. The assessment criteria for a technological project are limited to a calculation of the costs versus the profits, commercial opportunity and the like. One of the characteristic features of an economic model of technological development is its tendency to disregard any other type of values which could impose limits on the technology's development: the mantra is 'if something new can be done and there is a market for it, it should be done'. This view of technological development generally goes hand in hand with a business model based on continuous growth. Critics of this model tend to advocate alternative models based not on a technology's economic value in the market economy, but on the social value as a public or community asset, and, ultimately, on the containment of growth or even on the promotion of degrowth. The "intermediate technologies" model put forward by Schumacher (1973) is also a response to this economistic paradigm. It is a model of technologies that are accessible to poorer societies, whose social benefit, within the local economy, is greater than that of more mature technologies on the global scale.

Given these options, the engaging technology model is intended to be compatible with different value systems. In itself, it is not incompatible with an economy guided by the generic objective of increasing wealth, though this will never be the only value that is taken into account. Nor is it incompatible with respect for the general operating framework of a morally acceptable market economy, or with the operation of a cooperative social economy, provided that these respect the principles set out above.

It is too early yet to estimate all the potential consequences of the application of the engaging technology model. In the immediate future, we shall need to further revise and refine the model and, above all, develop application protocols to test how it will work in practice, in line with the proposal made by M. Parselis in his doctoral thesis (2016b). Meanwhile, though, we can consider that it represents one more option, amongst the many that currently exist; one which claims that a different world is achievable, in which technologies are not only better, but also less alienating. That is, a world more worthy of respect and care, one which is more engaging.

Acknowledgments I am indebted to Ben Engel for his professional contribution to the English translation of the original Spanish text.

References

Asimov, I. (2004). *I, Robot* (Vol. 1). New York: Spectra.
Bauer, M. W. (2009). The evolution of public understanding of science-discourse and comparative evidence. *Science, Technology & Society, 14*(2), 221–240.
Broncano, F. (1988). Las posibilidades tecnológicas. Una línea de demarcación entre ciencia y tecnología. *Arbor, 507*, 47–70.
Brundtland Commission, et al. (1987). *Report of the World Commission on Environment and Development*. United Nations. A/42/427.
Bunge, M. (1966). Technology as applied science. In F. Rapp (Ed.), *Contributions to a philosophy of technology* (pp. 19–39). Dordrecht: Springer.
Bunge, M. (1985). *Treatise on basic philosophy: Volume 7. Epistemology & methodology III: Philosophy of science and technology. Part II: Life science, social science and technology*. Boston: Reidel.
Bunge, M. (2014). *Ciencia, técnica y desarrollo*. Pamplona: Editorial Laetoli.
Correa Lucero, H. E., & González, J. E. (2014). El lugar de la estructura social capitalista en la concepción de las tecnologías entrañables. Apuntes críticos sobre la propuesta de Miguel Ángel Quintanilla. *Revista CTS, 27*(9), 207–213.
Dick, P. K. (1968). *Do androids dream of electric sheep?* London: Gollancz.
Dosi, G. (1982). Technological paradigms and technological trajectories. A suggested interpretation of the determinants and directions of technical change. *Research Policy, 11*(3), 147–162.
Ellul, J. (1954). *La Technique ou l'enjeu du siecle*. Paris: Armand Colin.
Ellul, J. (1964). *The technological society*. Translated from the French by John Wilkinson. With an introduction by Robert K. Merton. New York: Vintage Books.
FECYT. (2011). *Percepción social de la ciencia y la tecnología en España*. Madrid: FECYT.
Hall, M. B., Woodcroft, B., & Greenwood, J. G. (1971). *The pneumatics of hero of Alexandria*. London: Macdonald & Company.
Kline, S. J. (1985). Innovation is not a linear process. *Research Management, 28*(4), 36–45.
Lawler, D. (2007). *Las acciones técnicas y sus valores*. Salamanca: Ediciones Universidad de Salamanca.
London, B. (1932). *Ending the depression through planned obsolescence*. New York. https://catalog.hathitrust.org/Record/006829435. Accesed 1 Oct 2017.
López Cerezo, J. A., & Cámara Hurtado, M. (2014). Cultura científica y percepción del riesgo. In B. Laspra & E. Muñoz (Eds.), *Culturas científicas e innovadoras. Progreso social* (pp. 159–178). Buenos Aires: Eudeba.

Marx, K. (1844). *Manuscritos Económicos y filosóficos*. 2016. https://www.marxists.org/espanol/m-e/1840s/manuscritos/. Accessed 26 July 2016.

Mumford, L. (1967). *The myth of the machine [Vol. 1]. Technics and human development*. New York: Harcourt, Brace & World.

Mumford, L. (1970). *The myth of the machine [Vol. 2]. The pentagon of power*. New York: Harcourt, Brace & World.

Norman, D. (2013). *The design of everyday things* (Rev and exp ed). New York: Basic Books.

Ortega y Gasset, J. (1939). *Ensimismamiento y alteración. Meditación de la técnica*. Buenos Aires: Espasa Calpe.

Parselis, M. (2016a). El valor de las tecnologías entrañables. *Revista CTS, 32*(11), 71–83.

Parselis, M. (2016b). *Las tecnologías entrañables como marco para la evaluación tecnológica*. Doctoral thesis, Universidad de Salamanca, Salamanca.

Quintanilla, M. A. (1993). The design and evaluation of technologies: Some conceptual issues. In C. Mitcham (Ed.), *Philosophy of technology in Spanish speaking countries* (pp. 173–195). Dordrecht: Springer.

Quintanilla, M. A. (1996). The incompleteness of technics. In G. Munévar (Ed.), *Spanish studies in the philosophy of science* (1st ed., pp. 89–102). Dordrecht: Springer.

Quintanilla, M. A. (1997). El concepto de progreso tecnológico. *Arbor, 157*, 377–390.

Quintanilla, M. A. (2002). La democracia tecnológica. *Arbor, 683*, 637–652.

Quintanilla, M. A. (2005). *Tecnología: Un enfoque filosófico y otros ensayos de filosofía de la tecnología*. México: Fondo de Cultura Económica.

Quintanilla, M. A. (2009a). *Tecnologías entrañables*. Sin Permiso (Blog). http://www.sinpermiso.info/textos/index.php?id=2843

Quintanilla, M. A. (2009b). Recetas para hacer real otro mundo posible. In D. Lawler & J. Vega (Eds.), *La respuesta a la pregunta: Metafísica, técnica y valores* (pp. 95–104). Buenos Aires: Editorial Biblos.

Ribeiro, B. E., & Quintanilla, M. A. (2015). Transitions in biofuel technologies: An appraisal of the social impacts of cellulosic ethanol using the Delphi method. *Technological Forecasting and Social Change, 92*(March), 53–68. https://doi.org/10.1016/j.techfore.2014.11.006.

Schumacher, E. F. (1973). *Small is beautiful: A study of economics as if people mattered*. London: Blond and Briggs.

Stallman, R. (1986). *GNU emacs manual*. Cambridge, MA: Free Software Foundation.

Turkle, S. (2005). *The second self: Computers and the human spirit*. Cambridge, MA: MIT Press.

Willoughby, K. W. (1990). *Technology choice: A critique of the appropriate technology movement*. Boulder: Westview.

Willoughby, K. W. (2004). Technological semantics and technological practice: Lessons from an enigmatic episode in twentieth-century technology studies. *Knowledge, Technology, & Policy, 17*(11). https://doi.org/10.1007/s12130-004-1002-7

Winner, L. (1980). Do artifacts have politics? *Daedalus, 109*(1), 121–136.

Chapter 9
Re-thinking Innovation as a Lever for Development Taking into Account Inequality

Rodrigo Arocena and Judith Sutz

Abstract This chapter posits that development is no longer a "place" which anybody may reach by climbing similar types of ladders. In conjunction with the growing awareness of the socially exclusionary consequences of prevailing knowledge and innovation policies, this calls for alternatives in terms of both development and knowledge and innovation policies. Starting from the notion of Sustainable Human Development as a normative characterization, the chapter addresses knowledge-based inequality, analyzing the impacts of the weak structural demand for knowledge in underdevelopment and exploring an analytical tool to examine the relationships between social exclusion and innovation. It likewise explores the required characteristics for knowledge and innovation to become part of efforts towards equality and proposes alternative academic and governmental policies to foster their achievement.

9.1 Introduction

Development has been thought, both explicitly and implicitly, as a comparable process to that of climbing a ladder that reaches up to a place called development. This place has most usually been identified with the situation of the advanced capitalist countries.However, a new situation is now emerging, particularly because development understood as the place where people have access to what advanced capitalist societies provide on average is not sustainable in environmental terms even in these societies; even more so in terms of extending such access to the overwhelmingly poor regions of the world. Development has ceased to be a place as defined above; other conceptualizations, such as human and sustainable development, have now

R. Arocena (✉) · J. Sutz
University of Republic, Montevideo, Uruguay
e-mail: roar@fcien.edu.uy; jsutz@csic.edu.uy

B. Laspra, J. A. López Cerezo (eds.), *Spanish Philosophy of Technology*, Philosophy of Engineering and Technology 24,
https://doi.org/10.1007/978-3-319-71958-0_9

come to the fore. These conceptualizations refer more to processes than to places, as places like these have not yet been achieved anywhere.

The question thus arises as to which levers can enable the advancement of these processes. Some conceptualizations place innovation-based industrialization at the center of development processes; for instance, the conceptualization issuing from the Economic Commission for Latin America and the Caribbean. The main challenge is to narrow the productivity gap with the highly industrialized world, because this gap is responsible for the huge informal sector and for the poor quality of the jobs offered by the formal sector. This requires the incorporation of technology and innovation, key factors in the transformation of the productive structure. Providing this is achieved, social policies aimed at fostering the features that characterize human development can be implemented (ECLAC 2012). The sustainability part of this proposal is not clear. Other conceptualizations indicate that development has been pushed aside by innovation-led economic growth because, although it has been "smart", it has not been "inclusive" (Lazonik and Mazzucato 2012). However, the factor to blame for this lack of inclusiveness is not innovation; in fact, innovation is treated as neutral in this regard. Growing income inequality, so eloquently depicted by Piketty (2013), is attributed to socio-political arrangements that promote the concentration of wealth.

Nonetheless, there are also some conceptualizations that place the blame for inequality in its diverse manifestations on science, technology and innovation as they are currently carried out. "Knowledge-based inequality prevails in the contemporary world" (Tilly 2005, p. 121) is a particularly telling phrase in this respect. It is also a succinct way of explaining why our work has focused on innovation and the democratization of knowledge seen as a contribution to inclusive development (Arocena and Sutz 2014).

If we accept that diminishing inequality in its diverse expressions is a constitutive trait of human and sustainable development, we need to question the role of knowledge and innovation in development. However, even if the trend linking inequality to knowledge and innovation may refer to a universal phenomenon, the way it is expressed is highly idiosyncratic. By the same token, linking knowledge and innovation to the fight against inequality also requires idiosyncratic approaches.

This chapter addresses this issue by: (i) elaborating on previous statements that lead to the notion of Sustainable Human Development as a normative characterization; (ii) connecting this notion with the fact of knowledge-based inequality; (iii) analyzing the impacts of the weak structural demand for knowledge in underdevelopment on the difficulties science, technology and innovation face to contribute to human and sustainable development; (iv) providing an analytical tool based on Sen's work (Sen 2000) to examine the relationships between social exclusion and innovation as it is currently carried out; (v) explaining the specificity of the characteristics knowledge and innovation require to form part of the efforts towards equality in developing countries; (vi) proposing alternative academic policies to foster the achievement of this specificity; and (vii) proposing alternative governmental policies to this same end.

9.2 A Lesson from Experience

The very concept of development has changed rapidly in recent decades. Sixty years ago, development usually designated both a desirable *place* – defined by its occupants, the so-called developed countries – and a *ladder* that the other countries could – and should – climb in order to reach that place, a process known as *catching up*.

Alternatively, up to a few decades ago the place called development was characterized by some as the situation of the most advanced socialist countries. The ladder that was assumed to lead to that place was quite different in this case, but, as in the former case, industrialization was a sine qua non. In both cases, knowledge and innovation were needed to climb the ladder, though in different ways. In fact, the conceptualization of innovation as a requisite for development was reinforced by the crisis and fall of the socialist world, attributed in no small part to the inadequacy of its social relations in fostering technological innovation.

The "place and ladder" conception of development was quite comfortable: the normative goals were defined by the main traits of those countries situated in the place of development; their past was factual proof of the viability of climbing the ladder; their present showed the future to the rest, so prospective issues were not a big deal. As values, facts and trends seemed quite clear, attention focused mainly on proposals for climbing the ladder, which usually included fostering economic growth and industrialization, as well as major discussion concerning the roles of markets and states.

This notion is still widely endorsed, at least implicitly, yet it is also untenable. Let us assume that the title of the 2011 Human Development Report, "Sustainability and equity: a better future for all", is something that we all want. The aforementioned place looks neither very sustainable nor equitable, and climbing the ladder to get there is simply not feasible for the whole world. Prevailing types of economic growth foster environmental degradation – which threatens a climate catastrophe – and in many cases is associated with increasing inequality, particularly as regards income and environmental damage (UNDP 2011, pp. 7, 9).

The main example of development as a place has usually been the USA, while the most spectacular process of catching up is seen in China. Neither the USA nor China is characterized by either sustainability or equity. If every country in the world climbs that sort of ladder, the result will be a worse future for everyone.

9.3 The Move Towards New Paradigms

The conception of development proposed by Sen (1999) has contributed to a major change in development theory (Evans and Heller 2015) that can be seen as replacing previous dominant paradigm.

We suggest that the emergence of new paradigms could be helped by combining four approaches to development issues: (1) a normative approach, stemming from Sen's conception and its potential connections with the notion of Sustainable Development; (2) a (theoretical-)factual approach, focused on the dynamics of the capitalist knowledge society that has emerged in the (old or new) "central countries" and which fosters structural inequality in the world at large; (3) a prospective approach that explores major trends and the possibilities of overcoming inequality and underdevelopment in the new global context; and (4) a propositional approach aimed, as already stated, at democratizing knowledge and innovation.

9.4 The Normative Approach as the Starting Point

The history of development strongly suggests starting out not from a theory of social evolution or an overall strategy: values should guide the definitions of agendas for the study of facts and the elaboration of proposals.

This idea suggests that what is needed first is a solid, broad-ranging agreement about the normative goals of development, offering a plural space for more detailed conceptions and providing a general orientation for combining different concrete actions. This constitutes a major potential merit of the characterization of development as proposed by Sen (1999) which inspires the following synthetic concept: "Human development is the expansion of people's freedoms and capabilities to lead lives that they value and have reason to value." (UNDP 2011, p. 1). A fundamental clue to the propositional approach stemming from this characterization is Sen's assertion that the expansion of freedoms and capabilities is not only the normative goal of development, but also its main tool, and implies seeing people not as patients, but as agents.

The above formulation maintains the original commitment of development to improving the quality of human life. What has been achieved – for example, in health, life expectancy, and education – is closely linked to the quantitative and qualitative expansion of the production of goods and services, which is in turn deeply rooted in technological innovation. However, the prevailing types of production have damaged the environment to such an extent that we have witnessed an increasing demand in recent decades for sustainable development, understood as the set of processes that fulfill the needs of current generations while protecting the possibilities of future generations to fulfill their own needs. In fact, the self-sustainability of development depends not only on not undermining today the possibilities of tomorrow, but also on expanding in the present the individual and collective freedoms and capabilities that will help people to live valuable lives in the future. This concerns the environment, knowledge and skills, and social conditions.

In this respect, the normative approach stemming from Sen can be summarized by stating that Sustainable Human Development is: (i) the expansion of people's freedoms and capabilities, both individual and collective, (ii) in order to lead lives that they value and have reason to value, (iii) in ways that preserve such possibilities for future generations, (iv) assuming that the expansion of freedoms and capabilities is both the defining aim of development and its main tool, which (v) implies treating people not as patients, but as agents.

9.5 The Crucial Issue of Inequality

Levels of inequality are widely recognized today as being relevant factors in preventing many people from leading "lives they value and have reasons to value". The degree of consensus concerning equality according to Sen is worth recalling here: "Equality was not only among the foremost revolutionary demands in eighteenth-century Europe and America, there has also been an extraordinary consensus on its importance in the post-Enlightenment world." (Sen 2009, p. 291). We do not seem to be progressing in that direction, however.

The issue of (in)equality links the previously sketched normative approach with a factual approach which focuses on knowledge and innovation, seen as major factors for both fostering and hindering Sustainable Human Development.

Concerning the former aspect, we may simply recall that in the last 200 years, life expectancy at birth has risen on a global average from about 30 years to around 70 years (Hodgson 2015a, p. 1).

As regards the most technologically advanced country, the USA, it has been stated that "the main driver (of growing inequality) is exponential, digital, and combinatorial change in the technology that undergirds our economic system" (Brynjolfsson and McAfee 2014, p. 61).

Knowledge is related to inequality quite directly because it is a resource that increases with its use and decreases when it is not used. This benefits regions or social groups that are already strong in terms of knowledge and damages those that are weak. People with opportunities to study at an advanced level and work in conditions that foster permanent learning expand some of their capabilities, while the contrary happens to people without such opportunities. This is in some sense similar to what Merton (1968) termed, when speaking of the academic reward system, the "Matthew effect", because of a verse in the Gospel by Matthew where it is said that he who has plenty will be given abundance and he who has little will have it taken from him. A generalized Matthew effect seems to be acting in society at large in relation to knowledge and learning that probably compounds the relatively general trend towards rising inequality that has been observed in many places (Held and Kaya 2007; Milanovic 2011; Deaton 2013; OXFAM 2016).

9.6 A Factual Approach to (Under)Development

In the orthodox formulations of the "place and ladder" conception of development, the notion of underdevelopment did not exist. It was introduced by heterodox formulations to characterize the situation of non-developed (or "peripheral") countries that could not climb the textbook ladder of development because developed (or "central") countries used their power to prevent them from doing so. Therefore, the former had to climb alternative ladders to catch up with the latter. In a nutshell, this was the experience of Japan and other countries in East Asia.

Underdevelopment is virtually neglected by Sen (1999), but it is an actually existing situation that hinders the expansion of freedoms and capabilities. It is rooted in the "peripheral condition", characterized by specializing in productive activities with comparatively low added value stemming from advanced knowledge and high qualifications. In the 1950s, this usually meant producing primary goods, a specialization often induced from abroad by means that included forced deindustrialization (Rodrik 2011, p. 136, 141). Since then, several peripheral countries have become more or less industrialized while still being characterized by a barely diversified and low complexity production system, which can be seen as a source of lasting poverty (Reinert 2007). Meanwhile, a few countries have been able to greatly upgrade the complexity of their productive activities and thus escape their peripheral condition.

Today, the central countries are those where a knowledge-based and innovation-driven economy (de la Motte and Paquet 1996) is being consolidated. Where this does not occur, economic dynamics generate a weak commercial demand for advanced knowledge, usually not addressed to indigenous generators of knowledge. This situation inhibits knowledge generation and use, as well as the fostering of advanced education (Arocena and Sutz 2010). This in turn hinders the expansion of capabilities and constitutes one of the main examples of how peripheral countries are disadvantaged at an international level. This occurs in terms of economic power and also in terms of political, military and even ideological power. Conversely, central countries often use their power to "kick away the ladder" (Chang 2002); that is, to prevent peripheral countries from putting into practice today certain productive policies used yesterday by central countries. Thus, underdevelopment can be characterized as the combination of the peripheral condition with external subordination. It is not precisely favorable for the expansion of freedoms and is a fundamental cause of inequality, "apparent even in people's bodies" (Deaton 2013, p. 164).

9.7 Prevailing Innovation and Social Exclusion

Inequality and social exclusion are relational concepts associated with the capabilities people have to live worthy lives: "...the real importance of the idea of social exclusion lies in emphasizing the role of relational features in the deprivation of capability..." (Sen 2000, p. 12). Along these lines, Sen distinguishes two modalities of social exclusion, each with two expressions. One is related to impacts on people's lives: if the exclusion leads in itself to an important deprivation, it is termed *constitutive*; if it is not so determinant, but may lead to further deprivations, it is termed *instrumental*. The other modality of social exclusion refers to its origins: if there is a deliberate attempt to exclude, we have an *active* exclusion; if not, that is, if social exclusion occurs without purpose, we have a *passive* exclusion. These four expressions of social exclusion may be combined in a two-by-two matrix whose quadrants represent the following combinations: (I) active and constitutive; (II) active and instrumental; (III) passive and instrumental, and (IV) passive and constitutive. We posit that this matrix is analytically useful to explore the relationships between knowledge and innovation policies and inequality, particularly so when the latter is understood as multidimensional, that is, not only as income inequality. Examples of these relationships are, in each quadrant:

Quadrant I (active and constitutive exclusion): "In the Third World (the patent system) is a matter of life and death for people who cannot afford new brand-name drugs but might be able to afford generics" (Stiglitz 2007, p. 46). The war against generics, lobbied by "Big-Pharma", continues though, as the reinforced mechanisms of TRIPS-plus show. The patent system is an intrinsic part of innovation policy; its exclusionary effect is deliberate, and so the exclusion it may entail is active; its consequences in the case of medicines are constitutive in terms of direct and severe deprivation of opportunities and capabilities.

Quadrant II (active and instrumental exclusion): Big data is a relatively new phenomenon in its current development; research and innovation in ICTs have opened up possibilities of unprecedented accumulation and combination of information about people; the private appropriation of such data – by insurance companies or job consultants – may lead to decisions related to different types of exclusions that threaten further opportunities to live worthy lives. It is not only the use of data that is at stake; the philosophy of ICT's development is a main cause of the problem. Without normative guidelines, STI policies may be fostering technical developments that will augment the imbalance of the prevailing share of power.

Quadrant III (passive and instrumental exclusion): The uneven spread of technical progress is a major source of structural heterogeneity, leading to "two speed" economies, one mainly formal, with relatively well-paid jobs and prospects for upgrading through learning, and another, mainly informal, poorly paid, with bad working conditions and little prospects of progress by learning. Dual economies of this kind are present in most underdevelopment landscapes and are a major source of multidimensional forms of exclusion, even if such exclusion has not been voluntarily established and does not threat life directly.

Quadrant IV (passive and constitutive exclusion): What happens when some problems are not included in research agendas? If (i) such problems severely affect the opportunities and capabilities people have to live lives they have reason to value, (ii) research may provide workable strategies to address such problems and (iii) no research efforts are made in these directions, we can talk about social exclusion that is constitutive, due to the importance of the problems, as well as passive, given that "not researching" is not an active exclusionary decision. David Hess coined the term "undone science" to refer to unfunded or generally ignored areas of research which are identified by different social actors as important tools in the achievement of a given result. (Hess 2007) An outstanding example of undone science comprises the so-called neglected diseases, ignored precisely because the bearers of such diseases are people without market purchasing capabilities.

The strategies to fight these different types of "knowledge- and innovation-driven" social exclusion are quite diverse; all require purposeful policy actions. This constitute the thread linking the (theoretical-)factual approach with the propositional approach.

9.8 Elements of a Prospective Approach

The following constitute five frequently acknowledged global trends:

1. *The expanding role of knowledge in social relations.* For better or worse, "big science and large technological systems [...] transform the extrasocial world, nature and the environment – and *with it* the social world" (Schroeder 2007, 49).
2. *Growth and (even more so) diversification of production of goods and services.* Mainly a consequence of the previously noted trend, this second trend is beyond doubt in the long run, despite persistent recessions. It is in turn a major cause of improvements in average living conditions, particularly as regards health, life expectancy and education (Roser 2015a, b). It also fosters possibilities and aspirations related to consumption, a phenomenon with obvious economic consequences and closely related to political options.
3. *Degradation of the environment and the climate.* Advanced scientific and technological knowledge allows the accelerated expansion of production and consumption in ways that have already caused great environmental damage and have opened up the possibility of a catastrophe. Such a threat may be the decisive factor in the emergence of new combinations of knowledge generation, styles of production and social relations: "the people demand more and more economic growth in order to consume more, as a citizen right. Ordinary citizens will have to change their lifestyles to avert disaster" (Mann 2013, p. 95).
4. *The increasing influence of financial capital.* As stated in a methodical recent study, "financial markets have become the main drivers of the world economy" (Hodgson 2015b, p. 136). Income and other benefits bound for the commanding heights of financial capital do not seem to have been eroded in spite of the great recession of 2008, of which they were main culprits. This gives an idea of their power, new proofs of which can be seen daily, North and South.

5. *Rising inequality.* We briefly discussed this issue in Section (d), in particular, its links with knowledge. It follows from such connections that, if prevailing patterns of knowledge generation and use do not change, a strong drive towards increasing inequality will keep operating. A relation between the first, fourth and fifth trends of this short list can be highlighted as follows: "In recent decades, the combination of financial capital and scientific-technical knowledge has gained unparalleled potency in the production of inequality between those who control the combination and those who do not" (Tilly 2005, p. 115).

Thus, if "business as usual" were to continue, we would expect to see the expansion of production, of inequality (both knowledge-based), and of environmental damage. This seems difficult, because of the inherent social and, above all, environmental lack of sustainability. However, it also seems very difficult to overcome the contradiction between, on the one hand, increasing wealth and decreasing average poverty, such that economic growth becomes a major clue of political legitimacy almost everywhere, and on the other hand, the depletion of natural resources and the aggravation of climate changes.

"Not ideas, but material and ideal interests, directly govern men's conduct. Yet very frequently the 'world images' that have been created by 'ideas' have, like switchmen, determined the tracks along which action has been pushed by the dynamic of interest." (Max Weber, quoted by Gerth and Mills 1958, p. 280). Neoliberal individualism has been laying the tracks for the last four decades. It erodes cooperation and fosters consumerism. Perhaps increasing related risks and conflicts will lead to switching tracks, which could enhance the ideological power of solidarity and frugality. In such a case, more political support could be expected for alternative knowledge and innovation policies. Its first priorities would be social inclusion and environmental improvement. Both aims point to doing better with less; that is, the general orientation would be to find more affordable and less resource-consuming solutions for better coping with collective problems. This can be termed frugal innovation (Bound and Thornton 2012) or, even better perhaps, frugal innovation for social inclusion. A necessary condition for its success is its link with advanced learning and first-rate research. Whose "material and ideal interests" could back such alternatives to prevailing trends? Most underprivileged people, if suitable political articulations are carried on and results are obtained. Perhaps also some researchers and innovators, if related rewards systems are included in innovation policies for social inclusion. More generally, all those who would prefer to avert disaster than keep on increasing consumption. Maybe sheer prudence is the first cause for tempered optimism:

> I see regulation as likely to make a comeback, especially to combat climate change and if there is a populist reaction against the rising inequality and exploitation that neoliberalism brings. If climate change is effectively combatted, the regulation of capitalism and consumerism would be considerable. If it is not, the ensuing last-minute regulation of market forces amid disastrous times would be much more punitive domestically and probably often vicious geopolitically. (Mann 2016, p. 286)

9.9 Democratizing Knowledge and Innovation as an Element of the Propositional Approach

Fostering the normative notion of Sustainable Human Development, in a factual and prospective context that includes among its defining traits those previously outlined, requires quite a diversified set of policies with strong social and political backing.

As regards the knowledge aspects of such policies, a related contribution by a well-known specialist on scientific policies ends "by pointing to the need to revisit the fundamental question of 'What is science (or knowledge production) for?'" (Martin 2010, p. 44). At this point, some previous remarks can be summarized by stating that, in the global society shaped by capitalism, advanced knowledge constitutes the fundamental infrastructure of power relations in general. Moreover, as this will continue to be so in the foreseeable future, some of the most relevant and socially destabilizing changes will be driven by technological innovation. Knowledge and innovation greatly influence who the winners and losers of scientific expansion and economic growth are, as well as those who are damaged less or more by environmental degradation. Therefore, democratizing knowledge and innovation is an inescapable requirement for development in the normative sense. It may thus be stated that the prima aim of science (or knowledge) production should be that of contributing to Sustainable Human Development. This includes actual or potential research and innovation stemming from every discipline and from interdisciplinary work, in relation to the highly diverse material and spiritual aspects of lives that people have reason to value.

As a contribution to the propositional approach to development, we briefly consider some academic and governmental dimensions of democratic knowledge policies whose aims include promoting frugal innovation for social inclusion.

9.9.1 On the Propositional Approach (1): Alternative Academic Policies

In order to place the power of academic knowledge at the service of social inclusion, it is first necessary to be able to perform first-rate research, in all fields of science. The problems to be tackled are extremely complex and need to be addressed from several angles and with the help of diverse approaches. However, the research to be conducted in this direction has another characteristic. The problems to be analyzed and studied need to be recognized as problems by some social actor – the excluded populations in some cases, those that work with them in others – and not exclusively by the researchers. This characteristic calls for specific academic policies. Research problems arising from social exclusion are both difficult to formulate and to address. They are particularly time consuming, beside their intrinsic complexity, because what the problem is and what a solution may be needs to be understood in dialogue with the social actors that recognize the problem and with those that have the will to solve it. They are interdisciplinary by nature and the difficulties associated with interdisciplinary research are well known. They often refer to local situations, which jeopardizes the possibilities of future academic publications in journals of international circulation. The current logic of the academic evaluation system goes against this type of research, as has been abundantly acknowledged in recent literature. This designs one of the avenues of an alternative academic policy: the implementation of a research evaluation system that assures academic quality in ways that do not hinder research carried out in interaction with non-academic actors, of an interdisciplinary nature and with local emphasis. This point is worth stressing: there is no research without researchers able and willing to do it. However, if it is important not to discourage such research, it is no less important to promote it. This means that funds and specific policy instruments should be put in place to foster research concerned with promoting social inclusion. However, this is not sufficient. Researchers are not usually aware of the kinds of social problems they may be able to help solve; identifying them is not their main trade. Bridging the gulf between researchers and social problems requires specific research and a good deal of experimentation. This is important, because inclusive research and innovation holds the same logic as all the rest: "…one feature that invariably characterizes successful innovation is ongoing communication between the producers and users of knowledge" (Sarewitz and Pielke 2007, p. 7).

Many continue to profess the comfortable conviction that good science will always find ways to solve the problems that it is able to solve. This conviction is currently being challenged and the need for alternatives for prevailing research orientation is gaining voice, North and South.

9.9.2 On the Propositional Approach (2): Alternative Governmental Policies

One of the main impediments for using knowledge for development in the global South is the weak demand for the capacities countries have to produce and use knowledge. It is not that knowledge demand as such is necessarily weak: both business and the state may use sophisticated knowledge-based technology, but this is usually overwhelmingly acquired from abroad. Why this is so admits several answers: being dazzled by foreign and state-of the art technology, unawareness of what kind of solutions national capabilities are able to provide, or risk-aversion in terms of avoiding being blamed for confiding in untested local solutions. However, national problem-solving capabilities, both in terms of knowledge production and knowledge use, are extremely important to tailor solutions to local conditions and constraints. A major hurdle is that those that could rely on such capabilities for workable innovations capable of improving their lives – in health, housing, education, or productive conditions –, whether as individuals or as collective actors, are usually unable to generate market demand for their problems. However, if the knowledge and innovation capacities needed to satisfy their substantive demand could be mobilized, not only suitable solutions would be available, but such capacities would also be enhanced. For this to occur, social policies must involve knowledge and innovation; that is, their requirements in terms of knowledge and innovation should become part of the agenda of science, technology and innovation policies. In turn, such policies should also be understood as social policies, in the sense that their objectives include collaborating in the achievement of social goals through knowledge and innovation.

"Silo-like" policies in which each public policy pursues its goals in isolation usually constitute a powerful obstacle for placing national science- and technology-based problem-solving capabilities at the service of inclusive development processes. This is why alternative, integrated governmental policies such as those outlined above are important. If research and innovation can be linked in order to address problems of social inclusion both effectively (improving how certain problems are dealt with to a certain extent) and efficiently (achieving a better use of resources which are often erroneously spent on expensive solutions that are neither adequate nor able to expand the endogenous capabilities), this will represent a path towards democratization accompanied by the spread of knowledge.

9.10 Conclusion: On Knowledge and Agency

Emerging innovation policies considered as part of social policies have strong normative foundations and mobilize embryonic networks consisting of academics, policy makers and NGOs. A fundamental question that must be asked is whether these policies may be ideologically related to significant "popular actors", parties and/or social movements. This is a fundamental question because such emerging socially oriented innovation policies can be considered counter-hegemonic in the prevailing innovation policy landscape and hence politically fragile. This is why an affirmative response appears to require feedback between these policies and a more general ideological outlook concerning the democratization of knowledge. However, important social actors will need to foster these socially oriented innovation policies, the linking and articulating role of the state is paramount precisely because of the "anti-silo" turn of such policies. If there are sectors of the state apparatus willing to promote these tasks of linking and articulation within the state, at the same time as promoting agency among concerned social actors, we may affirm that the resulting innovation policies will belong to the set of democratic knowledge policies, specifically aimed at reducing certain aspects of inequality. If they can grow within the interstices of the dominant trends, they will be able to contribute to the emergence of economies in peripheral countries in which the goods stemming from the power of knowledge will be more evenly distributed, thereby advancing on the road to the democratization of knowledge.

References

Arocena, R., & Sutz, J. (2010). Weak knowledge demand in the South, learning divides and innovation policies. *Science and Public Policy, 37*(8), 571–582.

Arocena, R., & Sutz, J. (2014). Innovation and democratisation of knowledge as a contribution to inclusive development. In G. Dutrenit & J. Sutz (Eds.), *National innovation systems, social inclusion and development. The Latin American experience* (pp. 15–33). Cheltenham: Edward Elgar.

Bound, K. Thornton, I. (2012). *Our frugal future: Lessons from India's innovation system*. www.nesta.org.uk. Accessed 26 July 2016.

Brynjolfsson, E., & McAfee, A. (2014). *The second machine age*. New York: Norton and Co.

Chang, H.-J. (2002). *Kicking away the ladder: Development strategy in historical perspective*. London: Anthem Press.

de la Motte, J., & Paquet, G. (1996). *Evolutionary economics and the new international political economy*. London: Pinter.

Deaton, A. (2013). *The Great Escape. Health, wealth, and the origins of inequality*. Princeton: Princeton University Press.

ECLAC, Economic Commission for Latin American and the Caribbean. (2012). *Structural change for equality. An integrated approach to development*. Santiago de Chile: United Nations.

Evans, P., & Heller, P. (2015). Human development, state transformation and the politics of the developmental state. In S. Leibfried, E. Huber, M. Lange, J. Levy, F. Nullmeier, & J. Stephens (Eds.), *The Oxford handbook of transformations of the state* (pp. 691–713). Oxford: Oxford University Press.

Gerth, H., & Mills, W. (1958). *From Max Weber*. Oxford: Oxford University Press.

Held, D., & Kaya, A. (2007). *Global inequality. Patterns and explanations*. Cambridge: Polity Press.

Hess, D. (2007). *Alternative pathways in science and industry. Activism, innovation, and the environment in an era of globalization*. Cambridge, MA: MIT Press.

Hodgson, G. (2015a, May 7). Conceptualizing capitalism: How the misuse of key concepts impedes our understanding of modern economies. *Books&Ideas.net*.

Hodgson, G. (2015b). *Conceptualizing capitalism. Institutions, evolution, future*. Chicago: The University of Chicago Press.

Lazonick, W., & Mazzucato, M. (2012). *The risk-reward nexus. Innovation, finance and inclusive growth*. New York: Policy Network Paper.

Mann, M. (2013). The end may be nigh, but for whom? In I. Wallerstein, R. Collin, M. Mann, G. Derluguian, & C. Calhoun (Eds.), *Does capitalism have a future?* (pp. 71–97). New York: Oxford University Press.

Mann, M. (2016). Response to the critics. In R. Schroeder (Ed.), *Michael Mann's anatomy of the twentieth century and beyond* (pp. 281–322). Cambridge: Cambridge University Press.

Martin, B. (2010). Inside the public scientific system: Changing modes of knowledge production. In R. Smits, S. Kuhlmann, & P. Shapira (Eds.), *The theory and practice of innovation policy* (pp. 25–50). Cheltenham: Edward Elgar.

Merton, R. (1968). The Matthew effect on science. *Science, 159*(3810), 56–63.

Milanovic, B. (2011). *The haves and the have-nots. A brief and idiosyncratic history of global inequality*. New York: Basic Books.

OXFAM. (2016). *An economy for the 1%. How privilege and power in the economy drive extreme inequality and how this can be stopped*. Oxford: Oxfam GB.

Piketty, T. (2013). *Capital in the XXI century*. Cambridge, MA: Harvard University Press.

Reinert, E. (2007). *How rich countries got rich and why poor countries stay poor*. New York: Public Affairs.

Rodrik, D. (2011). *The globalization paradox: Democracy and the future of the world economy*. New York: W.W. Norton.

Roser, M. (2015a). *Life expectancy*. http://ourworldindata.org/data/population-growth-vital-statistics/life-expectancy/. Accessed: 12 Aug 2016.

Roser, M. (2015b). *Literacy*. http://ourworldindata.org/data/education-knowledge/literacy/. Accessed 12 Aug 2016.

Sarewitz, D., & Pielke, R. (2007). The neglected heart of science: Reconciling supply and demand for science. *Environmental Science and Policy, 10*, 5–16.

Schroeder, R. (2007). *Rethinking science, technology, and social change*. Stanford: Stanford University Press.

Sen, A. (1999). *Development as freedom*. Oxford: Oxford University Press.

Sen, A. (2000). *Social exclusion: Concept, application and scrutin* (Social Development Papers 1). Manila: Asian Development Bank.

Sen, A. (2009). *The idea of justice*. Belknap: Harvard.

Stiglitz, J. (2007). Pizes, not patents. Post-Autistic Economics Review 42, 48–49. http://www.paecon.net/PAEReview/issue42/Stiglitz42.htm. Accessed July 28, 2016.

Tilly, C. (2005). *Identities, boundaries, and social ties*. Boulder: Paradigm Publishers.

UNDP, United Nations Development Program UNDP. (2011). *Human development report 2011*. New York: United Nations.

Chapter 10
University, Technology and Development: Reflections from the South

Jorge Núñez Jover and Galia Figueroa Alfonso

Abstract In this chapter, we address the debate on development concerning knowledge and science, technology and innovation (STI). Specifically, we discuss the role of universities as knowledge institutions promoting sustainable and inclusive development. We highlight the need to review the underlying conceptual foundations from which university policies are usually analyzed and close the chapter with a brief outline of what has taken place in Cuba in recent decades.

10.1 Introduction

Andrew Feenberg (2012) questioned whether humans must obey the cold logic of machinery or whether technology may be substantially redesigned to better serve its creators "That is the underlying problem on which depends the future of industrial civilization" (Feenberg 2012, p. 13). In his view, which is also our own, a critical theory that rejects the idea of both the neutrality and determinism of technology. We share the concern about what J. J. Salomon called the "dance on the Titanic" (Salomon 2008, p. 29). Sustainable development requires changes in the paradigms of scientific and technological policies and practices.

We begin with the debate on development models and their relation to technology and innovation. We assume that these changes should involve universities. In Latin America and the Caribbean, universities concentrate most of the capacities for knowledge production, dissemination and use that maybe useful for development and play an important role in STI systems. They are responsible for a major part of

J. Núñez Jover (✉)
Department of Social Studies of Science, Technology and Innovation,
University of Havana, Havana, Cuba
e-mail: jorgenjover@rect.uh.cu

G. Figueroa Alfonso
Center for Development Research (ZEF), University of Bonn, Bonn, Germany
e-mail: galiafigueroa@gmail.com

B. Laspra, J. A. López Cerezo (eds.), *Spanish Philosophy of Technology*,
Philosophy of Engineering and Technology 24,
https://doi.org/10.1007/978-3-319-71958-0_10

the scientific production and training of human talent that should have a positive impact on development. It would appear appropriate to move towards a model of higher education that promotes sustainable development, including poverty reduction and social inclusion. This debate is currently taking place in Latin America and the Caribbean (LAC).

Rethinking universities and their social function can benefit from renewed STI approaches. Using these concepts, it is possible to rethink the role of the university in technological change. Studies in Cuba highlight different experiences in which higher education institutions, as technology developers, help solve social problems and connect directly to people's daily needs, fostering inclusion, justice and welfare dynamics.

10.2 Technology and Innovation for Sustainable Development?

In recent decades, it has been argued that the hegemonic development model is economically, socially and ecologically unsustainable (Cassiolato et al. 2013). Economic growth is substantially divorced from the social welfare objectives that it should be promoting.

Using abundant empirical data, Piketty (2014) has confirmed what seemed obvious: economic inequality is increasing and accelerating at a dangerous pace in the twenty-first century. Poverty and social exclusion are accelerating in equal measure in line with this trend. Hence, "economic spillover" should not be expected to end up spontaneously benefiting the whole population, particularly those most in need.

The current development model is a "cumulative regime characterized by mass production, mass consumption, and excessive use of natural resources and greater exploitation of labor" (Cassiolato et al. 2013, p. 65). The rise of neoliberalism, the erosion of the welfare state and the affirmation of the global power of transnational corporations, the dominant role of finance capital and the privatization of knowledge are all features of the current globalization process, one of whose consequences is "a dramatic increase in inequality" (Cassiolato et al. 2013, p. 65). More comprehensive visions of development have been proposed based on these findings: sustainable development, human development, and inclusive human development (PNUD 2010, 2011, 2013), among others.

The concern for sustainable development was recently reflected in the UN document "Transforming our world: the 2030 Agenda for Sustainable Development" which outlines 17 Sustainable Development Goals (SDG) and 169 tasks (PNUD 2015). This document supersedes the Millennium Development Goals (MDG) with the intention of achieving them. Objectives and goals combine the three dimensions of sustainable development: economic, social and environmental. They are expected to stimulate action in critical areas for humanity and the planet over the next 15 years.

It is obvious that knowledge and technology must play an important role in achieving these goals. Aligning itself with the SDG, UNESCO recently published its Science Report 2030 (UNESCO 2015) at the 38th Session of the General Conference in November 2015. This session debated an agenda of broad and ambitious universal education and approved the Framework for Action 2030. The intention is to ensure the contribution of science, technology and education to sustainable development.

This debate has reevaluated the role of public policies implemented by the State and their special relevance with respect to issues such as combating poverty and meeting the basic needs of the population, including priority concerns such as health and education. Nonetheless, many obstacles remain. At least since the 1980s we are witnessing the emergence of a "capitalist society of knowledge" (Arocena and Sutz 2013, p. 29) in which knowledge is the core of the technology base underpinning power relations.

The configuration of a "capitalist society of knowledge" strengthens the trend towards the privatization of knowledge and promotes the capitalization of the diverse activities related to higher education, research and their applications. Such privatization hinders the use of advanced knowledge to improve the quality of life of the poorest in the developing world to the extent that, in the field of healthcare, it has been compared with the effect of weapons of mass destruction (Arocena and Sutz 2013).

In the twenty-first century, knowledge and innovation are often considered sources of social welfare and sustainable development, thought he connection between knowledge, technology and human development is actually contradictory. Evidence suggests that STI, which are drivers of such growth, are based on the predatory, unsustainable model and do not necessarily solve major social problems. According to an OECD report:

> There is a broad consensus that the disconnection between economic growth and social welfare is increasing. At the same time, research and innovation have become one of the main engines of growth. However, these two trends could not be reconciled: there is a clear lack of exploitation of innovative solutions to address social problems, a fact that entails large opportunity costs for society. Social innovation provides a way to reconcile these two forces, generating economic growth and social value at the same time. (OECD 2011, p. 8)

According to UNDP, "Technology is created in response to market pressures, not the needs of the poor" (PNUD 2001, p. 3), which is evident in the priorities of global spending on health research, agriculture and energy.

Although the issue of poverty and inequality are at the heart of the debate concerning research agendas in the social sciences and in the pronouncements and policies of international organizations, a gap has arisen between social policies on the one hand and STI policies on the other, including the scientific policies of universities. Social policies have not often considered the role that STI could play in combating poverty and reducing insecurity in its different dimensions (Casas et al. 2013, p. 57).

The growing gulf between economic growth and social development and inclusion, in addition to the evidence that innovation has been a source of greater inequality, has led to greater emphasis being placed on the need for the kind of innovation that can act as a tool to promote social development and social inclusion, reduce inequalities, contribute to more welfare for marginal populations, and increase accessibility to and the use of knowledge for vulnerable populations, among others (Salazar et al. 2013, p. 135).

All the above has led to a debate on STI policies and their course correction and re-orientation with the aim of solving social problems (Dagnino 2012; Crespi and Dutrénit 2013). This critical view is presented as an alternative to the belief that such policies are a matter of strict technical rationality, ignoring the dense network of economic and political interests and values that underlie STI policy decisions (LALICS 2012).

10.3 University and Society

In view of the above, it is possible to draw the conclusion that the SDG demand new capacities to produce, disseminate and use knowledge. Furthermore, alternative ways to articulate knowledge and society are needed. LAC universities are known to concentrate most of the capacities of available knowledge and usually occupy – some more than others – a significant place in STI systems (Albornoz and López Cerezo 2010). They are responsible for much of the scientific production and training of human talent that should have a positive impact on the development process. If so, it would be desirable to move towards a model of higher education that promotes sustainable development, including poverty reduction and social inclusion. This debate, currently taking place in Latin America, is promoted by the Union of Universities of Latin America (UDUAL).[1] In this context, the role of the university as an agent of local and regional development is being analyzed (Núñez and Alcázar 2015).

However, the traditional performance of universities does not correspond to these objectives. Too often, the ideals of science underlying institutional practices and science evaluation systems (crucial in academic careers) are aimed more towards the production of articles in peer-reviewed journals than towards the solution of local problems, even if these may demand complex solutions from a technological and scientific point of view. The issues associated with local development are usually interpreted as less "universal" and less able to join in the development of cutting-edge knowledge. Perhaps for the same reasons, "outreach" (relation-building) activities are often assumed to be secondary to basic research. The Latin American scenario is characterized by both a relatively weak scientific and technological capacity and a low demand for knowledge. This context partly explains the behavior of academic communities. According to Thomas:

> The forms of academic legitimacy, evaluation mechanisms, financing forms, institutionalized habits, training mechanisms, explain the endogenous trend, self-centered, internationally integrated and locally isolated from the Latin American scientific communities. (Thomas 2011, p. 14)

[1] The Union of Universities of Latin America and the Caribbean (UDUAL) defends the idea of higher education as a public and social good. The Latin American and Caribbean Forum on Regional Economic Development was held in Quito, Ecuador, in May 2015. This forum saw the launch of the Latin American and Caribbean Network Linking Local Development to University (Spanish acronym, DELUNI). More recently, the issue was discussed at the International Congress on Higher Education, University 2016, held in Havana.

While communities in central countries research and publish papers on issues that are relevant to their own realities, peripheral communities display research agendas and quality and relevance criteria that have little to do with their own realities, with the aim of being admitted to the circuits of the former communities.

Hence, Latin American researchers, especially those working in universities, assume scientific and technological agendas generated outside the region. They internalize these criteria of quality and relevance and develop their careers responding to such issues, procedures, criteria, and financing requirements.

All this leads to advocating the importance of promoting public, economic and social policies that create demand for new forms of knowledge funding (Dutrénit and Sutz 2013). This could help redirect the science, technology and training agendas of universities.

10.4 Conceptual Renovations: New Images of STI

The eventual construction of university policies whose central focus is sustainable development requires the renewal of the conceptual arguments on which these policies are based. It is necessary to combine approaches that support an "epistemic movement" which will enable the legitimization of new definitions of knowledge, new scientific and technological practices, as well as new relevance criteria and alternative forms of legitimation of university work (Clark 2011, p. 167). Instead of the linear model of innovation and the exogenous orientation and legitimation of endogenous scientific work, this new approach should incorporate a more interactive view, focusing on the context, in which multiple actors shape the technological paths that are co-constructed with the participation of universities aimed at meeting important social needs in local contexts.

As seen above, the starting point for this conceptual renewal is that of overcoming the identification of development with economic growth and the placement of knowledge-based, socially inclusive sustainable development as the focus of university strategies. It is a development model not solely for academic groups. It constitutes a model in which people are not only beneficiaries, but also actors who collaborate in identifying problems and finding technological solutions to be resolved.

The idea of the importance of context in defining university agendas points to a different epistemology from that which is hegemonic in our universities. This alternative epistemology is based on the premise that the production, distribution and use of knowledge and technology always take place in particular contexts, with their economic peculiarities, culture and values. These latter factors should shape the priorities and development of such agendas.

By assuming this epistemology, universities can help to foster the local scope of knowledge and innovation, expanding its capacity to promote human welfare in different regions. Thus, a much more intimate encounter may occur at the municipal level between knowledge strategies, development priorities and resulting human wellbeing via the use of both internal and external resources.

Improving the quality of life of people implies achieving growth and economic development, food safety, education (coverage and quality), health, employment, housing, culture, environmental sustainability, water (access and quality), roads and transportation, services to the population, functioning legal systems and social participation in decision-making processes. In the words of Amartya Zen (Arocena and Sutz 2013), people should be seen as agents and not as patients.

The proposed "epistemic movement" can rest on concepts such as social technology (ST) and social innovation (SI). The argument for ST and SI is based on the alarming social and economic situation that characterizes Latin America and the Caribbean. Marginality, unemployment, poverty and social violence are tending to increase and deepen. A series of deficits can be perceived: housing, food, education, access to goods and services, among others. There is huge accumulated social debt. Overcoming this situation is a huge political and economic challenge (Thomas 2008).

This challenge can be translated into technological demands and different ways of conceiving the relationship between technology and development.

> The technological dimension of the problem is a challenge in itself ... It does not seem possible to meet the challenge by simply and uncritically multiplying the existing technological equipment ... it would imply a huge demand for energy, materials, natural resources, with high risks of environmental impact and new social misalignments ... simple multiplication of R&D budget will be insufficient to generate a significant change in social dynamics. (Thomas 2011, p. 13)

ST is based on technology approaches arising from contemporary theories and authors such as Callon (1987), Latour (1992), and Bijker and Pinch (2008). The metaphor according to which what is "social" and what constitutes "technology" forma "seamless fabric" (seamless web) emphasizes the social nature of technology and the technological nature of society (Hughes 1986). Technology is much more than just applied science. Technology is socially constructed by "relevant social groups", while society is technologically constructed. Technological development is neither unique nor inevitable: there is a process of co-construction between technology and society. All these ideas are key to rethinking the links between technology and development and to proposing technological alternatives.

ST refers to those technological efforts that are able to address social and economic problems and important deficiencies in housing, health, food or education. Creating such technologies is a major social and economic challenge. It is also a cognitive challenge, however, as it involves seeking technological solutions that allow access by these marginalized groups to goods and services, in addition to meeting the cognitive demand of the process of social inclusion (Dagnino 2009). It supposes converting exclusion and poverty into a scientific and technological challenge and generating "socio-technical adjustments" (Dagnino et al. 2004; Dagnino 2009; Thomas and Fressoli 2009) to help solve social problems. Social technologies can be knowledge-intensive; they can be produced in the market and generate income for communities and countries. Several authors consider that ST should contribute to "building another society" (Dagnino 2009).

Among the most frequently attributed features of ST is the orientation to solving social and environmental problems that benefit broader sections of the population, the search for alternative solutions to conventional technology, and the involvement of various social groups in the deployment of technological trajectories. To generate capacities for diagnosing the issues involved, planning, designing, implementing, managing and evaluating technologies are needed to implement ST. The participation of researchers and ST developers (from R&D institutions, universities, NGOs, businesses, etc.) is likewise required, as well as the participation of policy makers, decision makers and policy implementers (governmental institutions, international cooperation agencies, public agencies and representatives from the different business sectors). The participation of end users/beneficiaries in the design, production and implementation of ST is also essential, strengthening the role of grassroots communities in the processes of policy making, decision making and monitoring.

Efforts to develop social technologies are visible in countries like Brazil and Argentina through the "*Rede de Tecnologia Social*" [Social Technology Network] and the "*Red de Tecnologías para la inclusión social*" [Technology Network for Social Inclusion], respectively.

Theories on innovation systems (Lundvall 1992) are closely linked to this approach. Innovation is conceived in these systems as a social, interactive and systemic process. Innovation systems include organizations, institutions, and interactions between collective actors in such a way that the role of networks, interactions and actors are highlighted. An innovation system brings together various actors who interact to produce, disseminate and use knowledge, with the support of institutions, policies and regulatory systems.

The local dimension is crucial to the connection between knowledge, innovation, and development (Cassiolato et al. 2013). Consequently, policies should focus on the actors and their contexts, aiming to bolster their efforts and make them more effective through human resource training, information, design, computer services and consulting. Without ignoring the role of R&D activities and institutions, they emphasize the role of the interactions between a variety of actors, designed to generate, acquire, disseminate and use knowledge.

The idea of local innovation systems (LIS) (Lastres et al. 2005) is especially useful in this context. Traditionally, industrial development models, in which science and technology played a major role, were assumed to be indifferent to their contexts. Today, however, all theories on innovation attach great importance to the local dimension with its geographical, historical, cultural peculiarities, as well as its traditions and institutional identities.

The LIS approach aims to promote policies that spread the purview of the work of local actors (governments, cooperatives, farmers, businesses, the media, education sector, health institutions, credit institutions, universities, among others) in a way that closely links them with national and foreign actors with knowledge, technology, and financing capacities for local development. These LIS can be targeted to produce advances in education, health, employment, food safety, risk management, environmental care and other benefits via, let us stress once more, interactions, flows of knowledge, learning processes, and technology transfer.

According to some authors (Lundvall et al. 2009) there are two modes of innovation: the STI mode (science, technology and innovation) and the DUI mode (doing, using, and interacting). The former is called "narrow" and the latter, "comprehensive". The former model restricts innovation systems to R&D and their associated infrastructure. According to this model, the main mechanism is thus the promotion of R&D activities. This approach has been dominant up until now in Latin American STI policies and has generated what is called the "paradoxes of innovation" (Cassiolato 2012).

Scientific research can certainly be useful for certain types of innovations that serve local development, but it is important to possess capacities to absorb technology, adapt to local conditions, and learn to use them both efficiently and effectively. The university is an actor that can help build such local innovation systems by fostering links between actors, training them, contributing to the design of territorial strategies, identifying technological needs, and supporting the creation of networks that allow knowledge and technology transfer, among other actions.

In the last few years, a number of concepts related to these issues have appeared, namely "innovation systems for inclusive development" (Dutrénit and Sutz 2013), "socially inclusive innovation systems" (Sutz 2010) and "social innovation" (Bortagaray and Ordóñez-Matamoros 2012; Arocena and Sutz 2012), among others. These different formulations stress that governance schemes for social inclusion require greater participation in which knowledge users do not behave merely as passive agents or receivers of an unalterable information, but are involved in the generation of solutions to problems of health and housing, in addition to actively participating in creating opportunities and decision-making (IDRC 2011). Hence, the term "inclusive or democratic innovation" (Johnson and Andersen 2012) has likewise emerged. This inclusiveness is concerned, on the one hand, with the aims of meeting the needs of the excluded, and on the other, with the participation of not only scientists, engineers and administrators, but also citizens as consumers and workers in the process of innovation. Universities can likewise participate in these efforts.

The concepts of "integrated science" (Vessuri 2014) and "sustainability science" (González and Núñez 2014) have emerged from the growing pressure for relevant and credible science able to inform and influence public policies. They have arisen as a call to professionals and researchers to address real-world problems and collaborate in transformations. This demand requires the use of inter- and transdisciplinary approaches, in addition to the need to overcome natural barriers between social sciences, humanities and engineering.

10.5 Universities for Development

According to our proposed approach, the demands for sustainable and inclusive development directly question universities as knowledge institutions and the role they play within STI systems. Via their cognitive capacities, can universities help to fight poverty, exclusion and the environmental crisis, among other problems? In our view, the answer is yes, although the path is beset with conceptual and practical obstacles. In a well-known periodization, Etzkowitz (2004) postulates a first academic revolution to refer to the incorporation of research in universities, and a second academic revolution when universities established the relationship with the business sector, there by producing the capitalization of knowledge and the entrepreneur scientist, etc. Perhaps it is time to propose a third revolution, in which the commitment to development is the main goal. This is precisely the proposal of the so-called "developmental universities" (Brundenius et al. 2009). These authors propose that STI activities at universities should enhance social, sustainable and inclusive development. They rely on innovation system (IS) approaches and the ideals included in the proposals of social technologies and innovations and models of integrated science and sustainability.

Are there university policies that are moving in the direction proposed by the concept of "developmental universities"?

Previously, we referred to the debate concerning university/society relations and the importance attached to the role of universities in local development. A modest example of progress in this direction can be found in what has taken place in Cuba. Two networks maybe cited here in the form of example: the Program for University Management of Knowledge and Innovation for Development (Núñez 2014; Núñez and Alcázar 2015), and the Programme for Local Agrarian Innovation (Ortiz 2013). Both have emerged from higher education organizations and are committed to building local innovation systems. The former covers a wider spectrum of ST (housing, energy, health, etc.), while the latter focuses on food safety in different regions.

Both share a few key points:

1. They recognize the importance of local public policies and the role of local government in the governance of innovation processes.
2. They assume the interactive and systemic approaches of innovation systems, aiming to build institutional arrangements, strategies, and policies, as well as regulations that make knowledge an important resource for development.
3. They recognize that sustainable and inclusive development requires knowledge capacities that higher education possesses to a greater extent than other actors. They understand that higher education must work within models of "integrated science" and "sustainability". The traditions of scientific and technological research, the research group, high-level training, and the benefits derived from strong international ties can be combined with cognitive and technological capacities and the productive experiences of local actors.
4. They share a DUI innovation mode, in which participatory learning is a key element.
5. They encourage a strong link between technology, innovation, and social participation.

10.6 Conclusion

The SDG require new practices and scientific and technological policies, and particularly new university policies. The issue is now part of the Latin American debate. Rethinking these policies requires making use of different concepts. In this chapter, we have discussed some of these concepts emerging from the current discussion, which can help us think about new links between technology, innovation, and development. The brief reference to what has taken place in Cuba shows that higher education policies can be based on these concepts and ideals.

References

Albornoz, M., & López Cerezo, J. A. (2010). *Ciencia, tecnología y universidad en Iberoamérica.* Buenos Aires: OEI-EUDEBA.

Arocena, R., & Sutz, J. (2012). Research and innovation policies for social inclusion: Is there an emerging pattern? In H. M. M. Lastres, C. Pietrobelli, R. Caporalli, M. C. Couto Soares, & M. G. Pessoa de Matos (Eds.), *A nova geração de políticas de desenvolvimento productivo, sustentabildade social e ambiental* (pp. 101–113). Brasilia: BID/BNDES/SESI/CNI.

Arocena, R., & Sutz, J. (2013). Innovación y democratización del conocimiento como contribución al desarrollo inclusivo. In G. Dutrénit y J. Sutz (Ed.), *Sistemas de Innovación para un Desarrollo Inclusivo. La experiencia latinoamericana* (pp. 19–34). México: Foro Consultivo Científico y Tecnológico, A.C.- LALICS.

Bijker, W., & Pinch, T. (2008). La construcción social de hechos y artefactos: o acerca de cómo la sociología de la ciencia y la sociología de la tecnología pueden beneficiarse mutuamente. In H. Thomas, y A. Buch (Coords.), *Actos, actores y artefactos: Sociología de la Tecnología* (pp. 19–62). Buenos Aires.

Bortagaray, I., & Ordóñez-Matamoros, G. (2012). Innovation, innovation policy, and social inclusion in developing countries. *Review of Policy Research, 669, 29,* –671.

Brundenius, C., Lundvall, B. A., & Sutz, J. (2009). The role of the universities in innovation systems in developing countries: Developmental university systems-empirical, analytical and normative perspectives. In B. A. Lundvall, K. J. Joseph, C. Chaminade, & J. Vang (Eds.), *Handbook of innovation systems and developing countries. Building domestic capabilities in a global setting* (pp. 311–336). Cambridge, MA: Edward Elgar.

Callon, M. (1987). Society in the making: The study of technology as a tool for sociological analysis. In W. E. Bijker, T. P. Hughes, & T. J. Pinch (Eds.), *Social construction of technological systems* (pp. 83–106). Cambridge, MA: MIT Press.

Casas, R., De Fuentes, C., Torres, A., & Vera-Cruz, A. O. (2013). Estrategias y gobernanza del Sistema Nacional de Innovación Mexicano: Retos para un desarrollo incluyente. In G. Dutrénit y J. Sutz (Ed.), *Sistemas de Innovación para un Desarrollo Inclusivo. La experiencia latinoamericana* (pp. 35–64). México: Foro Consultivo Científico y Tecnológico, A.C.- LALICS.

Cassiolato, J. (2012). Aprendizaje y capacitación en innovación en América Latina Workshop, documento para discusión, Brasilia, 16–17 de Octubre, 2012.

Cassiolato, J. E., Lastres, H. & Soares, M. C. (2013). Sistema Nacional de Innovación de Brasil: Desafíos para la sostenibilidad y el desarrollo incluyente. G. Dutrénit y J. Sutz Sistemas de Innovación para un Desarrollo Inclusivo. La experiencia latinoamericana (65–95). México: Foro Consultivo Científico y Tecnológico, A.C.- LALICS.

Clark, B. (2011). *Cambio sustentable en la universidad.* Palermo: Fundación Universidad de Palermo.

Crespi, G., & Dutrénit, G. (2013). *Políticas de ciencia, tecnología e innovación para el desarrollo: La experiencia latinoamericana*. México: Foro Consultivo Científico y Tecnológico, A.C.- LALICS.
Dagnino, R. (2009). *Tecnología Social. Ferramenta para construir outra sociedade*. Campinas: IG/Campinas.
Dagnino, R. (2012). Why science and technology capacity building for social development? *Science and Public Policy, 39*, 548–556.
Dagnino, R., Brandão, F., & Novaes, H. (2004). A Tecnologia Social e seus desafios. In A. E. Lassance, C. J. Mello, E. J. Siqueira Barbosa, F. A. Jardim, F. Cruvinel Brandâo, H. T. Novaes, J. Rutkowski, J. De Oliveira Pena, J. S. Pedreira, L. Dowbor, M. Rillo Otero, P. Singer, R. Dagnino, S. Lianza, S. Caccia Bava, & S. M. Portella Kruppa (Eds.), *Tecnologia social. Uma estratégia para o desenvolvimento* (pp. 187–216). Rio de Janeiro: Fundação Banco do Brasil.
Dutrénit, G., & Sutz, J. (2013). *Sistemas de innovación para un desarrollo inclusivo, La experiencia latinoamericana*. México: Foro Consultivo Científico y Tecnológico, A.C.- LALICS.
Etzkowitz, H. (2004). The evolution of entrepreneurial university. *International Journal of Technology and Globalization, 1*(1), 64–77.
Feenberg, A. (2012). *Transformar la tecnología. Una nueva visita a la teoría crítica*. Bernal: Universidad Nacional de Quilmes.
González, P., & Núñez, J. (2014). Desafíos emergentes en los vínculos entre ciencia, naturaleza y sociedad: La Ciencia de la Sostenibilidad. *Humanidades médicas, 14*(2), 522–546.
Hughes, T. (1986). The seamless web: Technology, science, etcetera, etcetera. *Social Studies of Science, 16*, 281–292.
IDRC (2011). *Innovation for inclusive development. Program prospectus for 2011–2016*. http://es.slideshare.net/uniid-sea/october-2011-innovation-for-inclusive-development-program-prospectus-for-20112016. Accessed 15 June 2016.
Johnson, B., & Andersen, A. D. (2012). *Learning, innovation and inclusive development: New perspectives on economic development strategy and development*. Aalborg: Aalborg University Press.
LALICS (2012). *Declaración LALICS: Aportes desde la ciencia, la tecnología y la innovación a la inclusión social*. http://lalics.org/index.php?option=com_content&view=article&id=180:declaracion-lalics&catid=8&Itemid=108&lang=es. Accessed 15 June 2016.
Lastres, H. M. M., Cassiolato, J. E., & Arroio, A. (Orgs.). (2005). *Conhecimento, sistemas de inovação e desenvolvimento*. Rio de Janeiro: Editora UFRJ.
Latour, B. (1992). Where are the missing masses? The sociology of a few mundane artefacts. In W. Bijker & J. Law (Eds.), *Shaping technology-building society: Studies in sociotechnical change* (pp. 225–259). Cambridge, MA: MIT Press.
Lundvall, B. A. (Ed.). (1992). *National systems of innovation: Towards a theory of innovation and interactive learning*. London: Printer.
Lundvall, B. A., Joseph, K. J., Chaminade, C., & Vang, J. (2009). Innovation system research and developing countries. In B. A. Lundvall, K. J. Joseph, C. Chaminade, & J. Vang (Eds.), *Handbook of innovation systems and developing countries* (pp. 1–32). London: Edward Elgar.
Núñez, J. (2014). *Universidad, innovación, conocimiento y desarrollo local*. La Habana: Editorial Félix Varela.
Núñez, J., & Alcázar, A. (2015). ¿Universidad y desarrollo local/territorial? Argumentos conceptuales y sugerencias para las políticas institucionales. In J. Núñez Jover & A. Alcazar Quiñones (Eds.), *Universidad y desarrollo local: Contribuciones latinoamericanas* (pp. 191–204). La Habana: UDUAL-Editorial Universitaria Félix Varela.
OECD (2011). *Fostering innovation to address social challenges—Workshop proceedings*. https://www.oecd.org/sti/inno/47861327.pdf. Accessed 15 June 2016.
Ortiz, R. (2013). *La biodiversidad agrícola en manos del campesinado cubano*. Mayabeque: INCA.
Piketty, T. (2014). *El capital en el siglo XXI*. México: Fondo de Cultura Económica.

PNUD. (2001). *Poner el adelanto tecnológico al servicio del desarrollo humano*. México: Ediciones Mundi Prensa.

PNUD. (2010). *Informe sobre el Desarrollo Humano 2010. La verdadera riqueza de las naciones: Caminos al desarrollo humano*. Madrid: Mundi-Prensa. Retrieved May 18, 2016, from http://hdr.undp.org/sites/default/files/hdr_2010_es_complete_reprint.pdf

PNUD. (2011). *Informe sobre el Desarrollo Humano 2011. Sostenibilidad y Equidad: Un mejor futuro para todos*. Madrid: Mundi-Prensa Retrieved May 18, 2016, from http://hdr.undp.org/sites/default/files/hdr_2011_es_complete.pdf

PNUD. (2013). *Inclusive development*. Retrieved June 15, 2016, from http://www.undp.org/content/undp/en/home/ourwork/povertyreduction/focus_areas/focus_inclusive_development/

PNUD. (2015). *Transformar nuestro mundo: la Agenda 2030 para el Desarrollo Sostenible*. Retrieved Nov 27, 2017, from http://www.un.org/ga/search/view_doc.asp?symbol=A/70/L.1&Lang=S

Salazar, M., Lozano-Borda, M., & Lucio-Arias, D. (2013). Ciencia, tecnología e innovación para un desarrollo inclusivo en Colombia: Programas piloto desarrollados por Conciencias. In G. Dutrénit y J. Sutz (Ed.), *Sistemas de Innovación para un Desarrollo Inclusivo. La experiencia latinoamericana* (pp. 125–154). México: Foro Consultivo Científico y Tecnológico, A.C.- LALICS.

Salomón, J. J. (2008). *Los científicos. Entre poder y saber*. Bernal: Universidad Nacional de Quilmes.

Sutz, J. (2010). Ciencia, Tecnología, Innovación e Inclusión Social: Una agenda urgente para universidades y políticas. *Psicología, Conocimiento y Sociedad, 1*, 3–49.

Thomas, H. (2008). Estructuras cerradas vs. procesos dinámicos: Trayectorias y estilos de innovación y cambio tecnológico. In H. Thomas & A. Buch (Eds.), *Actos, actores y artefactos. Sociología de la tecnología* (pp. 217–262). Bernal: Universidad Nacional de Quilmes.

Thomas, H., & Fressoli, M. (2009). En búsqueda de una metodología para investigar tecnologías sociales. In Renato Peixoto Dagnino (Org.), *Tecnologia Social. Ferramenta para construir outra sociedade* (pp. 113–138). Campinas: Editora Kaco.

Thomas, H. (2011). Tecnologías sociales y ciudadanía socio-técnica. Notas para la construcción de la matriz material de un futuro viable. *Revista do Observatório do Movimento pela Tecnologia Social da América Latina Ciência&Tecnologia Social. A construção crítica da tecnologia pelos atores sociais*. Volume 1 – número 1 – jullho de 2011. http://periodicos.unb.br/index.php/cts/article/download/3838/3295. Accessed 15 June 2016.

UNESCO (2015). *UNESCO science report: Towards 2030*. http://unesdoc.unesco.org/images/0023/002354/235406e.pdf. Accessed 15 June 2016.

Vessuri, H. (2014). Los límites del conocimiento disciplinario. Nuevas formas de conocimiento científico. In P. Kreimer, H. Vessuri, L. Velho, & A. Arellano (Eds.), *Perspectivas Latinoamericanas en el estudio social de la ciencia y la tecnología* (pp. 31–43). México: Siglo XXI.

Part IV
New Technological Frontiers

Chapter 11
The Philosophy of Technoworlds and Technopersons

Javier Echeverría

Abstract After first reexamining Ortega's philosophy of technique and Quintanilla's philosophy of technology, this article will propose a philosophy of technoscience that distinguishes between different scales in present-day technological worlds. There are macro-, meso-, micro-, and nano-cosmic scale technoworlds whose properties and structures are still being researched by contemporary technosciences. The philosophy of technology needs to be specific for each kind of world under study, including the social worlds, not only those of nature or the biosphere. An important example in this respect comprises technopersons, artificial entities that superimpose themselves on physical and legal persons and interact in today's major technoworld: the digital world. At the end of the article, some of the possible consequences and risks of the expansion of the information and communications technologies (ICTs) to microcosms and nanocosms, specifically, to the human brain, will be highlighted.

11.1 The Philosophy of Technique: Ortega y Gasset

In his *Meditación de la técnica*, Ortega y Gasset defined technique as "the reform that man imposes on nature in order to satisfy his needs" (Ortega y Gasset J. 1939, p. 63). This definition was a starting point for the European philosophy of technique in the first half of the twentieth century, although some points need to be modified in the twenty-first century:

1. A distinction should be made between techniques, technologies and technosciences
2. I propose "transformation" instead of "reform"
3. Where the word "nature" appears, "worlds" would be preferable

Regarding the third point, today's technosciences not only modify nature, but also societies, people, and the relations between them. This is the case for the infor-

J. Echeverría (✉)
Santamaria Center, University of the Basque Country, San Sebastián, Spain
e-mail: javier.echeverria@ehu.es

B. Laspra, J. A. López Cerezo (eds.), *Spanish Philosophy of Technology*, Philosophy of Engineering and Technology 24,
https://doi.org/10.1007/978-3-319-71958-0_11

mation and communications technologies (ICTs). First of all, they transform relations between people and make tele-relations possible via television, telephones, the Internet, the social networks, and electronic banking, for example. However, with the consolidation and daily use of the ICT system, people themselves have been transformed. In this article, I hold that *techno-persons* are emerging at the beginning of the twenty-first century. I will try to specify this concept, which, if valid, would be the key to the philosophy of technosciences. We can even speak of the progressive emergence of *technosocieties*, as I will suggest at the end of the article.

I am taking the first change for granted and shall not refer to it because it has already taken place: radio-television, the telephone, the Internet, and social networks permit different kinds of tele-relations and interactions among physical and legal persons, regardless of their geographical location and recurrently in time. This is because the ICTs have made it possible for a new kind of space-time to emerge, a space-time that was initially called cyberspace, but which I, for my part, call the third environment (the digital world), as opposed to the first and second environments (*physis* and *polis*), in which different human cultures have developed. I will focus on the emergence of this new social and technological space-time and, within it, the progressive constitution of *technopersons*, who superimpose themselves on physical and legal persons. As for technosocieties, for now they only exist as such in specific areas and on certain scales. However, different technosocieties will foreseeably develop and become consolidated throughout the twenty-first century and their technological component will be substantive, not only adjectival. For this reason, instead of talking about technological or scientific-technological society (Rescher 1999; Agazzi 1998; Queraltó 2008), I propose to reflect upon technosocieties, the societies whose social relations and interactions are only possible via the mediation of the ICTs. In these societies, there are technopersons, who, in turn, can be characterized as persons whose capacities and actions are determined by the ICT system.

Having presented the main proposals that I intend to make, I will now return to the second point regarding Ortega's definition in order to qualify and extend it. Some technologies not only reform, but also transform nature, sometimes radically. Ortega himself fostered this idea when he stated that, thanks to techniques, human beings have generated *supernatures*:

> man, thanks to his technical gift, manages for everything that he requires to be found in his surroundings – he creates, thus, a new, more favorable circumstance, segregating, we could say, a supernature by adapting nature to his needs. (Ortega y Gasset 1939, p. 65)

Some technological devices and constructions from the industrial period transform nature in such a way that they even conceal it, to the point that many people have lived and continue to live their lives in practically artificial worlds. We can take the city of New York as an example, insofar as this "supernature" and its multiple skyscrapers are superimposed on the island of Manhattan. Nuclear power plants are another good example, because, in order to work in them, the human body must be completely enclosed in anti-radioactive clothing. I thus propose substituting "reform" by "transformation" and, where "nature" appears, I suggest using the terms "worlds", in general, including natural microcosms and macrocosms, as well as social and urban mesocosms, artistic spaces, and, of course, scientific microcosms, whose main canon are the physics and chemistry laboratories of the modern epoch, with their corresponding handcrafted or industrially produced instruments. In short, technologies and their derived effects transform the different kinds of worlds, including the *Lebenwelten* (worlds of life) of each person, as occurs with any person who has to travel to work in a metropolis each day by car, subway or bus. The philosophy of technology must first specify the kinds of artifacts which concern it: whether only artisanal techniques or industrial technologies as well. Subsequently, it should not limit its concern to the artifacts, but should analyze the different kinds of technical and technological actions that they make possible, analyzing not only the results, but also the derived effects.

1. I accordingly propose a consequentialist philosophy of technology which not only analyzes artifacts, but above all the actions, consequences, and risks derived from their use. In this sense, Quintanilla's distinction between common techniques and industrial technologies seems to me a very important one (Quintanilla 1989, p. 50).
2. On the other hand, I find it indispensable to clarify, right from the start, the scale of the world on which these technological actions are produced, as well as their effects: macrocosmic, mesocosmic or microcosmic, and, since the beginning of the twenty-first century, nanocosmic.

Investigating galaxies using space probes or the Hubble telescope is not actually the same as gaining access to nanocosms using an atomic force microscope or a scanning electron microscope, which in fact should be called nanoscopes. The reason is clear: Einsteinian physical laws rule on the galactic scale, but they are not the same as the Newtonian laws that hold in our habitual mesocosms, and they coincide even less with the quantum mechanics laws that are in force on the nanometric scale. As scientists know, certain physical-chemical properties change depending on scale. In contrast to artisanal and commonplace techniques, technologies and technosciences are based on scientific knowledge. Therefore, their derived results and effects must be analyzed in the corresponding nomological context. The consequences of technological actions are quite different depending on the kind of cosmos we are talking about and depending on the scale on which this world is observed and measured. In the case of the neurosciences, which I will address at the end of this article, action is on the microscopic and nanoscopic scale, with all of the resulting ontological, methodological, and instrumental implications. The properties of material are different on the nanometric scale, one reason why the philosophy of nanocosms and of the technologies operating within them is significantly different from the philosophy of mesocosms and macrocosms.

On the mesocosmic and microcosmic scale, it is also necessary to take into account the existence of social technologies and social technosciences. This chapter, in fact, states the need to create a *philosophy of social and human technologies and technosciences*, with the cognitive sciences and neurosciences as relevant examples, although many others could be given. The majority of the philosophers of technologies have reflected upon the technologies that transform nature (such as nuclear energy), biotechnologies, and the impacts of technological actions on the environment. Now in the midst of the twenty-first century, it is also necessary to analyze the social transformations made possible by certain technologies, such as the ICTs, as well as the transformations that human beings themselves experience when they become technopersons. It is even necessary to consider the possible emergence of artifacts that go beyond the capabilities of the human species, with the multiple consequences that this would have in the future.

Social technosciences can indirectly affect the biosphere, for example, when it comes to eliminating the waste materials that televisions, computers, cellphones, and other ICT artifacts generate, but, above all, they transform societies, people, and the relations between them. This is especially true in the case of the information and communications technologies, to which authors such as Manuel Castells (1996–1998) have attributed the appearance of a new era, the *Information Era* (Castells, 1996–1998), called the *Digital Revolution* by others. I am not personally in favor of talking about New Eras, quite the contrary, but if I had to do so, I would allude to the *Scientific-Technological Era* (Rescher 1999) or, even better, the *Technosciences Era*, as there are several of these technosciences.

1. Returning to Ortega's definition, in my opinion the expression "satisfaction of needs" seems too restrictive. I prefer to speak of the "satisfaction of values," including within the term 'value' different kinds of needs, demands, and objectives, both individual and collective. In order to do philosophy of technosciences, therefore, I adopt an *axiological approach*, as do other authors (Rescher, Agazzi, Olivé, Álvarez, etc.)
2. Finally, I believe it is necessary to distinguish between techniques, technologies, and technosciences, generating three modalities of philosophy, each of which has its own field of study: the philosophy of techniques (artisanal techniques, for example), the philosophy of technologies (industrial technologies), and the philosophy of technosciences, to which I will devote the following section.

11.2 From the Philosophy of Technology to the Philosophy of Technoscience

In his book, *Tecnología, un enfoque filosófico* (1989), Quintanilla defined techniques as "systems of human actions oriented toward transforming specific objects to achieve a valuable result in an efficient way" (Quintanilla 1989, 34). This definition is precise and rigorous and can hence be taken as a starting point for creating a serious philosophy of technique. After Ortega, Quintanilla can be said to have become the second major author of reference on issues pertaining to the philosophy of technique, within the Spanish context.

In this article, I will only mention four of Quintanilla's contributions, those that seem to me to be most important for my own line of research. First, he stated that techniques are human actions and that they cannot be reduced to artifacts. Second, he applied systems theory to the philosophy of technology, referring above all to *technological systems*, rather than to isolated artifacts. Third, he made a distinction between techniques and technologies, arguing that technologies are based on scientific knowledge, in contrast to techniques, which exist in all human cultures. Fourth, he stated that technologies have a strong industrial component, whereas most techniques are artisanal.

Paraphrasing Quintanilla, in 2003 I proposed a "definition" of technology that reexamined several of the points that he had highlighted and which included some additional ones: "technologies are systems of action that are regulated, industrial, and linked to science, carried out by agents, with the aid of instruments, and intentionally oriented toward the transformation of other systems with the purpose of achieving valuable results and avoiding unfavorable consequences and risks" (Echeverría 2003, p. 58).

Clearly, a definition of this kind does not attempt to trap the "specific difference" of technologies compared to techniques or sciences, for the simple reason that sciences, techniques, and technologies are not species belonging to the same genus, and neither are the technosciences to which I will refer later. More than defining these concepts, the philosophy of technology must attempt to throw light upon them. In order to do so, I usually use a methodology of distinctive features, which I will not go into here (see Echeverría 2003, Chap. 2).

The technosciences were born in the USA at the time of World War II, thanks to the macro research projects (*Big Science*) that the federal government and some of its agencies (NSF, NIH, NASA, Brookhaven) promoted. These projects were developed by military institutions, universities, industrial businesses, scientific policy offices and, last but not least, scientists, engineers, and technicians. The war saw the emergence of technophysics (Radiation Laboratories, the Manhattan Project), technomathematics (the ENIAC project, computer science, artificial intelligence), and technochemistry (chemical weapons, nylon, other synthetic materials). In the following decade, technomedicine (created by the NIH in the United States and by large pharmacological companies) and technoastronomy (Sputnik, NASA, Hubble, etc.) appeared. Starting in 1980, technosciences as such arose, particularly biotechnologies (the Genome Project, Monsanto) and the information and communications technologies (Internet, IBM, Microsoft, Apple, Linux, Netscape, etc.). All of this took place in the USA, although other countries began to create some relevant technoscientific agencies and companies, as was the case in Europe (CERN, European Space Agency, Nokia, Ericsson, etc.).

In general, I think that the second half of the twentieth century produced a major *technoscientific revolution* that has affected scientific practices above all, transforming them radically by introducing ICTs in research and subsequently in the administration, communication, and teaching of science. This technoscientific revolution gave rise to the constitution of national and regional systems of innovation (Lundvall 1992; Nelson 1993) from 1990 on. Its clearest manifestation comprises the ICTs, but it should be borne in mind that there are other technosciences to consider, not only the ICTs. Foreseeably, this technoscientific revolution will continue to develop throughout the twenty-first century. At the end of the twentieth century, the revolution had also reached some social and human sciences, which is why we can talk about human and social technosciences: one of the main examples is the *Big Data* industry, as well as advertising, social networks, artistic visualizations and simulations, digital archives, music, and e-books. The technoscientific revolution has transformed *the worlds* (natural, social, artificial, cultural, artistic, political, business worlds, etc.) into *technoworlds*. In these technoworlds, technological systems are constituent, not only instrumental, because the events and phenomena that happen in them are a result of technological actions, without the mediation of which they would not happen. There are still facts (physical, chemical, biological, social, and economic facts) in the world, but techno-facts, resulting from technological actions for whose execution ICT devices are indispensable, are superimposed upon them. Technoscientists work in front of their computer screens and the subject/world relationship is mediated by technologies.

Technoworlds exist on different scales (macrocosmic, mesocosmic, microcosmic, and nanocosmic scales) and are the main object of study of the philosophy of technosciences. Consequently, different kinds of technopersons have appeared in these technoworlds: cyborgs, artificial limbs, avatars, virtual worlds, photographs on social networks, videos on YouTube, "intelligent" artifacts, etc.).

In addition to technoworlds, the philosophy of technosciences must deal with some technoscientific macroprograms that are particularly decisive for the future of people and societies, such as *Converging NBIC* (nano-bio-info-cogno, NSF 2001) *Technologies,* and the *BRAIN* (USA 2009) and *Human Brain* (EU 2009) programs. The latter two are derived from NBIC and their objective is to carry out in-depth research on the brain and the central nervous system. Macroprograms such as NBIC have defined the main *technoscientific agenda* at the beginning of the twenty-first century. Apart from their long duration and the high economic investments required to carry them out, these macroprograms tend to create high expectations not only among scientists, but also among investors and companies, as they promise important innovations. Contemporary technosciences must pay great attention to the sources of private funding and, in order to do so, it is necessary to publicize the results and innovations that may potentially be derived from them in advance. This is the case in the field of artificial intelligence (think of *Deep Blue* or current visionary literature on the singularity of technology, Kurzweil 2005, 2012). There is a reason behind the attempt to improve human perceptive and cognitive capabilities by means of different technological devices whose construction and functioning is very expensive. This social, political, and financial component of the technosciences is undoubtedly important, but the philosophy of technoscience should focus on the structure of different scientific practices and the changes they undergo: research, application, evaluation, dissemination, and learning.

The technosciences are hybrid practices, being both scientific and technological, but they are also economic, legal, political, advertising and, sometimes, military practices. Not only do scientists and engineers intervene in these practices, but other kinds of agents do, too, contributing different cultures, interests, objectives, and value systems. The technosciences do not only attempt to describe, explain, and understand the world; their main objective is to transform it and generate innovations. Scientific knowledge has value in the technosciences, but it is worth more if it generates technological developments and innovations. Therefore, in order to reflect on the technosciences, it is necessary to start out from the different kinds of actors and actions that comprise them, as well as from the stakeholders involved. The philosophy of technosciences is not based on a conception of being, but on one of doing. Stated categorically: *it is not what the technosciences are, but what they do that matters*. My approach accordingly focuses on the actors and technoscientific agendas, as well as on the interests, objectives, and values that guide their actions and agendas. My starting point is the idea that the interests and objectives depend on the values, as it is possible to *assess the objectives and goals of the different actors, as well as their interests*. The philosophy of technoscience that I expound (Echeverría 2003, 2014) is grounded on an axiological approach, based on the existence of several actors who are relevant throughout the process of technoscientific innovations, whose values can be shared but which can also enter into conflict. Consequently, the philosophy of technology must pay particular attention to *technoscientific conflicts*, which sociologists usually analyze from the viewpoint of interests, and philosophers from the perspective of the values applied by technoscientific actors, in their different actions. Axiological conflicts seem to me to be a priority research subject for the philosophy of technoscience.

The neurosciences and *cognitive technosciences* are particularly important because they not only concern the philosophy of nature or society, but also the philosophy of the mind. Going beyond *cyborgs* (Haraway 1985) and actor-network theory (Latour 2005), I feel it is preferable to talk about technopersons and technoactors, in the manner that I will now explain.

11.3 Technopersons and Technosocieties

The construction of technopersons can be carried out in two main ways: (1) by implementing previously existing persons technologically, and (2) by giving machines and robots capabilities similar to those of human beings, and even improving these capabilities. The latter path is the path of artificial intelligence, which I shall not address in this chapter. The former focuses on technoscientific macroprograms such as those previously mentioned, part of whose objectives consist in providing artificial limbs and artifacts that make it possible to improve people's perceptive, cognitive, and motor capabilities. Industrial technologies improved human capacity for movement enormously by producing bicycles, automobiles, trains, and airplanes that people have used massively. This all generated different technological landscapes, typical of the industrial epoch: airports, train stations, maritime ports, and cities around these features (New York, Buenos Aires, etc.).

The ICTs make up a different technological system which has also superimposed itself on human beings, but which creates other kinds of techno-landscapes: websites, simulations, screens, and digital sounds, among others. The ICTs do not implement capabilities for physical movement, but rather other capabilities of action and interrelation, particularly expressive, communicative, and mental capabilities. Thus, new personal masks and therefore technopersons appear. Etymologically, 'person' means mask: images on television and on YouTube, avatars and photographs on Flickr and Instagram are all concrete examples of these techno-masks, which are superimposed on persons and provide a third kind of identity which is superimposed on physical identity and identity as a citizen, as I shall show.

In order to interpret this major transformation made possible by the ICTs, I hold the hypothesis of the third environment: the ICT technological system has made it possible for a new social space-time to emerge, the third environment, which has superimposed itself on the two main environments in which human beings have lived historically, the physis and the polis. According to this hypothesis, human macrocosms and mesocosms have three main dimensions in the twenty-first century: (1) the biosphere and the rural world, (2) States and cities, and (3) the digital and electronic world. Human cultures, techniques, and technologies have traditionally developed in the first and second environments. The contemporary technosciences have facilitated the emergence of a technoculture, i.e., a culture that develops in a social space-time whose structure is not Euclidean but reticular, where physical proximity is not a requirement for human and social interaction. The appearance and development of the third environment has multiple consequences, including the emergence of technopersons. Simplifying greatly, I will say that the first environment is inhabited by physical persons (bodies); the second, by legal persons (citizens); and the third, by technopersons. In the first environment, people communicate among themselves through speech; in the second, through writing; and in the third, through technolanguages, which are possible thanks to the machine language of computers (digitalization) and the subsequent languages of programming and the processing of texts, images, and data. In the third environment, memory is made up of large databases and digital memory repositories; while in the first, it is accumulated in the brain; and in the second, in books, archives, and libraries. Other additional structural differences between the three environments could be furnished.

The distinction between physical and legal persons and technopersons, which works on both the individual and collective level, derives from the hypothesis of the third environment. Technopersons can be organizations, not only individuals. Just as States assign a legal identity that one must bear in mind (name, surname, date of birth, place of residence, passport number, nationality, etc.) to physical persons, information and communications technologies attribute signs of identification to each technoperson (URLs for computers, e-mail addresses or user names on the social networks, credit and debit card numbers, secret passwords, etc.). These technological signs of identity, whose structure varies depending on the telematic service providers, are indispensable in order to be a technoperson in the third environment. The important issue is that this new kind of identity is very different from physical identity and from citizen identity, even though it superimposes itself upon them. Once admitted to a social network or to the Internet, technopersons can build a techno-house in a digital world (a website, space in *Second Life*, a Facebook wall, etc.), which they subsequently decorate according to their taste, though strictly following the format of the corresponding software application, which is imposed upon them when they sign the "Accept" agreement. Starting out from there, technopersons can do many things in the digital domain where they have installed themselves, while at the same time continuing to be legal persons in their corresponding city or country and, of course, while still being physical persons who breathe, eat, and reproduce as flesh-and-blood bodies. This is the triple dimension of the contemporary world, although the separation between these three dimensions is not strict and admits mixtures. The third environment superimposes itself upon the other two, similarly to the way that cities superimpose themselves on the country and territories, making new ways of organizing and developing social life possible. The Orteguian idea of how technical systems superimpose themselves also works for the third environment. However, there are many more technopersons than persons, as the same person can create several technological masks of him or herself. This is what usually happens and is the reason why the "demographic" growth of the third environment has been exponential in recent decades. Of course, technopersons tend to form groups, generating larger or smaller technosocieties. A social network is an example of a technosociety.

The main difference between persons and technopersons is that of plurality. A single (physical or legal) person can support several technopersons, not just one. This is also true for techno-objects, which can have many locations on telematic networks, even if they have a DOI (digital object identifier). The second difference was also mentioned earlier and is ontological in nature: what is important is not what the technopersons are, but what they do. Technopersons are characterized by the relations and interrelations they have in the third environment, most of which leave traces in the super-computers of the "Cloud." On the other hand, I should point out that I am not only referring to the Internet. Credit and debit cards and cash machines provide access to the financial "Cloud." Using these paths of access to the third environment, physical-legal persons manage part of their economic resources. Capital migrated to the third environment many years ago, as did savings and investment funds. In short, the third environment is not limited to the Internet, but includes many other ICT networks, including military and financial networks, as well as disseminated Intranets and large databases (Big Data), which require authorizations and additional identification passwords. Contrary to what is generally believed, most areas of the digital world are not open to the public at large. There are passwords and private digital domains everywhere.

I myself think that the present-day techno-political and techno-economic structure of the digital world is based on the existence of several *informational fiefdoms* (*techno-fiefdoms*), which are not territorial, but reticular and technological in nature. Google, Apple, Facebook, Twitter, Microsoft, Amazon, etc. are business technopersons that give their names to these techno-domains. Despite its major relevance, in this chapter I will not address the emergence of a new kind of power in the third environment, a kind of power that we should call *techno-power* in order to distinguish it from the political power of the States in the second environment, or the physical-chemical, energetic, and biological power of bodies in the first environment. I will only say that the identity of technopersons does not depend on States (whereas their identity as legal persons does). Nor do many of the regulations that reign in the digital fiefdoms depend on States. When a legal person (whether individual or collective) subscribes to some service in the "Cloud" and clicks on the "Accept" button, he or she becomes a technoperson in this specific domain and must follow the norms that the owner of the domain has established. This does not prevent him or her from being a techno-person in other digital techno-worlds. Today, we human beings are responsible for several technopersons, who are *ourselves*, even though each one has its own signs of technological identity. This is an important difference, which can be summarized as follows: *in the third environment, the identity of technopersons is plural*. On the other hand, a technoperson that I have generated can be managed by another person, as we well know. I not only refer to identity theft, as can happen with zombie computers or when someone alters photographs, images, or texts that someone else has put online, but above all to the possibility any network administrator has to manage our technological artifacts from his or her control post and, ultimately, from the "Cloud." This is one of the keys of contemporary techno-power and it involves a major structural change in human techno-relations, which are strictly hierarchical, even though they do not seem to be. External cyber-manipulation of the actions of technopersons is always possible, precisely because the corresponding persons have become technopersons.

Up to now, I have referred above all to individual persons, but what I have said also goes for collective legal technopersons. When any organization decides to become a technoperson, it has to go through the basic technological mediations that shape the new social space. This is precisely what turns it into a technoperson. We usually only think about hardware and software, but there are many other mediations to take into account, particularly the social technologies that intervene in the networks, which are numerous. What can or cannot be done socially on a telematic network is previously established by the owners, managers, and administrators of these networks. The States are not the ones who set these norms; it is the owners and designers of the telematic networks who do so. I call them the *Lords of the Air* or *Lords of the Networks* (Echeverría 1999), to compare them with the feudal lords of medieval Europe. Today, they could be called the Lords of the Clouds. What is important is that the structure of power in the digital world is neo-feudal. The control network administrators can exert over the actions of their users and their capacity to expel them from their domains without trial is the source of this neo-feudal techno-power, a power which is not only technological, but above all techno-social and techno-political.

11.4 Nanoworlds and Technopersons

To end, I wish to insist on one of the reflections that I have presented in this article: the importance of taking into account the scales of the worlds which are the object of philosophical study. Previously, I stated that technological actions have different consequences depending on different spatial scales; now, I would add that this is also true for time scales. Doing philosophy of technology based on ephemeral technological phenomena, no matter how intense at any specific moment, is not the same as proposing hypotheses for interpreting long-lasting technological changes. The hypothesis of the third environment is a change of this second kind and, up to now, I have only presented it on the mesocosmic scale. However, in the present state of the art, nothing seems to prevent the third environment from expanding toward the microcosms and the nanocosms of the body, as the NBIC program itself suggested in 2001. ICT nano-devices that facilitate or implement relations, functions, and interactions between different organs in the human body, including the central nervous system, can be imagined, even if they seem very hard to achieve at present. Should they be achieved, we would be faced with a major disruptive innovation, which I would dare to call "techno-neurons," although they do not yet exist. In fact, the genome sequence of the 1990s can be interpreted as the creation of a new kind of entity, techno-genes, which superimpose themselves on biological genes. Techno-genes are computational sequences of information and can be patented, generating economically profitable innovations. They are, therefore, typically technoscientific entities, which is why it is a good idea to call them techno-genes, to distinguish the technological implementation from the biological genes that come from natural evolution and are subject to Darwinian laws of variation and mutation. Techno-neural networks would be a similar technological implementation of neural networks and, when the time comes, as artificial products in contrast to neurons, they could be patented.

Techno-perceptions (high-resolution digital screens, virtual reality helmets, stereophonic headphones, etc.) are, at present, frequent, though on the mesocosmic scale, i.e., as a superimposition upon certain external organs of the human body. There are also hearing aids, data gloves, electronic tongues, electronic noses, and other ICT devices that make possible or enhance sensory perceptions by means of different technological implementations. Artificial intelligence experts even talk about building "machines that think", using Turing's test as the criterion for defining what thinking is with regard to a computer. I will not enter into this debate, which lies beyond the scope of this chapter. I personally distinguish between thoughts and techno-thoughts in order to introduce a semblance of conceptual order into the terminological chaos that the conceptual marketing of technoscientific companies generates. In any case, lines of research have been opened up that point toward the possibility of there being techno-cognitive processes on the microcosmic scale, regardless of whether these processes are carried out by "intelligent chips" or by technologically improved neural networks. Bypasses to restore lost connections already exist, allowing us to palliate some of people's perceptive and cognitive insufficiencies. This all involves the appearance of a new kind of technoperson; in this case, on a microscopic and perhaps even nanoscopic scale. Thus, I think that the philosophy of technology should remain very aware of these lines of technoscientific research being developed in the BRAIN and Human Brain programs. The possibility of establishing ICT connections among neural networks or between neural networks and internal or external sensory organs is of major philosophical importance, because it affects a kind of nature that is very human: the body and its brain.

The mental techno-landscape that I am suggesting seems very distant. Yet it is no less certain that both artificial intelligence and the aforementioned technoscientific macroprograms point in this direction. A minimally serious philosophy of techno-science must take an interest in the transformations that the ICTs can generate in the microcosms and macrocosms, not only in the mesocosms (local) and the macro-cosms (global). To use Ortega and Gasset's terms, nanocosms are an issue of our times, as they were discovered at the end of the twentieth century. The expansion of the third environment and the ICT system to the interior of the body and of the human brain – even without reaching Matrix-like dystopias – will be a decisive step in the construction of technopersons.

References

Agazzi, E. (1998). *La technoscience et l'identité de l'homme contemporain*. Fribourg: Ed. Universitaires Fribourg.

Echeverría, J. (1999). *Los Señores del Aire, Telépolis y el Tercer Entorno*. Barcelona: Destino.

Echeverría, J. (2003). *La revolución tecnocientífica*. Madrid: FCE.

Echeverría, J. (2014). *Innovation and values: A European perspective*. Reno: UNR- CBS.

Haraway, D. (1985). Manifesto for Cyborgs. *Socialist Review, 80*, 65–108.

Kurzweil, R. (2005). *The singularity is near*. New York: Penguin Group.

Kurzweil, R. (2012). *How to create a mind: The secret of human thought revealed*. New York: Viking Books.

Latour, B. (2005). *Reassembling the social. An introduction to actor-network theory*. Oxford: Oxford University Press.

Lundvall, B. A. (1992). *National systems of innovation: Towards a theory of interactive learning*. London: Pinter.

Nelson, R. R. (1993). *National systems of innovation*. Oxford: Oxford University Press.

Ortega y Gasset, J. (1939). *Meditación de la técnica* (p. 2015). Madrid: Biblioteca Nueva.

Queraltó, R. (2008). *La estrategia de Ulises, o Ética para una sociedad tecnológica*. Sevilla: Dos Ediciones.

Quintanilla, M. A. (1989). *Tecnología: un enfoque filosófico*. Madrid: Fundesco. 2ª ed., 2005. México: FCE.

Rescher, N. (1999). *Razones y Valores en la Era científico-tecnológica*. Barcelona: Paidós.

Chapter 12
Ethical and Political Delusion in the Model of Cloud Computing

Javier Bustamante Donas

Abstract Cloud Computing (CC) is not just a revolutionary development in information technologies (IT), but also a powerful metaphor and new sociotechnical paradigm, one that is, at least theoretically, intended to empower users. Transformations on this scale raise serious ethical and political dilemmas that need to be addressed. This chapter argues that such dilemmas have to do mainly with the very nature of cloud computing as an inherently political technology (a technology that is strongly compatible with certain models of social organization). It is argued that CC is an inherently political technology in the strong sense (it requires a set of political and social requirements to function properly). Once fundamental decisions are taken, changing the influence and public understanding of its social impact will be extremely difficult to achieve.

12.1 Definition of Cloud Computing, Characteristics, Service Models, and Implementation Models

A widely accepted definition is needed to provide a reasonable baseline for discussing the nature of Cloud Computing (CC), forecasting its social impact, and discussing how to best use it. According to a Working Definition of Cloud Computing published by the U.S. Government's National Institute of Standards and Technology (NIST), "Cloud computing is a model for enabling ubiquitous, convenient, on-demand network access to a shared pool of configurable computing resources (for example, networks, servers, storage, applications, and services) that can be rapidly provisioned and released with minimal management effort or service provider interaction" (Mell and Grance 2011).

J. Bustamante Donas (✉)
Department of Philosophy of Law, Moral and Politics, Complutense University of Madrid, Madrid, Spain
e-mail: jbustama@ucm.es

B. Laspra, J. A. López Cerezo (eds.), *Spanish Philosophy of Technology*, Philosophy of Engineering and Technology 24,
https://doi.org/10.1007/978-3-319-71958-0_12

The NIST document document includes other defining elements of CC, namely characteristics and implementation models. CC presents five specific characteristics:

1. *Services under automatic user demand* (without interaction with each service provider),
2. *Broad network access* (access to network resources can be obtained from a wide range of devices, including thin clients like tablets, smartphones, netbooks, etc.),
3. *Resource pooling* (memory, storage, bandwidth, processing power, and virtual machines are assigned dynamically according to customer demand)
4. *Rapid elasticity* that make resources to be available for multiple users simultaneously – each user experiences a sense of unlimited availability of resources (scalability of systems)–,
5. *Measured service* (continuous monitoring of resources usage providing transparency to both users and providers of the provided service).

CC services can be deployed according to four implementation models:

1. *Private cloud* (the infrastructure is provisioned exclusively for a single organization serving multiple users, and the cloud is operated by this organization or a third party).
2. *Community cloud* (resources are provisioned exclusively for a community of users sharing concerns or interests. It can be managed by one organization within the community or by a third party).
3. *Public cloud* (cloud infrastructure is provisioned for open use, and can be ruled by an academic institution, government agency, non-profit organizations, business, etc.)
4. *Hybrid cloud* (This cloud is a mix of two or more private, community, or public infrastructures. They are bound together by protocols and technology that enables data and application sharing).

CC can offer different models of services according to the way it simulates tasks ordinarily performed by software applications, operating systems, or hardware. The most accepted taxonomy segregates three levels (Yoo 2011; Foster et al. 2008; Vaquero et al. 2009; Mell and Grance 2011):

1. *Software as a Service* (SaaS) includes applications that the end user can access through thin clients or web browsers. The end user does not have any control over networking infrastructure or the design of the applications. Gmail and Google Docs are examples of these services.
2. *Platform as a Service* (PaaS) offers an operating environment and software tools that users can employ to customize their own applications. Microsoft Windows Azure and Google App Engine are within this category. PaaS allows customer control at the application level, but not at the level of the system's physical infrastructure.
3. *Infrastructure as a Service* (IaaS): end users have direct access to processing, storage, and networking resources. They can configure those resources and run operating systems and applications on them as they wish. Examples: Rackspace, Amazon Elastic Compute Cloud (EC2), and IBM Computing on Demand.

12.2 Cloud Computing as an Emerging Socio-technical Paradigm

The history of computing demonstrates that technical decisions usually have social consequences, since design patterns look for much more than resource optimization or efficiency: they are tools for reengineering everyday life. It is almost impossible to find areas of human life immune to the influence of IT, and there are no strong reasons to believe nowadays in the neutrality of technical decisions. Langdon Winner (1986) has demonstrated how certain technologies can embody models of life, ideological forms, power-related positions, and how technologies can be ways to solve power disputes.

From time to time, a new sociotechnical paradigm arises. The shift from centralized to distributed models in the 1990s represented to some extend a democratization of the access to information, as well as a change of power from the center of the system to periphery. The Web 2.0 carried out a citizen Internet, and more opportunities for a social appropriation of technology. Smartphones, tablets, *phablets*, portable computing, VR devices, thin clients, and now cloud computing. How do they relate to a cultural context?

There are currently a great number of services that take advantage of CC technology, from Gmail and Yahoo electronic mail systems and the controversial Google Glasses to cloud robotics and computer games like Pokémon Go by Nintendo. In that light, CC can be considered one of the most powerful and suggestive paradigms within the field of information and communication technologies, particularly when its ubiquity is combined with the power of *big data*. For some authors, it will help bring about social changes as radical as those raised by the invention of the Personal Computer (Carr 2008). CC will have a major political and ethical impact on global society, although these *external* impacts have been frequently neglected, along with other non-technical parameters of implementation. Some serious thought on CC development is needed, because CC strongly implies a post-political model, a blend of information technology and human relationships, that could eventually become a *de facto* political constitution.

12.3 Is Cloud Computing as an Inherently Political Technology? Cloud Computing as a Form of Life

From an STS (Science, Technology and Society) perspective, *Cloud Computing* is much more than a revolutionary development in IT. It can be seen as a new socio-technical paradigm that theoretically empowers users and facilitates many kinds of information-related activities, but it also presents serious ethical and political problems that have to be addressed. This chapter argues that such problems have to do mainly with the very nature of cloud computing as an inherently political technology (a technology that is strongly compatible with certain models of social organization), one that discourages or even prevents other particular uses of CC. Thus, there is a need to specify how it is that *CC is an inherently political technology in the strong sense*: a technology that requires a set of political and social requirements to function properly and flawlessly). For once fundamental decisions are taken at the stage of design and implementation, changing the character and meaning of its social impact would be extremely difficult.

In the world of CC, all kinds of data will be at hand, ready to be used in our thin clients, cloud resources will be available, storage capacity will be virtually unlimited, and connectivity will be extended to any area of leisure and work. However, in order to take advantage of these possibilities, our way of proceeding must be strictly submitted to the rules and protocols of CC. A lack of full control over our data, and the need to be permanently connected are two of the requirements that end users of CC are compelled to accept. The great oligopolies of information are particularly happy with this proposal. There is no better way to promote a given technology that creating a social perception of an unacceptable shortage. Before long we will witness new legislation defending universal availability of access to cloud resources, a step seen as a new human right that needs to be protected. Social activities will be more efficient, but also more dependent than ever on a highly coupled and complex technology; all elements have to function flawlessly and timely, and must be perfectly synchronized to meet the operating standards of a CC system. Since a failure of any strategic element of the system soon spreads to the whole service with serious consequences, people will willingly accept enormous investments in fault-tolerant and redundant infrastructures essential to ensuring an uninterrupted access to cloud services. Yes, here will be benefits for citizens. But this technology is clearly more profitable for the corporate sector. The balance between corporate power and civil society will be increasingly precarious.

CC is not politically neutral. It encourages a concentrated and centralized techno-social system. Certain technological decisions can carry on a polarization of political relations. Far from being a problem of malicious use or unfortunate slip-ups, power relations and operating requirements of CC arise independently of any political system or social organization. The Cloud imposes its material conditions of production and distribution on society as a whole. More centralized and concentrated informational power frequently leads to more authoritarian forms of social life. Regardless of later stage design modifications, CC will maintain a set of essential characteristics: synchronization, normalization, concentration, and central control. We will have to ponder the kinds of social institutions that would be compatible with such a powerful technological paradigm as CC spreads over to ever-widening areas of political life and economic activity.

Back in the 1980s, when Langdon Winner elaborated his methodological critique of socio-technical systems as inherently political realities, technology had already become an intense factor modeling our societies, as is particularly evident in both micro and macro scales of our lives. Such a trend has now deepened at a dizzying speed to a point where technologies are no longer instrumental entities that modulate our relationship with the world we live in, challenging conventional understandings of social life, human rights, and life in democracy. Taking Winner's perspective as a starting point, we can begin to explore CC relationships within social structures and our culture overall. This powerful paradigm is driving a shift in the current forms of power, and laying the foundations for a new techno-political culture based on a return to centralized and controlled computing. Traditional distinctions about relationships among infrastructure, structure and superstructure, become meaningless since means of production, tools for data processing, communication channels, protocols and software involved in CC are by no means simply neutral, ductile entities. When combined, elements of this kind are organized in ways that embody the ends and strategies of those groups who promote them. Technologies can be understood as forms of life, or even as political constitutions, a sociotechnical order well defined and established. Some of the characteristics of these technological constitutions can be applied to the model of CC.

First, CC facilitates the control of an unprecedented amount of information, particularly when combined with *big da*ta structures, from a limited number of centers. After years of decentralized Web 2.0, the return to highly centralized models of services is becoming a dominant trend. Given the current faith in deregulation within the business world, the scale and power of CC companies make them highly immune to effective compensatory influences. Second, CC fosters a world-scale model of business. Large bureaucracies and corporations are the ones that benefit most from the model of CC, and the Cloud makes them very efficient at a gigantic scale. Local initiatives can be easily stifled by global institutions that seek to provide global solutions to social needs. Such solutions often tend to eliminate cultural differences and homogenize all possible distributed, decentralized and smaller-scale alternatives. Third, CC implies its own rationality, which tends to produce its particular forms of hierarchical rearrangement. Because of the need for a flawlessly uninterrupted service, new stiffer protocols and standards will likely be created as ways to define the range of communicative and informational forms permissible within the system. Fourth, those competitors that can benefit most form CC can readily displace and marginalize smaller and less efficient competitors, including social services associated with non-global models of conviviality. Fifth, CC will facilitate a new concentration of power in corporate hands. This will increase corporate capacity of controlling political institutions that should regulate them. Academic institutions, parliaments, and regulatory agencies that could control business operation are often subject to manipulation by these powerful new agents (Winner 1986; Bustamante 2013).

12.4 Ethical and Political Dilemmas in Cloud Computing

Several authors have discussed ethical problems that arise with the adoption of CC. According to Timmermans et al. (2010), these problems can be situated within a set of distinctive categories. First are questions about privacy and loss of control over access and rights of use of personal information stored on a cloud. Second, there are dilemmas related to intellectual property rights due to the offshoring of data and their dispersion throughout different jurisdictions. De Bruin and Floridi (2016) take the point of view of business ethics, arguing that the way to deal with cloud services providers, corporate clients and final users have a mutual epistemic obligation to request and provide information about sensitive items such as user privacy, reliability of services, data ownership and data mining. They develop the concept of *interlucency* as an epistemic virtue that should control ethically effective communication in cloud computing. According to this principle, cloud services providers should provide users with accurate information about what the technology really does, and actively track whether those users understand what it tries to communicate. Users must have a sufficient knowledge about the pros and the cons of CC in general, as well as a specific knowledge about the risks involved in the services they are willing to buy or use. Due to a lack of interlucent interaction with the providers, many users are unaware of the real costs of outsourcing their personal data to a CC service. Interlucency is defined in this context as an epistemic virtue aimed to create a shared knowledge that is necessary for an efficient coordinated action and social cooperation (De Bruin and Floridi 2016).

Another classification of risks involving CC distinguishes among *privacy issues* (loss of control over access and use of personal information stored on the cloud), *intellectual property issues* (due to delocalization of servers and multiple legislations), and *problems related to the inherently nature of CC*. Since technical and political decisions on CC implementation are interconnected, they are by no means independent to one another (Bustamante 2013). The following problems and dilemmas are among those frequently encountered:

1. There is a new factor of digital divide: Fast and universal network access and a guaranteed broad bandwidth are more urgent than ever, and sociotechnical infrastructure becomes a critical factor.

Because CC demands much greater levels of infrastructural network resources, its usefulness is limited to those places equipped with fast, reliable, and cheap Internet access. Those elements that deepen the digital divide in traditional environments will multiply their effect when efficiency and productivity depend on CC services. Social integration and territorial policies will achieve a more crucial role, and more coherent and ambitious public policies towards the development of a digital citizenship will be needed. Since nothing can be done without Internet access, any infrastructural weakness will multiply its effect on digital exclusion.

2. Reliability becomes a critical element: Fragility of vital subsystems for sustaining many social functions.

As a consequence, there is an increasing collective perception of vulnerability. Uninterrupted functioning is a must, since CC services cannot be accessed without Internet connection. As we return to a new breed of centralized network topology, fear of cyber-terrorist attack soars. This could be a problem for civil liberties because reliance upon a complex technology involves many critical connections between social life and technical systems. What should be protected first, civil rights and constitutional guaranties, or the cloud?

3. Internet will need a more democratic architecture to prevent CC from becoming a new tool for social control.

The universalization of CC services, spreading their use to all social classes, will require a reengineering of network access and use. Yoo (2011) emphasizes that Border Gateway Protocol (BGP) determines the flux of data packages between domains. It is controlled by centralized routing systems, and it is not especially flexible. He advocates a more democratic and decentralized multi-path intelligent pattern of routing as a much needed way to democratize Internet and to improve the global performance of CC, a way to avoid informational traffic congestions and to facilitate faster recovery from system failures.

4. Delocalization of information, extraterritoriality of laws.

CC is ubiquitous, whereas communication infrastructures belong to specific countries. CC services operate in cyberspace. However, servers for data storing and processing are located in physical spaces regulated by the legislation of specific states. In this light, the legal framework for CC becomes a confusing matter at the very least. In fact, national legislative frameworks become extraterritorial laws affecting CC users, independent of their nationalities.

5. Risks for privacy and integrity of personal information.

CC implies massive data traffic outside both personal and corporate firewalls. To keep the integrity of cloud information is not and easy task. Shared storage and shared channels, shared resources also create new vulnerabilities for privacy. Data from different corporations will share storage space in the same servers, and adequate sealing must somehow be provided. Another vulnerability affects information processing, since the same data can be processed at once across a number of virtual machines under different operating systems, yet it must also preserve its integrity. Data flow throughout many countries with different data protection laws, but the user is ultimately responsible for any violations of particular provisions in these laws.

6. Challenges to social empowerment. "Transparency" has its own risks.

In the model of applications services (Saas), end users hardly need any skill, since they do not need any expertise to run cloud applications. Intelligence resides in the Cloud itself, not in the capabilities of the users. End users can concentrate on their own targets without paying attention to the technical details of the tools they are using, but this benefit has a tradeoff. One possible price to be paid is the risk of shutdown of a cloud service. Technical autonomy of the user, an important element of empowerment, vanishes. According to Richard Stallman, creator of the GNU/Linux system and defender of free software, this *transparency* – CC makes our life much easier – implies a high cost: "One reason you should not use web applications to do your computing is that you lose control. It's just as bad as using a proprietary program. Do your own computing on your own computer with your copy of a freedom-respecting program. If you use a proprietary program or somebody else's web server, you're defenseless. You're putty in the hands of whoever developed that software" (Johnson 2008).

7. Menaces to network neutrality.

Since thin clients are enough to make use of CC services, the quality of cloud services relies upon network access. Network neutrality and universal access become critical elements. The break-up of network neutrality is by itself a menace to the democratic structure of Internet. However, there will likely be strong pressure to provide privileged access to those willing to pay more in order to get a more efficient service. The resulting technological gap would tend to generate new social inequalities.

8. Need for a higher decentralization of Internet.

As the Internet is used to further centralize computing power and user data, the pendulum seems to swing away from individual autonomy and toward more concentrated power in fewer hands (Miller and Voas 2010). Within the domain of an inherently political technology once the fundamental decisions are taken, changing the character of its social impact will be extremely difficult. Since CC will extend its model of functioning to an increasing number of areas of human activity, a true democratization of the Internet will become and urgent need. Citizens should be considered as stakeholders.

9. Dominant corporate agents are vulnerable to the influence of central states.

Even dominant corporate agents are vulnerable to a power shift in favor of central States (those nations whose decisions over technological policy strongly affects beyond their frontiers). CC services providers and big data companies are under pressure for several reasons (national security, law enforcement, war on terrorism, defense of national values, protection of free trade, etc.), required to play a key role in implementing these policies. They can also act as cultural and political battering rams (convergence of interests between the state and the Companies). Corporations like Google, Microsoft and Apple have admitted that they accepted subsidies for the US Government in exchange for user and content identification. They have also yielded general access without explicit judicial orders to files that have been uploaded by users of the cloud.

10. Unauthorized spreading of information (function creep).

One of the most pervasive problems created by CC involves data collected for a certain purpose, but which can be later used for unauthorized, unanticipated, unwanted purposes. With the global prominence of CC systems, with its relaxed control and lack of interest in the specific purposes for which data is used, function creep is a serious risk (Timmermans et al. 2010).

CC email services like Gmail or Yahoo regularly scan email in order to target advertisement to users. Of course, the collected information can be combined with other personal data for other purposes. In a lawsuit against Google, one of its attorneys declared that non-Gmail users who exchange emails with a Gmail user should have no legitimate expectation of privacy:

> Google, which has more than 425 million Gmail users worldwide, said in its submissions: "Just as a sender of a letter to a business colleague cannot be surprised that the recipient's

> assistant opens the letter, people who use Web-based email today cannot be surprised if their emails are processed by the recipient's ECS [electronic communications service] provider in the course of delivery." Citing other case law, it adds: "Indeed, a person has no legitimate expectation of privacy in information he voluntarily turns over to third parties." CW spokesman John Simpson [Privacy Project director of Consumer Watchdog (a consumer pressure group)] said: "Google's brief uses a wrong-headed analogy; sending an email is like giving a letter to the Post Office. I expect the Post Office to deliver the letter based on the address written on the envelope. I don't expect the mail carrier to open my letter and read it. Similarly when I send an email, I expect it to be delivered to the intended recipient with a Gmail account based on the email address; why would I expect its content will be intercepted by Google and read?" [...] In its motion to dismiss the case, Google said the plaintiffs were making "an attempt to criminalize ordinary business practices" that have been part of Gmail's service since its introduction. (Calahan 2013)

11. Ethical problems related to Cloud robotics, cloud automation, and driverless cars.

Cloud robotics is a new breed of CC technology applications. It is called upon to play a major role of technological innovation in the next decades. There is an enormous potential in CC for improvements in robotics, automation and other areas of contemporary life including drones and autonomous vehicles. John Markoff (2016) considers the case of self-driving cars, and shows how they raise many ethical issues related to everyday situations. According to Asimov's robotic laws, autonomous vehicles should take decisions in order to not harm any human being and doing their best to achieve the greatest good. However, things are not so simple. These cars rely heavily for a safe and efficient operation on cloud computing resources like GPS-positioning information, weather reports, information about the condition of the roads, etc. Within particular circumstances a vehicle needs to make decisions involving the wellbeing of its passengers, other vehicles and their drivers and passengers, pedestrians in a crosswalk, etc. Car makers will have to create algorithms that should choose what should be protected first in case of a potential crash or in a situation involving the risk of running over a pedestrian or person on a bicycle. A set of six online survey conducted in the United States between June 2015 and November 2015, using the Mechanical Turk public-opinion tool, showed that most people wanted cars programmed for the greatest good (Bonnefon et al. 2016). However, when confronted with practical situations that could imply personal sacrifices in order to avoid a major damage to others, most respondents would rather save their own lives first. Autonomous cars, drones and automated weapons take classic dilemmas of robotic morality to another level with more complex equations. It represents also another version of *the trolley problem* introduced by Philippa Foot in 1967, in which a trolley out of control is a about to run over five workers on the track, but a lever can be operated that would divert the trolley onto another track on which there is just one workman instead of five. The question is: What is the right decision to take?

12. Cloud Computing and the New Meaning of Privacy.

There is a two-way causal relation between CC technology and the society that is ready to adopt and implement it. CC involves a substantially new definition of privacy. In fact, De Bruin observes that cloud computing is usually most welcomed by individuals who adopt a deflated view of the value of ownership and an inflated view of freedom. Particularly, Generation X or the Millenials do not care very much about who owns a personal picture stored on a cloud service, and care more about the freedom to share it and show it at will (De Bruin 2010; De Bruin and Floridi 2016). As a member of what can be called *Generation Cloud*, you do not exist if you neither broadcast your life almost in real time nor immortalize every moment of everyday life and share it with your friends. The right to keep control over your own life has been replaced by the right to live in a global cybernetic showroom. CC is also reflects sweeping social changes, as it mirrors our dreams and our quest for an utterly interconnected world. Social networks and other social media are an *open book* in which who we are and who we want to be can be read. Processes of identity construction emerge from the *mediascape* created by the technologies that surround us. In some sense there is a *personal soundtrack* composed by the music we listen, the TV-series we watch, and the web pages we trust. It is a *life soundtrack* we share with our partners, friends and associates. It is a new landscape we live in, that goes always with us, a mobile and evolving landscape, perhaps a most realistic *umvelt* upon which our very identity builds. This phenomenon explains the existence of cultural ghettos, ideal nesting sites for nurturing radical ideologies: the world we live in is not the physical reality around us, but the set of cultural elements meaningful to us that transmits and reinforces shared values of a given group of which we are parts. On the cloud, personal information of any kind is more vulnerable to unwanted access and prone to be well beyond the owner's control.

12.5 The Salvific Rhetoric of Technology: Cloud Computing as a God Term

Beyond the ethical and political delusions created by the powerful techno-social paradigm of CC, the task of inserting it into a democratic culture is not a theoretical question, but a matter of political prudence. Stronger social control is required if we are to avoid pungent, direct applications of market logic. Social-defined values and methods of adequate supervision should promote socially relevant ends, as well as regulatory mechanisms for the social appropriation of this technology. Otherwise, the very dynamics of CC, which are supported by the interests of corporate and economic agents frequently hidden to public scrutiny, will drive the pace and direction of its development and implementation. CC has entered our collective imaginary not just as a new ensemble of applications, procedures and devices, but also as a set of powerful metaphors. CC is the paradigm of a new interconnected individual, a domestic cosmopolitan living the world enclosed within the walls of his or her home. The very image of the cloud conjures up dreamlike, symbolic resonances. If information is on the cloud, what could go wrong? Information is close to heaven, closer to God than ever, wisely safeguarded and protected by angels and archangels.

Imagery of this kind is a clever and subtle way to create an atmosphere of trust that encourages the adoption of a given technology and the political apparatus it may bring along. If something resides on the cloud, it belongs to all of us, it is beyond human frontiers, outside the miseries of this world. CC as a metaphor of powerful resonances arises as a one of the great icons of the mainstream faith in a democratic Internet, a truly social computing technosystem that we should not be afraid of.

The rhetorical strength of the term CC can be understood from Richard M. Weaver's distinction between *god terms* and *evil terms*, two expressions coined in his book *Ethics of Rhetoric* (Weaver 1953). *God terms* are words or expressions with deep roots. They can be vague or generalist, but always demonstrate an inherently powerful meaning. For example, the *last frontier* belongs to this category. Langdon Winner borrows these concepts in order to elaborate an acute criticism on how for decades *the last frontier* has been a constant reference in American politics, an appeal to the quest for an achievable dream (Winner 2014). In a time of American history, it refereed to the conquest of the Far West. In the worst times of the Cold War, those of the Soviet lead in military technology, the space was christened as *The last frontier*. The *Space Race* had an almost mythological value for the dignity of the American people. After the great disappointment over the successful launching of Soviet first artificial satellite, the *Sputnik I*, and the first man in the outer space, Yuri Gagarin, American self-esteem was elevated by the success of the Apollo project and the placing on the Moon of the Stars and Stripes. When the battle against cancer or against autoimmune diseases is called "the last frontier", it takes the character of a selfless and messianic enterprise, almost a salvific one. In Spanish language, the term *crusade* (*cruzada* in Spanish) has a similar symbolic value. The so-called *crusade against communism* eventually became synonymous with political repression. Other expressions such as *the crusade against hunger* (development plan), *the crusade against ig*norance (literacy campaign), etc., lend religious dignity to what are often highly problematic undertakings. We should not forget that, historically speaking, Papal Bulls exonerated those involved in the *crusades* from any liability for crimes that their actions entailed.

Within the project of Enlightenment, another powerful *god term* can be identified. Committed to carrying humans out of the shadows of myth and religious obscurantism, the Enlightenment applied science and technology as levers for projects of *progress*. There are few expressions as powerful and untouchable as *progress*. Anyone who dares to criticize it, and rejects the rightfulness this fight of man against nature will be regarded in a suspicious light. Any kinds of atrocity in the name of progress are justified in the end. For they are surely due to the imperfect nature of man, never a consequence of an essential failure of the project itself. Science is not immune from a rhetorical closure of problems, and often *god terms* play this role called *blue-sky effect*. Some words unmistakably give a positive meaning to any sentence. Terms like *artificial intelligence* or *electronic brain* create a sense of hope and confidence. How can we distrust the behavior of a military drone or a driverless car that is thought to be *artificially intelligent*? Weaver suggests that we should be cautious with a rhetoric based on such terms. We need an ethical rhetoric in order to remove a diffuse terminology laden with a heavy load of propagandistic significance. According to Weaver and Winner, the way we use language reflects the good or bad health of our society. Another variety of rhetorical closure is a widespread strategy in today's politics: a lie that is constantly repeated as a mantra ends up becoming an accepted truth.

12.6 Conclusions: Cloud Computing as an Inherently Political Technology

Our final conclusions can be summarized in six points: First, CC is a revolutionary development in IT, and deserves a specific political analysis. Second, it has to be considered as a new sociotechnical paradigm, and a strong source of metaphors involving a strong faith in technology. Third, there is a strong financial pressure to invest in infrastructures that are crucial for deploying a global CC system. Fourth, CC is an inherently political technology in the strong sense (it requires a set of political and social requirements to function properly). Fifth, we don't have to wait for its global implementation to forecast its impact on societies. Those impacts stem from its very nature, not to any particular use, and they are independent of the context in which they are situated. And finally, CC is in fact a political constitution: The discussion on CC has to do with which values we are willing to support, and the society we want to create. As CC extends the power of technology to ever more areas of human activity, citizens should be considered as stakeholders. We agree with Floridi (2013) and Zakhem et al. (2007) in that an ethical approach to CC should start with the identification of the individuals and groups confronted with the effects of CC activities, affecting their rights and their interests. In this sense, most people on this planet ought to be, sooner or later, considered as stakeholders.

Acknowledgements This chapter has been sponsored by the Spanish government-funded research project Science, Technology and Society: Political and ethical problems of cloud computing as a new socio-technical paradigm (FFI2013-46908-R).

References

Bonnefon, J. F., Shariff, A., & Rahwan, I. (2016). The social dilemma of autonomous vehicles. *Science, 352*(6293), 1573–1576. https://doi.org/10.1126/science.aaf2654.

Bustamante, J. (2013). Ética en la nube: Dilemas éticos y políticos en el modelo de Computación en nube. *Argumentos de Razón Técnica, 16*, 37–54.

Calahan, P. (2013). Google: Gmail users can't expect privacy when sending email. *Independent*, August 14, 2013.

Carr, N. (2008). *The big switch*. New York: Norton.

De Bruin, B. (2010). The liberal value of privacy. *Law and Philosophy, 29*(5), 505–534.

De Bruin, B., & Floridi, L. (2016). The ethics of cloud computing. *Science and Engineering Ethics*. https://doi.org/10.1007/s11948-016-9759-0.

Floridi, L. (2013). *The ethics of information*. Oxford: Oxford University Press.

Foot, P. (1967). The problem of abortion and the doctrine of the double effect. *Oxford Review, 5*, 5–15.
Foster, I., Zhao, Y., Raicu, I., & Lu, S. (2008). Cloud computing and grid computing 360-degree compared. In *Proceedings grid computing environments workshop: GCE 2008* (pp. 1–10). https://doi.org/10.1109/GCE.2008.4738445.
Johnson, B. (2008). Cloud computing is a trap, warns GNU founder Richard Stallman. *Guardian. co.uk*, September 29, 2008.
Markoff, J. (2016). Should your driverless car hit a pedestrian to save your life? *New York Times*, June 23, 2016.
Mell, P., & Grance, T. (2011). *The NIST definition of cloud computing* (version Sept 2011). http://nvlpubs.nist.gov/nistpubs/Legacy/SP/nistspecialpublication800-145.pdf. Accessed 25 July 2016.
Miller, K., & Voas, J. (2010). Ethics and the cloud. *IT Pro, 12*, 4–5. Sept/Oct 2010, IEEE Comp. Soc.
Timmermans, J., Stahl, B., Ikonen, V., & Bozdag, E. (2010). The ethics of cloud computing: A conceptual review. *IEEE Second International Conference on Cloud Computing Technology and Science, 2010*, 614–620.
Vaquero, L., Rodero-Merino, L., Caceres, J., & Lindner, M. (2009). A break in the clouds: Toward a cloud definition. *ACM SIGCOMM Computer Communication Review, 39*, 50–55.
Weaver, R. M. (1953). *Ethics of rhetoric*. Davis: Hermagoras Press.
Winner, L. (1986). *The whale and the reactor*. Chicago: University of Chicago Press.
Winner, L. (2014). *Thinking outside of the box is the new box*. http://technosciencepeople.wordpress.com/2014/05/17/outside-is-the-new-box/. Accessed 17 July 2014.
Yoo, C. (2011). Cloud computing: Architectural and policy implications. *Review of Industrial Organization, 38*(4), 405–421.
Zakhem, A., Palmer, D., & Stoll, M. (Eds.). (2007). *Stakeholder theory: Essential readings in ethical leadership and management*. New York: Prometheus.

Chapter 13
The Promises of Synthetic Biology: New Bioartefacts and Their Ethical and Societal Consequences

Jorge Enrique Linares Salgado

Abstract In this chapter, we analyze the current production of bioartefacts derived from synthetic biology, and their ethical and societal consequences. In addition to the socio-economic problems and growing technological gap between North and South, synthetic biology raises new environmental and biological risks, triggering a global challenge. Such risks are also beginning to worry some scientists, philosophers and civil society organizations, because the use of new synthesized organisms extends their long-term effects and increases the risk of bioterrorism and 'bioerrorism'. However, the principal and most profound implications of synthetic biology will occur in the global economy, in biomedical and pharmaceutical research, and in the food and biofuels industries, fomenting and intensifying social and ethical controversies regarding the development of biotechnology, its capacity to determine the future of technological civilization, and its essential relationship with all the ecosystems on Earth.

13.1 Introduction

Synthetic biology is a new techno-science[1] that may bring about an entire industrial revolution that expands the human capacity to transform living beings and manufacture new kinds of bioartefacts, altering the ecological and ethical relationship that humans have developed throughout history with all forms of life, perhaps irreversibly.

The ethical and societal issues raised by this new bio-technoscience, which represents the latest step in biotechnology, are not only limited – as many authors

[1] For the concept of technoscience, see Echeverría (2003) and Linares (2008).

J. E. Linares Salgado (✉)
School of Philosophy and Literature, University Program of Bioethics, National Autonomous University of Mexico (UNAM), Mexico City, Mexico
e-mail: lisjor@unam.mx

B. Laspra, J. A. López Cerezo (eds.), *Spanish Philosophy of Technology*, Philosophy of Engineering and Technology 24,
https://doi.org/10.1007/978-3-319-71958-0_13

think – to biosafety and biosecurity issues in the lab. The creation of new life forms (Boldt 2013) such as synthetic genomes and synthetic cells – probably new species – raises a broad range of ethical and societal issues: the question of patents and private property, legal ownership controversies, biosafety and biosecurity, 'bioerrorism' (Kaebnick and Murray 2013), biological weapons, and bioterrorism. But it also reactivates former challenges related to the management of risk and its social perception, the matter of the fair distribution of technological benefits and the increasing responsibilities of scientists and biotechnologists, governments and civil society regarding new technological effects – both positive and negative – that are difficult to foresee.

The development of biotechnology in the twentieth century began with molecular biology, genetic engineering, and recombinant DNA technology. In the twenty-first century, synthetic biology, enhanced by bioinformatics, bio-nanotechnology, and the development of new genetic editing techniques such as CRISPR may give rise to a whole range of innovations which will revolutionize the industry with several kinds of biological products and new types of bioartefacts. However, at this stage of its development, synthetic biology is more a promise and a magnification of our technical capability than a new real technological power under our control. Nevertheless, the environmental dilemma already posed by biotechnology is whether humans can: (a) adapt these techno-scientific innovations to the structures and limits of ecosystems and natural resources, or (b) adapt living organisms and ecosystems to the conditions and development of the contemporary bio-technosphere.

It is clear that the path followed by synthetic biology is based on the second alternative. Its productions may thus radically change ecosystems in the near future and even the evolution of species in the long term. Combined with already existing transgenic organisms, the products of synthetic biology certainly could generate new technological possibilities to solve diverse challenges, including pharmaceutical, environmental and energy problems. However, this could also create a new form of biosynthetic matter capable of interacting with different species and natural ecosystems, with unforeseeable consequences. This great leap in the history of technology may hence lead to social conflicts and controversies arising from disputes regarding the regulation of these new artefacts, while at the same time generating emerging factors of social and economic inequality between developed and developing countries and precipitating environmental problems during the current century.

13.2 The Philosophical Foundations of Synthetic Biology

In his *Meditation on Technics* (1939), a pioneering text in the philosophy of technology, José Ortega y Gasset anticipated the rationale and consequences of the current scientific and technological revolution. According to this philosopher, *technics* is "the reform that man imposes on nature in view of the satisfaction of his necessities" (Ortega y Gasset 1939, p. 63). It is, therefore, a reverse movement running contrary to the natural process of biological adaptation. According to Ortega y Gasset, *technics* may be characterized in the following way:

1. It is a universal attribute in all societies; no human being exists without technics.
2. Technics varies to the same extent as it changes the content of human well-being, according to the idea that the humans have of their own well-being.
3. The purpose of technics is to create new possibilities, producing what nature cannot produce.

Thus, synthetic biology embodies Ortega's conception of technics as an essential characteristic of human existence.

> [...] Technics, appearing, on the one hand, as capacity, unlimited in principle, makes man's life empty, because he lives by faith in technics and only in it. Since becoming a technical being and only a technical being means only to be able to be everything and hence to be nothing determinate. Full of possibilities, technics [...] is unable to determine the content of human life. (Ortega y Gasset 1939, p. 81)

For Ortega y Gasset, as Carl Mitcham points out, "technics may even be thought of as a kind of human projection, but not on strictly natural or organic foundations [...] There is a break or a rupture between the human and the world. [...] The reason, he argues, is that the human being is not part of nature but has an *idea,* an *interpretation* of nature" (Mitcham 1994, p. 47).

In our time, synthetic biology is fostering a new idea of human well-being and of a bio-technological world which involves the engineering of life. This is the radical 'reform' of nature, according to Ortega's program for technology. However, Carl Mitcham claims that Ortega y Gasset argued that the predominant modern technology leads us to a new existential problem:

> In the past people were mainly conscious of things they were unable to do, of their limitations and restrictions. After willing some project, a person had to expend years of energy in solving the technical problems involved in its realization. Now, however, with the possession of a general method for discovering the technical means to realize any projected ideal, people often lose the ability to will any ends at all. (Mitcham 1994, pp. 48–49)

Synthetic biology apparently has a clear idea about its goals, yet it cannot explain why we desire this radical transformation of nature. Is it just because we can do so? Or because it is absolutely necessary to solve our socio-economic issues? Is it necessary, perhaps, to boost the production of food, drugs and medical technologies? Or is it really indispensable to imagine a new idea of human well-being? In his time, Ortega inferred what the transcendent aim of technics was:

> [...] Besides, these are the technical, specific acts of man. All of them are the technics that we can certainly define as the reform imposed by man on nature in view of satisfying his needs. [...] These, we have seen, were impositions of nature on man. Man responds in turn by imposing a change on nature. It is, therefore, technics, the strong reaction against nature or circumstance, that leads to creating a new nature, in between nature and man, a supernature. (Ortega y Gasset 1939, p. 63)

Thus, synthetic biology embraces Ortega's program to design and develop a new natural world, as defined by Church and Regis: a veritable *Regenesis* (Church and Regis 2012) that reinvents nature as an entire biological system through the engineering of life. Moreover, biotechnologist and leader in the bio-industry J. Craig Venter believes that, with the creation of the first synthetic living entities (synthetic genomes and synthetic bacteria), the scientific team at his institute had put to rest any remaining notion of vitalism once and for all. For the first time, the achievements of biotechnology demonstrate to us that life is the phenomenon of an orderly composition of genetic material, i.e., a biochemical structure that can be modified and manipulated.

13.3 The Digital Era of Biology

As Venter explains, the quest to manipulate life in the laboratory "has come a long way since the early days of recombinant DNA, in the 1970's, when Paul Berg, Herbert Boyer and Stanley Cohen began to cut and splice DNA" (Venter 2013, p. 83). Since then, biotechnologists have assumed that living beings are 'machines' (not in a metaphorical sense) and that their molecular processes are 'robots' that code the genomic script of every cell. According to Venter, we have initiated the "digital era of biology". This conception has been possible since the invention of recombinant DNA, which consists in the transference of DNA from one organism to another.[2] Subsequently, the first transgenic animal (a genetically modified mouse) was created in 1974. In response to the rapid development of this novel recombinant DNA technology and its bioethical implications, Paul Berg promoted a scientific moratorium on these biotechnological experiments at the historical Asilomar Conference in 1975. Regardless of public unease, this DNA technology launched a new era in the industrial biotechnological production of bioartefacts. The first company of this new industry was Genentech,[3] founded in 1976 by Bob Swanson and Herb Boyer. Their goal was "to create a business based on manipulating the genes of microorganisms to produce vital new medicines" (Genentech 2016). Its first and most famous product was transgenic insulin via a joint venture with the Eli Lilly Company in 1978 (Venter 2013, vid. 33). This was the first biotechnological product to be placed on the market. Nevertheless, the possibility of patenting different kinds of biotic matter or even entire organisms arose following the granting of the patent (after a long legal process in the United States) for the first genetically modified bacterium, produced by Ananda Chakrabarty in 1980.

[2] Recombinant DNA was the invention of Paul Berg, Herbert Boyer and Stanley Norman Cohen, in Stanford 1971 (Venter 2013).

[3] Nowadays, Genentech has obtained nearly "20,000 patents worldwide covering the innovations made by its researchers, as well as its engineering and manufacturing operations." Genentech is one of the most prolific patent owners in the biotechnology industry (Genentech 2016).

This is now the era of digital biology, in which the proteins and other interacting molecules in a cell can be viewed as its hardware and the information encoded in its DNA as its software. All the information needed to make a living, self-replicating cell is locked up within the spirals of its double helix. As we read and interpret that code, we should, in the fullness of time, be able to completely understand how cells work, then change and improve them by writing new cellular software (Venter 2013, p. 47).

However, this industrial use of 'DNA software' to produce intended performances in 'living hardware' of organisms does not suppose the full understanding of the complexity of life and its evolutionary processes, as Richard Feynman formulated in his famous phrase, "What I can create, I do understand." This is a fallacy, for the reason that the bioartefacts that biotechnology can create are not necessarily understood in all the dimensions of the complexity of life. The biotechnical ability to modify and 'create' a new life form (whether a genome or an entire organism) is necessary, but not sufficient to fully understand the processes of living organisms (Schmidt 2009). Moreover, Markus Schmidt adds that the "difference between having enough knowledge to create a new bio-system and having enough knowledge to fully grasp all possible interactions and its complete set of behavioral characteristics, is exactly what makes the difference for a sustainable and safe development" (Schmidt 2009, p. 81).

In 2010, the Venter Institute published in *Science*[4] that it had produced the first living organism with a synthetic genome: a synthesized *Mycoplasma mycoides* bacterium with the minimum DNA needed to survive (473 genes), called *JCVI-syn3.0.* This synthetic genome was inserted in a natural bacterium, *M. capricolum,* thereby changing its genomic structure. Venter stated, "we could now be absolutely confident that we had a new and novel mechanism to transform the genetic identity of a cell that does not involve DNA recombination or natural transformation mechanisms. […] we now knew that we had the first cells derived from the deliberative transplantation of one species' genome into a host cell from another species. In doing so, we had effectively changed one species into another" (Venter 2013, p. 109).

For Venter, this scientific success teaches us that "genome transplantation not only provided a way to carry out a striking transformation but would also help prove that DNA is the software of life" (Venter 2013, p. 96) Nevertheless, it has not been synthesized 'from scratch' in a chemical sense, but from a computational model based on the original DNA sequence of a natural bacterium. Despite the recognition of this achievement, many scientists and the media stated that it was not the "creation of artificial life". Creating life from scratch would imply knowing how "to write" genomes chemically, in such way that the genomic text make sense; i.e., so that it works biochemically within an organism to make it live and reproduce by itself. For the time being, it has only been able to transfer the genome from one bacterium to another, though certainly changing its genomic identity. This operation was rather a genomic transplant between two bacteria.

[4] Vid. *Science,* http://science.sciencemag.org/content/351/6280/aad6253. Accessed August 20, 2016.

Nevertheless, regarding this achievement, the US Presidential Commission for the Study of Bioethical Issues replied in its 2010 report on synthetic biology:

> The technical feat of synthesizing a genome from its chemical parts so that it becomes self-replicating when inserted into a bacterial cell of another species, while a significant accomplishment, does not represent the creation of life from inorganic chemicals alone. It is an indisputable fact that the human-made genome was inserted into an already living cell. The genome that was synthesized was also a variant of the genome of an already existing species. The feat therefore does not constitute the creation of life, the likelihood of which still remains remote for the foreseeable future. (Presidential Commission for the Study of Bioethical Issues 2010)

Venter has confidence in the possibility of producing synthetic microorganisms on a mass industrial scale. For him, the second major impact of the first genome transplant between two different bacteria is that it provided a new and deeper comprehension of the biochemical structure of life. From the viewpoint of synthetic biologists, a new epistemic paradigm of biology has been achieved. For them, DNA is the software of life – this is not a metaphor – and if we humans can change that software, we will be able to change the natural species and hence evolution, altering the entire hardware of cells and organisms. Nonetheless, some sort of ancient vitalism still pervades scientific and largely philosophical thinking, despite this empirical evidence of the biochemical composition and structure of life, as Venter maintains:

> This is precisely the result that those yearning for evidence of some vitalistic force feared would come out of good reductionist science, of trying to break down life, and what it meant to be alive, into basic functions and simple components. Our experiments did not leave much room to support the views of the vitalists or of those who want to believe that life depends on something more than a complex composite of chemical reactions. These experiments left no doubt that life is an information system. (Venter 2013, p. 109)

Thus, the advance of the new engineering of life encompasses the ability to operate chemical synthesis in order to change DNA code and transplant it into a cell. Synthetic biology can now start fabricating biological circuits, BioBricks or genomic chassis to design and rebuild new types of bioartefacts, similar to the emergence of the production of electronic circuits during the second half of the last century, which promoted the development of electronic technology and its myriad applications. Venter rightly points out the similarity between the informational revolution and the biosynthetic revolution, because synthetic biology is capable of designing new biological 'operating systems', which, like computers, may run instructions for driving the performance and function of manufactured organisms.

As his team had modified the natural genome of *M. mycoides*, Venter claims that the new synthetic cell has no direct ancestors and that is the main reason, moreover, for requesting the patent of this first synthetic living organism. Venter argues, "we now know how to write that code *de novo,* with the help of computers, which opens up the potential for designing almost any type of living thing, as we discover more details of the machinery of life" (Venter 2013, p. 129). Synthetic life is defined by Venter as "self-replicating biology based on a synthetic genome, a synthetic code-script" (Venter 2013, p. 129). Thus, synthetic biology has the potential to modify the genetic structure and functions of any cell and to create a new one.

Venter recognizes the ethical issues of synthetic biology such as the 'bioerrorism' caused only by biohackers or amateur biotechnologists, according to him. However, he overlooks the fact that experienced scientists also commit major errors that could be dangerous, in spite of biosafety measures and rigorous methods of verification. Nonetheless, in terms of the most problematic risks, Venter's major fear is expressed in the following statement, a typical example of technological determinism, which is the dominant ideology not only in the scientific realm in the USA, but also worldwide:

> [...] my greatest fear is not the abuse of technology but that we will not use it at all, and forgo a remarkable opportunity at a time when we are overpopulating our planet and changing environments forever. If we abandon a technology, we abandon the means to use it to improve and save lives. The consequences of inaction can be more dangerous than the improper use of technology. (Venter 2013, p. 157)

Venter uses a well-known and very widespread fallacy of 'all or nothing' in the development of technology. If we read Ortega y Gasset carefully, we realize that humanity is a technical species. That means we will never set aside our technical devices or tools, as there is no other way to survive. No one, not even '"the Luddites" (as Venter puts it), have advocated abandoning technologies or even the new biotechnologies. The ethical and political problem is rather one of adequately assessing their repercussions and risks in order to ensure beneficial products for all human beings and to ensure that they never cause damage to the environment. The question is not whether we use this or that new technology, but rather what kind of technological world we want to build, what kind of benefits and acceptable environmental risks we are willing to take. A model in which bioartefacts are concentrated in private enterprise with benefits for the few, but risks for everyone? Or a model of common, fair access, sharing benefits, while controlling the risks for society as a whole.

Biotechnology companies will attempt to monopolize and control the world market for these synthetic bioproducts in the same way that companies like Monsanto and Syngenta have attempted to monopolize and control the world market for transgenic seeds, or as Microsoft and Apple have monopolized operating systems for computers. These products thus constitute the new projects of this biocapitalism that has begun to emerge in the twenty-first century (Sunder Rajan 2012).

13.4 What Is a Bioartefact?

Bioartefacts[5] are technically modified living organisms with a specific function assigned by human agents. They are therefore technical biological objects or *biotic artefacts*, as defined by Keekok Lee (2003). They have origin, structure or natural-

[5] In the Spanish philosophy of technology, Ana Cuevas Badallo (2008) first formulated the concept of the bioartefact.

biological consistency, but with different degrees of bio-artefactuality or technical control and modifications of their biological functions, because they have been modified intentionally in their genetic, genomic, physiological, teleological or functional natural properties. They thus pose multiple philosophical and scientific issues due to their intermediate or ambivalent status: they are living organisms belonging to some evolutionary branch (so far, but they may not have biological ancestors in the future); yet they are also 'living artefacts', i.e., technical objects intentionally placed in the world via human ingenuity, to which humans have materially (biochemically) assigned technical functions, and have been adapted, altered or modified by human agents. As Lee puts it, the biotic artefact "is the embodiment of human intentions and ends, and would not have come into existence or continued to exist but for such human intervention and manipulation" (Lee 2003, p. 2).

The commencement of bioartefactuality actually began thousands of years ago during the long historic process of domesticating plants and animals, artificial selection and breeding, but has now achieved a higher level thanks to genetic engineering and synthetic biology. Keekok Lee distinguishes three historic levels of biotic artefacts, in ascending order of *artefacticity*, or, as we define it, bioartefactuality:

1. Produced through the trial and error method of traditional selection and breeding by farmers and horticulturists.
2. Produced through the combined contribution of Mendelian genetics and hybridization technology by scientists.
3. (a) Produced through the combined contribution of molecular genetics and genetic engineering; (b) to be produced through the combined contribution of molecular genetic engineering, molecular nanotechnology as well as microcomputer technology and the various sciences which underpin them (Lee 2003, p. 11).

We may also add as type 3(c) the current productions of synthetic biology, which means that the techno-scientific development of our time has achieved more efficient forms of technical control over living things than in the past. It has also opened up a new era of synthetic bioartefactuality based on the mass production of bioartefacts.

Humans have been modifying and controlling abiotic materials such as stone, wood and iron to make artefacts and, produce electricity and oil fuel as energy sources. To a certain extent, we could say that we have even recently been able to 'control' power generated by the nuclear fission of radioactive material. Technically, however, we still cannot control many viruses, microorganisms, weather events such as hurricanes, or abnormal biological processes such as cancer and various genetic mutations. The concept of technical control implies that human agents are capable of intentionally producing, for a sustainable period of time, technical changes in natural objects, artefacts or processes as well as anticipating, preventing or avoiding unintended consequences by correcting or altering their designs, artefacts or technical actions. We can hence deduce that some systems and things created by human technical action are beyond our control, e.g., vehicular traffic in cities or the flow of information over the Internet. However, bioartefacts seem to be *controlled artefacts*. Nevertheless, the problematic nature of technical controllability arises if we consider our artefacts from a more extended temporal duration. In the case of bioartefacts, their complexity and evolutionary adaptability, in addition to potential genetic interactions and genetic exchanges with other organisms, makes their controllability a major challenge.

13.5 The Dual Nature of Bioartefacts[6]

Bioartefacts have a dual natural and artefactual character. We cannot completely control the design and operation of the bacteria, cells and living creatures that we may transform technically. In bioartefacts, in particular, intrinsic structures and unknown biotic interactions remain, highlighting the indomitable character of nature and its relentless causality and necessity. Thus, bioartefacts are rather a boundary line between the natural and the artefactual, between the controllable and the uncontrollable, as they have a life of their own. Moreover, as Aristotle states in his *Physics,*[7] they have the cause of their existence, growth and movement by themselves and for themselves. Bioartefacts are revealed as dual natural/artefactual organisms and for that reason, they now require a dynamic, evolutionary ontology, as well as a set of epistemological and axiological demarcations.

As already stated, it exists continuity from the ancient domestication of plants and animals, which began more than ten thousand years ago, to present-day biotechnologies. After a long period of artificial selection, hybridization, and cross breeding to produce several generations of bioartefacts with effective, but unintentional genetic modifications,[8] biotechnological manipulation of genes has emerged in the twentieth century, leading to the production of recombined bioartefacts. The latest stage is the advent of synthetic biology, which initiates a new era of bioartefactual production through synthesis, which ventures to design biological entities previously not existing in nature.

This major biotechnological transformation is driven by the deployment of capitalism and the hegemonic global model of contemporary technoscience. These factors explain the fundamental difference between ancient and contemporary biotechnologies, revealing that it is not only an epistemic and technological, but also a socio-political and economic mutation in human history.

From a diachronic perspective, therefore, the discourse of many biotechnologists seems fallacious because they believe that what they do represents the simple continuity of a technical transformation of nature carried out by our species for thousands of years (since the emergence of agriculture in the Neolithic). They consider that we

---

[6] The next two sections are based on the chapter written in Spanish by María Antonia González and myself in Linares & Arriaga (Eds.) (2016, in press).

[7] Vid. Aristotle (2008), *Physics,* B, 1, 192b13-24. We can attribute, in part, the same characteristics of natural beings to bioartefacts.

[8] Keekok Lee denominates this period of the production of biotic artefacts the "Mendelian genetics and hybridization". Vid. K. Lee (2005, chapter 3).

always had the intention and ability to radically alter organisms and ecosystems. Such as vision is anachronistic because it takes into account only the genetic changes resulting from agriculture and farming after thousands of years of adaptation of cultivated and bred species. However, to understand the entire history of the technical transformation of living organisms, we must regard this temporal process in a complex historical network of environmental, technical and cultural relationships.

Whatever the case may be, the process underlying this radical mutation of the technical transformation of nature is the development of bioscience and biotechnology, as the latest manifestation of a modern project of planetary human domination of nature. The main goal of this project is technological intervention in nature. The ultimate aim of this project of technical colonization of nature is to bend natural forces, entities and processes to human ends to provide abundant means for the material well-being of growing humanity and its intensive use of the planet's resources, as we have seen in Ortega y Gasset, and as we can read in the conception of modern technology in Martin Heidegger's *The Question Concerning Technology*. This project, which is characteristic only of contemporary technology, not of ancient biotechnics, is driven by the development of industrial capitalism in order to produce new merch for a modern technological way of life: biomaterials, biofuels, bioreactors, new generations of synthetic drugs and, now, synthetic genomes and organisms.

From the current standpoint of biotechnology, looking back in retrospective anachronism, all natural entities and, in general, nature as a whole is conceived as available raw material for producing bioartefacts. Or as Heidegger argued, as a *Bestand* [boundless stock] for technical production and transformation; i.e., natural beings should be transformed, altered and adapted to the conditions and functions that humans need, imagine or desire. This is a very anthropocentric *Weltanschauung* [world conception]. Thus, all natural entities in this modern technological conception are reduced to a single instrumental value because their intrinsic value as part of evolution is irrelevant to the techno-scientific understanding of the world. And that, precisely, is another source of persistent ethical and political controversy.

13.6 A Preliminary Typology of Bioartefacts

We can establish a preliminary typology of the degrees of artefactuality involved in the recent production of bioartefacts. The natural/artefactual ambivalence of bioartefacts never disappears, but it may be feasible to identify degrees of technical construction, by means of which we can highlight the salient features of bioartefacts and in turn conjecture how they relate to other living organisms and ecosystems. Moreover, the degree of artefactuality is pertinent for making conceptual distinctions and social regulations. The degrees of *artefactuality* and/or *artificiality*[9] in

[9] A degree of *artefactuality* implies that an entity has been modified by human technical action; a degree of *artificiality* implies that an entity has been produced or manufactured by human technical

bioartefacts can be analyzed in the classical Aristotelian causes as four interconnected factors:

1. The *material cause* of which the bioartefact is composed or made. If the biological material is not a processed raw material, but only selected, the device is very close to its natural counterparts and at first glance does not possess distinctive characteristics.[10] If it has been made with transformed biotic matter (secondary, genetically recombined or synthesized matter), it falls within a second degree of artefactuality, but retains its natural and essential biological properties. However, if the biotic material of which it is manufactured or built is fully synthetic or even *artificial*, it would be a biological (and biochemical) material produced de novo; this type of matter does not yet exist, however. Hence, in terms of material composition, a completely artificial bioartefact would be one that has been designed and fabricated entirely and whose material has been synthesized (synthetic biochemical matter) to create a nonexistent biological material in nature. For now, what bio-technosciences are capable of producing is recombined and transplanted biological material, i.e., secondary biological matter.
2. The *formal cause* that structures the bioartefact and determines many of its functions, capabilities and potentialities. If the form has not been modified, it retains its intrinsic and natural properties and hence its aims and capabilities of inherent movement, growth and reproduction. If the form has been altered to some extent, new functions and possibly capabilities are introduced.[11] However, if the organism's genome, i.e., its formal structure, has been generated entirely by design and built gene-by-gene in the laboratory, we may say that such a bioartefact has a synthetic form, which is somewhat the case of Venter's *Mycoplasma*. Its genetic structure would be novel and unprecedented in nature, though not artificial. This is one of the most ambitious goals of synthetic biology. The synthetic form of a living organism implies a new form and a high level of bio-artificiality, in the sense postulated by Massimo Negrotti, although a *bioartificial* bioartefact would not be a *naturoid* as Negrotti defines it, because of its natural biotic matter (Negrotti 2012).
3. As the *final cause*, bioartefacts maintain the intrinsic ends or *telos* of a living, unmodified organism from which they originate or with which they are related. However, all bioartefacts have at least one new non-natural purpose or end in their capabilities and features, a technical purpose or function (assigned extrinsi-

action to mimic or replicate some other natural (unmodified) object or process. Up to now, artificial things have been made with non-biotic materials; e.g., artificial hearts, artificial intelligence, 'artificial respiration', artificial satellites, and artificial flavors. However, *bioartificial* plants do not yet exist, nor do living artificial organisms or living artificial cells. Vid. Negrotti (2012).

[10] This is the case of bioartefacts resulting from assisted reproductive techniques.

[11] Transgenic corn, for example, has had its genome modified to perform in a new way: it produces a toxin that defends it from plagues. A transgenic animal may have a(n) (unnatural) technical capacity, or new modalities of action or function. For example, a rabbit may be fluorescent to certain types of light because of jellyfish genes transplanted into its genome; such as the rabbit *Alba* created by bioartist Eduardo Kac in collaboration with French geneticist Louis-Marie Houdebine.

cally by human intelligence) that appears as a result of the amendment made from the design and human production, which has been introduced from an external, efficient cause. A fully artificial bioartefact could develop new purposes that are non-existent in nature or provided by the technical design that brought it into existence. We do not know what new goals could be developed in more complex degrees of bioartefactuality that combine material and deeper formal changes in modified organisms.[12]

4. The *efficient cause*, which in the case of natural living organisms is inherent and intrinsic to their matter-form-purpose (*hylé, eidos, telos*), becomes an external cause in the case of bioartefacts. The external efficient cause is the design, manufacture and operation performed by at least one human agent (or perhaps a robot in the future). The efficient cause introduces formal alterations and material recompositions and recombinations within the organism and thus produces other biological features, functions and capabilities that meet the technical interests of humans.

13.7 The Synthetic Biology Project and Its Philosophical Issues

Synthetic biology aims not only to modify existing genomes, but aspires to go one step further by creating or designing new genomes and hence new organisms without immediate biological ancestry. The ultimate goal is to construct new genomes from a minimal genetic chassis, on which can be installed the necessary genes for the new living organism to function. The second primary goal of synthetic biology is to fabricate BioBricks, i.e., standardized, interchangeable sequences of genes, to manufacture other genomes and produce certain biological performances.

The leading doubts and philosophical questions concerning synthetic biology – as well as biotechnology as a whole – involve the question of whether we have gained sufficient knowledge of the substantial structures and regulatory mechanisms of living beings. How far have our knowledge and our ability to control synthetic genomes and DNA sequences advanced? If this biotechnical capability is effective, then it implies a radical transformation of the relationship and cognitive experience derived therefrom, between humans and the rest of nature and living organisms.

As Gregory E. Kaebnick argues in *Synthetic biology and morality*, the technical capacity of manipulating living organisms has already (conceptually) involved the

[12] In principle, it would be possible to design, e.g., an organism with a predetermined lifespan. Although mortality is substantial to all living entities, a precise lifespan is not predetermined. A new end could be assigned as scheduled, precisely specified mortality, with an 'expiration date'. Like the *Nexus replicants,* bioartificial cyborgs, in the cult science fiction film *Blade Runner* (Ridley Scott 1982), from the original novel of Philip K. Dick: *Do androids dream of electric sheep?* (1968).

idea that we can reduce nature to a technical product of our intervention. This is the old philosophical problem between externalist realism and internalist constructivism. For that reason, it is convenient to hold the idea, even for our own survival, that there exists an independent natural world of our concepts, desires and values about nature. Because nature is not reducible to our scientific cognitions and biotechnical operations. Kaebnick states, "One might argue that nature is a category of things that are and should be independent of human control (as a matter of eternal fact, not merely of human moral conventions)" (Kaebnick and Murray 2013, p. 56).[13]

However, synthetic biology aims to achieve a new stage of bioartefactual transformation of living organisms and that goal may have further-reaching consequences than we can imagine today. Bioartefactuality has thus reached a new phase of unknown ecological effects and conjectural risks, which, although unpredictable, will require new social debate regarding their ontological, ethical and political dimensions. This process of techno-scientific development, which is taking place within the specific context of the deployment of new lines of production of global capitalism, is driven by the economic and political forces of the global market, known as 'biocapital' (Sunder Rajan 2006). Bioartefacts constitute new types of merchandise and capital goods due to their reproductive biological capacity, transformation and adaptation to diverse industrial applications and interactions in ecosystems.

13.8 Ethical and Socio-economic Issues Derived from Synthetic Biology

Biotech companies have obtained patents for synthetic organisms, their digitized genome maps and DNA synthesizer machines, as well as for the actual set of genes from each of them, and any version derived from them that may further be produced. Patents for synthetic biology products are granted based on the argument that these bacteria or synthetic organisms do not previously exist in nature; i.e., they are bioartefactual inventions of possible social and economic utility. However, these patents have raised socio-economic issues in the global economy because they contribute to increasing the technological gap between developed and developing countries.

From the perspective of Latin American countries, the future development of synthetic biology is of utmost importance. Some of the most biologically diverse countries in the world can be found in America, namely Brazil, Mexico, Colombia, Ecuador, Peru, Venezuela and the United States. This import is not only due to the enormous biodiversity of flora, fauna, microorganisms, and bacteria existing in the Amazon rainforest or in the rainforests of southern Mexico and Central America,

[13] K. Lee argues at the beginning of her book *The natural and the artefactual*, "The threat then posed by modern *homo faber* is the systematic elimination of the natural, both at the empirical and ontological levels, thereby generating a narcissistic civilization" (Lee 1999, p. 2).

but also on account of the cultural riches that traditional knowledge and non-industrial common practices in these countries present, a heritage that has been developed for centuries among indigenous people, farmers and even traditional healers (who base their treatments on their vast knowledge of the therapeutic effects of plants and natural substances).

The Americas are the center of origin of vital crops such as corn, potatoes, beans, tomatoes and thousands of plants with culinary and/or medicinal uses. This rich biodiversity and biocultural heritage comprise the raw material of new biotechnological productions. It is well known that many companies have carried out the genetic sequencing of valuable sources of biodiversity for several years now in order to eventually produce synthetic organisms and patent them for use for various industrial purposes. In principle, highly productive and profitable industrial innovations may emerge, of course. Yet the fundamental problem is that there could be bioprospecting or out-and-out biopiracy in order to produce certain biotechnical modifications in commonly used living organisms, converting them into new private industrial property, for which developing countries will have to pay royalties.

Areas of great potential exist for the development of new biosynthetic technologies; e.g., the region of Cuatro Ciénegas in the middle of the Coahuila Desert, Mexico. This is a unique place in the world where its pools contain water organisms and microorganisms that have probably continued to exist there for millions of years. The greatest significance of this region as a source of scientific knowledge and raw material for new biotechnological production is based on the existence of living stromatolites and other microbial communities, as described on the National Commission of Natural Protected Areas of Mexico website.[14]

The biodiversity of microorganisms and Cuatro Ciénegas stromatolites will produce very fruitful research for understanding the development of life on our planet and possibly to understand how it could have developed on other planets, such as Mars. However, the generation of innovations in synthetic biology derived from knowledge of unique microorganisms in the world may follow the course of the system of private intellectual property. That is not fair for the satisfactory dissemination of scientific and technological knowledge about this worldwide natural heritage.

In this case, as in many others, a suitable proposal of international regulatory policy would mean favoring a system of global scientific collaboration, in addition to a licensing scheme for shared use of biodiversity, which is owned by the nations where is located, thereby promoting the production of industrial innovations free of royalties. Synthetic biology can generate a new lucrative industry based on the effi-

[14] "The geological characteristics of […] the chemical composition of the water bodies of Cuatro Cienegas have created a niche that has allowed the survival of life forms that were once common on earth and now survive only in few places in the world. In the waters of this valley there are still archaic organisms whose main function in the planet's history was the modification of the primitive atmosphere by the first known form of photosynthesis and subsequent production of oxygen […] In the current context, stromatolites are a component of vital importance to the pools of Cuatro Cienegas, because they are true oxygen tanks for living beings that exist there" (Comisión Nacional de Áreas Naturales Protegidas 2016).

ciency of its products, but should not rely on patents of exclusive use and intellectual property. Nonetheless, this is not the predominant scheme of ownership in biotechnology (Oye and Wellhausen 2009).

Unlike what Keekok Lee proposes (Lee 2003), bioartefacts (especially those based on natural genomes and previous living organisms) should not necessarily be patented, because they are not equivalent to the abiotic artefacts of conventional chemical and physical technologies. We need a new global policy with international jurisdiction regarding the ownership of biotechnological products in order to harmonize the different regulatory national frameworks. It is crucial for the regulatory ethical and political framework of biotechnology not to focus solely on biosafety, biosecurity and risk prevention; it should also be intended to create better conditions for distributive justice and world equity in development.

References

Aristotle. (2008). *Physics*. Oxford: Oxford University Press.

Boldt, J. (2013). Creating life: Synthetic biology and ethics. In G. Kaebnick & T. Mrray (Eds.), *Synthetic biology and morality: Artificial life and the bounds of nature*. Cambridge, MA: MIT Press.

Church, G., & Regis, E. (2012). *Regenesis. How synthetic biology will reinvent nature and ourselves*. New York: Basic Books.

Comisión Nacional de Áreas Naturales Protegidas de México. (2016). *Área de protección de flora y fauna Cuatrociénegas*. http://cuatrocienegas.conanp.gob.mx/biodiversidad.php. Accessed 28 June 2016.

Cuevas Badallo, A. (2008). Los bioartefactos: viejas realidades que plantean nuevos problemas en la adscripción funciónal. *Argumentos de Razón Técnica, 11*, 71–96.

Echeverría, J. (2003). *La revolución tecnocientífica*. Madrid: Fondo de Cultura Económica.

Genentech. (2016). *Genentech a member of the Roche Group*. http://www.gene.com/40th/patent-then-publish. Accessed 15 July 2016.

Kaebnick, G. E., & Murray, T. H. (Eds.). (2013). *Synthetic biology and morality: Artificial life and the bounds of nature*. Cambridge, MA: MIT Press.

Lee, K. (1999). *The natural and the artefactual. The implications of deep science and deep technology for environmental philosophy*. Lanham: Lexington Books.

Lee, K. (2003). *Philosophy and revolutions in genetics: Deep science and deep technology*. Basingstoke: Palgrave McMillan.

Linares, J. E. (2008). *Ética y mundo tecnológico*. México: UNAM-FCE.

Linares, J. E., & Arriaga, E. (2016). *Aproximaciones interdisciplinarias a la bioartefactualidad*. México:UNAM (in press).

Mitcham, C. (1994). *Thinking through technology: The path between engineering and philosophy*. Chicago: Chicago University Press.

Negrotti, M. (2012). *The reality of the artificial. Nature, technology and naturoids*. Dordrecht: Springer.

Ortega y Gasset, J. (1939). *Meditación de la técnica* (p. 2015). Madrid: Biblioteca Nueva.

Oye, K. A., & Wellhausen, R. (2009). The intellectual commons and property in synthetic biology. In M. Schmidt, A. Kelle, et al. (Eds.), *Synthetic biology: The technoscience and its societal consequenes* (pp. 121–140). Dordrecht: Springer.

Presidential Comission for the Study of Bioethical Issues. (2010). New directions: The ethics of synthetic biology and emerging technologies. In *Presidential comission for the study of bioethical issues. june de* 2010. www.bioethics.gov. Accessed 20 Aug 2016.

Schmidt, M. (2009). Do I understand what I can create? In M. Schidt, A. Kelle, et al. (Eds.), *Synthetic biology: The technoscience and its societal consequences* (pp. 81–100). Dordrecht: Springer.

Sunder Rajan, K. (2006). *Biocapital. The constitution of postgenomic life*. Durham: Duke University Press.

Sunder Rajan, K. (Ed.). (2012). *Lively capital: Biotechnologies, ethics, and governance in global markets*. Durham: Duke University Press.

Venter, J. C. (2013). *Life at the speed of light: From the Double Helix to the dawn of digital life*. New York: Penguin Books.

Chapter 14
Matters of Concern Regarding Nanotechnology

José Manuel de Cózar-Escalante and Andrés Manuel Núñez-Castro

Abstract Due to its potential to take advantage of novel properties of matter at the atomic level, nanotechnology promises to provide innovations that will have a tremendously beneficial economic and social impact. However, there are also risks and problems associated with it that must be properly assessed and responsibly dealt with, not just by researchers, businesses and public authorities, but also by society as a whole. The creation of matters of concern related to nanotechnology and of the different publics surrounding those concerns is an open process. To clarify the process, the epistemological status of nanotechnology as a technoscience should also be examined.

14.1 Introduction

From the start, nanotechnology has promised to provide innovations that will transform our lives, thanks to its capacity to manipulate and alter matter at the atomic level, thereby taking advantage of new properties that arise at that scale.

This transformative potential has also led the public to demand that nanotechnological innovations be carried out responsibly. To do so requires a democratic assessment of how, and to what end, these improvements will be carried out and, above all, how to manage possible risks that arise. In order to define the *matters of concern* and publics associated with different nanoinnovations, it is first necessary to determine the epistemological status of nanotechnology.

The investigation of the matters of concern regarding nanotechnology must also include its ethical and social implications. One of the main problems is how to regulate it, to figure out the best way for stakeholders and other interested and affected parties to take part in its governance. This debate on the ethical controversies that

J. M. de Cózar-Escalante (✉) · A. M. Núñez-Castro
University of La Laguna, La Laguna, Tenerife, Spain
e-mail: jcozar@ull.edu.es; ancastro@ull.edu.es

B. Laspra, J. A. López Cerezo (eds.), *Spanish Philosophy of Technology*, Philosophy of Engineering and Technology 24,
https://doi.org/10.1007/978-3-319-71958-0_14

arise from nanotechnology is vital because it has the potential to physically and figuratively turn our world upside down.

The classic pragmatism of John Dewey, updated by Noortje Marres and Bruno Latour, proposes a democratic *socio-ontological* focus on the issues that create the publics affected by the possible indirect and undesired effects of nanotechnology. By applying this approach, we stress the role of experts in anticipating possible nanotechnological scenarios through proactive assessment, which will allow them to provide support and guidance to the publics that might arise from nanotechnology.

14.2 The Promise of Nanotechnology

Nanotechnology refers to the manipulation of matter at the atomic and molecular level. The term itself is derived from the incredibly small scale at which nanotechnological research and applications are carried out: a nanometer is one billionth of a meter. Working at the nanoscale generally refers to between one and one hundred nanometers. In order to better visualize this scale, keep in mind that a human hair has a thickness of approximately 80,000 nm. Therefore, the nanoscale is invisible to the human eye and even beyond the capability of many of the most powerful microscopes.

Over the past few decades the development of nanotechnology has been a political and economic priority. Research in this area has been promoted in a wide range of fields, from pharmaceuticals to agro-food and electronics. The potential benefits of nanotechnology for healthcare, the environment or telecommunications have been enthusiastically publicized by governments and companies (European Commission 2004; NSTC 1999).

Nanotechnology is being developed in a very broad, complex and heterogeneous technoscientific field in which structures, components and systems are designed, characterized, produced and applied. Therefore it cannot be understood as a specific technology, but rather as a highly varied group of technological approaches and applications (Hodge et al. 2010). Furthermore, there is a distinction between nanoscience and nanotechnology. Nanoscience studies the properties and phenomena related to material at the nanoscale. At this scale the physical properties of material changes drastically, presenting new mechanical, optical, chemical, magnetic and electronic properties that the same substance does not display at larger scales; many of these properties are due to quantum effects.

On the other hand, nanotechnology refers to the production and manipulation of objects, materials, instruments, structures and systems at the nanoscale (for convenience we refer to it in the singular, but in fact there are many nanotechnologies). However, the separation between nanoscience and nanotechnology is not so clear because there is a close interaction between scientific knowledge and a complex series of technological innovations, such as the latest generation of scanning tunneling microscopes, among others, which are microscopes capable of viewing and manipulating nanoscale objects. Without nanotechnology we could not speak of nanoscience, but the opposite is also true: scientific research is a fundamental part of nanotechnological development.

Although this field of study was conceived by the physicist Richard Feynman at the end of the 1950s, research at the nanoscale is relatively recent because the advanced instrumentation and processes necessary to carry out such research have only been available for around three decades. The first companies dedicated to commercializing nanotechnology applications appeared at the end of the twentieth century. New nanomaterials began to be produced, such as carbon nanostructures (the most well-known were fullerenes, also known as *buckyballs*, and carbon nanotubes) and possible applications in industrial, energy and environmental processes were studied. The main properties of nanomaterials are, among others: fire resistance, durability, enhanced properties for polymers, bactericides, fungicides, etc. The majority are multifunctional; that is, a single product can offer two or more of these properties. They are already being used in aeronautics, the car industry, microelectronics, neurosurgery, hydrogen storage, adsorption or catalysis, among other fields.

One of the most popular nanomaterials over the past few years is graphene, an allotrope of carbon in the form of a two-dimensional, atomic-scale, honey-comb lattice, in which one atom forms each vertex. This extremely light structure is no more than a single atom thick. It can potentially be used to improve, for example, light panels or touch screens, and may become the material of the twenty-first century. And graphene is only part of a large family: more than a hundred stable, two-dimensional compounds with very promising properties are already being designed and produced. These new compounds will make it possible to produce amazing advances, ranging from invisible camouflage suits, to self-charging batteries for smartphones, improved water treatment and the absorption of radioactivity. The long term possibilities are limitless, but two imminent *technological revolutions* are expected to come in consumer electronics and aeronautics, as tiny devices are already being designed to be integrated in clothing or worn on the body, and airplanes can be made more resistant, less vulnerable to turbulence and hundreds of times lighter than current models.

It is believed that not only will nanotechnology make products more efficient, but also the processes used to produce them, due to its capacity to create multifunctional objects with less raw material and to replace some of the materials used today with others that have better characteristics, although this may disrupt many industrial sectors and raw material producers (Foladori 2010).

The business prospects derived from nanotechnology have been very promising from the start, and huge amounts of money (tens of billions of dollars) have been invested in its research and development by a large number of private corporations from all industrial sectors, but, above all, by the public sector in Japan, the United States, the European Union and China (Sargent 2014). Industrial applications of nanotechnology include the automotive industry, the medical and pharmaceutical industries, biotechnology, textiles, information technologies and telecommunications, among others. Therefore, nanotechnology is very present in society through different products already on the market. Nanomaterials can be found in impermeable textile products, drugs, sun screen lotions, electronic components, cosmetics, contact lenses, paints, etc.

It must be emphasized that a large part of nanotechnology's potency lies in its convergence with other emerging technologies, such as biotechnology, cognitive science, information technology or robotics. Interdisciplinarity, already characteristic of nanotechnological research, has increased thanks to the knowledge and abilities being shared between different scientific and technical specialties, from quantum physics to biochemistry, microelectronics and materials engineering, among many others. The repercussions of nanoscale research are potentially more significant due to the implementation of inclusive strategies. Nanotechnology has become a component of many emerging technologies such as artificial intelligence, biotechnology or robotics.

For all these reasons, nanotechnology is often considered the transformative technology of this century: it promises to produce materials with extraordinary characteristics that can be applied in many fields, as well as to revolutionize production processes in practically every industry. However, although nanotechnology has been and continues to be widely supported, researched and disseminated, it has yet to produce a technological revolution; what it has produced are some concerns.

14.3 The Epistemological Status of Nanotechnology

In order to carry out a democratic evaluation and promote responsible innovation, the publics and matters of concern associated with different nanotechnological innovations must be identified. To do so, we should first determine the epistemological status of nanotechnology. Is it a science? Is it technology? Is it a mixture of both? Or is it something different?

First of all, as mentioned earlier, a distinction must be made between *nanoscience* (scientific research) and *nanotechnology* (technological development, innovation). In practice, the *nanotechnology* tag has been placed on the immense majority of everything related to the nanoscale. That is, the term is used to group all kinds of scientific research and technological programs.

This is because much of the research in this field is applied research. Obviously, a large part of the public and private financing in this area is interested in practical goals and, therefore, research is oriented toward achieving practical objectives for economic, strategic or social benefits. However, another significant reason why nanoscale research is often classified as *nanotechnology* is that the way it is being developed makes it difficult to place in traditional categories that clearly distinguish pure scientific research from technological development (that later lays the groundwork for further innovation). This is where another concept enters the scene: technoscience (Hottois 1984).

In fact, today's most promising technologies – such as nanotechnology, information technology, synthetic biology and cognitive science – can be also described as technosciences. That means that they have a special epistemological status, being neither purely science nor purely technology (engineering), but a *hybridization*, a sui generis combination of both, developing in a specific institutional and economic context. This context is often described by means of labels like *big science*, *knowledge economy*, *bioeconomy*, *fourth industrial revolution* and, of course, *globalization*. What can initially be seen as something of mere philosophical interest, can potentially have highly worrying social and environmental repercussions.

Today technosciences (including nanotechnology) often have the following features (see, e.g., Nordmann 2010):

1. Instead of a single theoretical paradigm, they are based on a combination of knowledge from several scientific disciplines and technologies, operating at several levels of abstraction.
2. They focus on intervention and making, rather than representing and knowing (research is oriented to practical ends and not to knowledge in itself).
3. They emphasize (predictive and technical) controlling and developing capabilities at the expense of understanding.
4. When possible, they use reductionist methods.
5. They search for radical novelty, rather than incremental improvement.
6. There is a strong possibility that they will be disruptive (potential to displace existing technologies and industries).
7. They are based on dynamics of convergence, especially at the nanoscale.
8. Their research and commercialization is carried out with a high level of risk and uncertainty.
9. They generate high expectations using a calculated, unrealistic rhetoric.
10. They are increasingly dependent on financial resources and speculative capital.

The combination of these characteristics gives rise to a worrisome ethical, economic and sociopolitical scenario. Shortly, we shall see why.

At its origin, modern science was conceived mainly as the production of knowledge through observation, experimentation and theory. Knowledge was based on the epistemological authority of the scientists. Of course, it was also hoped that the knowledge acquired would have an impact on socioeconomic development through its technological applications, but only as a result of previous, independent research. However, this situation has changed. The technoscience of today is incapable of providing a definite representation, in theoretical terms, of the world. The traditional distribution of roles, in which scientists act as experts advising public authorities who, in turn, take (or at least tried to take) the best decisions for the welfare of the community based on those reports can no longer be maintained without tension. There is no sharp distinction – conceptual, practical or simply temporal – between the area of generating expert knowledge, with its epistemic and technical values, and the social, political and economic interests and values borne by their application. Technoscience has an obvious interest in minimizing risk and uncertainty in terms of the possible damage caused to humans and the environment, if only because the contrary would be detrimental for businesses and public research bodies. But it also nearly always prioritizes economic profitability, given that a product meeting the requisite of efficiency and the other technical demands, but that does not provide economic benefits, will obviously have little chance of thriving in a market economy.

The crucial idea that must be understood here is that by using exclusively technological means to control risk and uncertainty, while simultaneously trying to maximize economic profits, technoscience itself has become a source of serious risk and uncertainty (Beck 1998). The convergence of nanotechnology with biotechnology, information technology, neuroscience, robotics, etc., creates such a high level of complexity that it would be naïve, or cynical, to assure citizens that it is occurring under reasonable supervision. Moreover, many of the proponents of these technologies are not interested in submitting their development to any kind of serious oversight by the general public, whether for economic reasons (launching successful products on the market) or because they firmly believe the benefits that will be obtained (for instance, human enhancement) far outweigh any complications that may arise. The history of technology, especially over the past few decades, shows that it is impossible to reduce the problem of controlling and evaluating technology to purely technical parameters, nor can it be left solely in the hands of its promoters, whether they are scientists, technologists, businesses or politicians.

So, does the special status of nanotechnology facilitate or impede its social evaluation and a meaningful public participation in decisions taken about the present and future of nanotechnology? On the one hand, the technoscientific reality of nanotechnological research, development, and innovation seems more *open*, in a manner of speaking, than the traditional approach, which leaves assessment in the hands of experts and decisions in the hands of public authorities. It is true that this traditional approach is still largely used in many countries, although counteracted by the growing power of business concentration in a globalized economy. But, at least in public discourse, it is increasingly common to hear that it is necessary to consider the opinion of the general public. In fact, it is necessary to consider it, even if only instrumentally; that is, to avoid consumers rejecting a technological innovation. However, in such a complex and changing scenario, with so many factors in play, it is difficult to clearly present the state of the issue and the available alternatives, or to provide a plausible approximation of the scenarios that could arise. A continuous real-time assessment (Guston and Sarewitz 2002) of nanoinnovations has been called for, while at the same time the public still hasn't perceived any problems about which to debate with any reliability. Furthermore, the margin of decision left to the public is in many cases more fictional than real, given that there are increasingly large political and economic conglomerates that, in defense of their interests, offer a somewhat veiled opposition to true democratic assessment.

14.4 The Debate on Its Ethical and Social Implications

As mentioned, the public already has access to numerous nanotechnological products on the market, a situation that has been allowed by governments, industry and scientific institutions, despite an almost complete absence of public debate or regulatory frameworks. For example, in Europe there is currently no specific regulation or legislation regarding nanomaterials, although over the past few years there has been a push to modify the European Union's REACH (Registration, Evaluation, Authorization and Restriction of Chemicals) regulation to better identify nanomaterials.

Meanwhile concerns regarding the possible effects of nanotechnological products on the environment, health and society have been growing, leading to all kinds of reactions, from demanding a moratorium on implementing nanotechnological innovations, to the moderate recommendations by the European Union (European Commission 2008) calling for increased research on risks.

Possible ethical, social and economic impacts of nanotechnology have been identified and analyzed. In the environmental and energy fields, there is the possibility of contamination by nanoparticles or the spread of nano-bio entities on a large scale (also known as the *green plague*). At the same time, other potential dangers appear, such as the undetermined borders between natural and artificial, which, among other consequences, could generate problems with *patents on life*. In the domain of health and quality of life, negative impacts refer to the possible toxicity of the nanoparticles, genetic discrimination, discrimination in access to treatments due to their prohibitive costs or diverse threats to human nature and dignity (eugenics, creation of cyborgs, etc.), among others.

In the areas of security and military technology, privacy and liberty could be threatened through abusive surveillance and control of citizens. Furthermore, nanotechnology can offer disproportionate military advantages, while simultaneously posing a risk of serious terrorist threats – for example, by deliberately releasing highly dangerous nanoproducts in the environment.

The disruptive effects on the economy (obsolescence, unemployment, etc.) are also a genuine risk, as well as the greater concentration of power in companies increasing inequality in poor countries and minimizing the political power of citizens everywhere. In other words, there is a threat of creating a nano-division, namely the unequal and unfair distribution of wealth among citizens. The liberty and privacy of social institutions may also be threatened. There is also the threat of new forms of life emerging, with traumatic consequences.

The development of materials and production processes could lead to accidents caused by the abusive or uncontrolled implementation of nanotechnologocial production and the distribution of nanoproducts. Other consequences of the development of nanotechnology could include the public rejecting scientists *playing God*, the rise of difficulties inherent to scientific-technologic illiteracy (lack of public understanding of nanotechnology) or the emergence of legal problems related to property rights.

These circumstances and foreseeable risks have created doubts about the current governance of nanotechnology, which is often perceived as insufficient or inadequate to deal with the challenges involved in this technoscience. This criticism has favored a shift toward renovating nanotechnology governance in two major ways: the adoption of mild regulatory instruments and the growing importance of involving all stakeholders (Arnaldi 2014). The milder the regulatory frameworks are, the more prominent the stakeholders become.

Graeme Hodge et al. (2010), among others, suggest refocusing the regulatory problem to reflect the complexity of the challenges being faced. They propose that new approaches to dealing with issues should be innovative, providing all stakeholders with clear and accessible information, which is essential in ensuring a well-informed regulation.

The debate is essential because technoscience conceptually and physically disrupts the world and what we refer to as its natural order. In fact, we are experiencing a gradual dissolution between what is natural and technical, in that natural elements are incorporated in the artificial and vice versa. There is a genuine risk that if living and artificial material merge, the result could be hybrid organisms that are not easily controlled or that behave in unpredictable ways, affecting the environment and, by extension, the life and health of human beings. In addition to the simple and straightforward pollution of the environment by the growing production and emission of nanoparticles. Therefore, the main ethical dilemma concerns health risks that may arise from the new nanomaterials, because there has already been irresponsible behavior as occurred in the past with other toxic industrial material, such as asbestos.

However, the most important issue is still the lack of social control over nanotechnology and its applications, the lack of a debate that spells out the controversies implicit in all new technologies. It is not enough to ask the scientific community to regulate itself, since it is too closely tied to professional and academic interests, which are increasingly dependent on the businesses and financial groups that support their research financially. Of course advances cannot be made without assuming some risks, but it is necessary to find a way to minimize them or at least to make them socially acceptable.

The expectations generated around the nanorevolution has stimulated a productive ethical debate, given that transformation the world through nanotechnology has also been associated with a broader project to improve human capacities. The ethical and social aspects of nanotechnology refer to an ethical approach to and reflection about an emerging and powerful new technology (Palm and Hansson 2006). The idea is to democratically assess what is being researched and how it is applied; that is, how the results of such research are transferred to society. This constructive, real-time assessment must be as complete, multidimensional and objective as possible. In other words, it must include ethical as well as technical aspects.

The ethical debate on nanotechnology (nanoethics) has already been around for almost two decades. It has been an extremely heterogeneous debate, due to two main factors: the lack of a universally accepted definition of nanomaterials and uncertainty regarding an appropriate method to analyze and classify the issues at stake (Ferrari 2013). Moreover, nanoethics should not be limited to nanoscale objects or processes, as this would exclude some of the most important aspects, which transcend such object-categories and are not reducible to them. Nanoethics should be founded in questions about human relations, imaginations, meanings, commitments and normative visions of valuable ends and needs to which human knowledge and technology should be dedicated (Wynne 2006). Rather than focusing on specific innovations, the problems posed by this new technoscience should be approached from a broader perspective, represented by the convergence of nanotechnology with other technologies (mainly the NBIC convergence between Nanotechnology, Biotechnology, Information Technology and Cognitive Science) and with other dominant global economic patterns of the *knowledge economy* or the *knowledge society*.

It is necessary to adopt an approach that is more appropriate for nanotechnology, because for the most part, the technologies in question do not yet exist in material reality, at least in terms of the promised *revolutionary* innovations, although they do carry metaphysical and ideological dimensions. Nanoethics should force us to raise uncomfortable questions about aspects of the human condition we normally take for granted (Dupuy 2007).

With this in mind, looking at the benefits and risks of the new technologies, that is, their possible consequences, is a rumination on the future. It is a reflection of expectations and promises about events that, in many cases, may never take place. An ethics that precedes the future evolution of the new technologies is, therefore, a reflection in the present of a possible future, since it deals with promises, expectations and imagined trajectories that may or may not occur. In other words, it is an analysis on whether a common paradigm exists among the majority of the nanoinnovations being developed today that orients them toward certain technical trajectories, instead of others. What are the criteria that lead toward a certain design, fabrication or application rather than one of the many other alternatives? All innovation incorporates values, but in our plural societies, those values can differ and manifest in different ways. However, there are global values, worldviews and cultural and economic standards that probably have a greater influence on guiding the specific trajectories of innovations in all fields and therefore nanotechnology as well. The technical methods and standards that we use on a daily basis were once adopted in a process that included discursive reflection by intellectual and political elites on the nature of the values they wanted to promote and then instilled in the technical codes accepted today as the norm. The experiences that the artifacts produce in the actors are *included* in their design (their *technical code*) (Feenberg 2002). Thus the political aspect becomes preeminent over the ontological dimension, offering the possibility for citizens to act upon the design of these technical artifacts. Thus, it is our responsibility as citizens to take charge of the decisions regarding technological designs through broad democratic participation: how technology is developed, based on what criteria, under what values.

14.5 Publics of Nanotechnology and the Role of Experts

John Dewey's (1927) *classic pragmatism* proposal to study the democratic participation of citizens can be updated with the contributions by Noortje Marres (2007) and Bruno Latour (2007). The resulting reflection can be useful in order to facilitate democratic and responsible innovation in nanotechnology. This proposal is a *socio-ontological* approach to the issues – the matters of concern – that examines the formation of publics that may be affected by indirect and undesired effects of a complex, emerging technology, such as nanotechnology. A certain group becomes a *public* through indirect and undesired consequences of an issue that affects it. However, in the case of nanotechnology, the matters of concern that could form nanotechnological publics are still not precisely defined because it is an uncertain technoscience.

The democratization of science has been a fundamental concern of Science and Technology Studies (STS) from its inception. The most recent STS have arrived at the conclusion that to achieve the much desired democratization of the development of nanotechnology and other complex technologies, it is necessary to consult the public on the issues surrounding them. Marres argues that formulating issues is not just an important democratic process, but also it is what those processes are all about. That is, the STS have developed a *socio-ontological* sensitivity that has caused them to take an ontological turn in the conception of democratic processes. However, the STS still have to investigate what an issues-oriented perspective – referred to as the *turn toward the issue*– can contribute to and reveal about public participation. According to Marres, finding an answer to this matter becomes a necessity in order to assert that public participation in technoscientific policies can be explained in terms of the articulation of these issues and their solutions.

But how are publics formed in nanotechnology when responsibilities and consequences for these actions are not clearly defined? Dewey suggests that indirect consequences generate a public with a common interest in controlling those consequences. In other words, a public is a group of citizens brought together by a problem that has been detected; which means that for a public to exist, there first has to be an issue. In situations of uncertainty, it is the members of the public who define the problems and issues that concern them, and through democratic participation they can reach a consensus on how to manage the risks related to new technologies. Thus, a public for nanotechnology will only arise when citizens perceive a potential risk. Although there is no record of significant problems occasioned by nanotechnology on to date, if problems do appear they will be accompanied by controversy. The likelihood of such an event eventually occurring is worthy of consideration. Nanotechnology is considered to be one of the key technologies of this century, but its potential to improve multiple aspects of our lives can be seriously jeopardized if the potential users of this emerging technology feel threatened by it. Fear of nanotechnology becoming a political battlefield is growing along with the millions of dollars being invested in it. There are passionate debates among stakeholders and experts when analyzing and evaluating the underlying risks and ethical dangers of nanotechnology, just as happened some years ago in the case of genetically modified organisms in Europe. Environmental and health risks, as well as how to manage and communicate them, are being discussed. Questions about the definition of risk, the impartiality of the scientific information provided and the role played by political decisions are also at the center of the debate between some actors. However, from a Deweyan, pragmatist perspective, democratic public participation in the constructive assessment of nanotechnological developments and the management of controversies regarding the opportunities and possible risks of nanotechnology can only occur after publics affected by the indirect consequences of the diffusion of nanotechnological products appear.

This state of affairs has nearly paralyzed all action. Citizens are not coalescing into publics because they are overwhelmed by a technoscience whose causes and effects cannot be completely followed even by the experts. In order to escape this situation, an approach inspired by Dewey's analysis of the public could improve the methods and conditions involved with nanotechnological controversies. According to Dewey, it is not necessary for the great majority of the citizens to have enough knowledge to do scientific research; on the other hand, it is necessary for them to be able to judge the value of the knowledge offered by others regarding public affairs.

Dewey believes that while it is necessary to listen to expert knowledge, in the end it should always be the public that takes the political decisions. So, what can experts in STS contribute to public debate on nanotechnological controversies? How can they orient their work to generate a broad debate that facilitates public participation in a new and uncertain technoscience like nanotechnology?

Nanotechnological development will continue to be promoted by national and supranational institutions for the social and economic wellbeing it promises to provide, unless serious ethical problems arise, such as threats to the autonomy, liberty or dignity of people (European Commission 2012). The possible benefits of nanotechnology must be safeguarded from an opaque technology transfer process. The role of experts in this debate is extremely important. They must formulate the incipient issues, because most citizens possess very little and diffuse knowledge regarding what nanotechnology represents in their lives. The matters of concern can be manipulated because the groups have still not clearly defined nanotechnological issues, due to their inherent complexity. Experts must anticipate possible nanotechnological scenarios by proactively assessing and developing instruments that allow them to advise nascent nanotechnological publics when they ask for assistance (de Cózar-Escalante 2010).

Complementarily, Marres supports the value of STS when they help citizens to participate in public debates. The knowledge accumulated by STS when designing and articulating participatory events can be enriched by adding this socio-ontological and pragmatist perspective on the association of the actors. Public participation is shaped by institutional methods, and it should be approached from the perspective of the issues that concern all actors. It should be focused on developing a large set of structured events in which issues are articulated as potentially widespread matters of concern.

With the help of those initiatives, matters of concern can be transformed, if everything functions properly, into opportunities to carry out innovative and responsible research. For example, the toxicity of nanoparticles can be assessed together with technical innovations that can counteract the risk of toxicity, thereby providing social and economic benefits.

Therefore, we propose a kind of continuous technological assessment, broadened and extended. Continuous over time; broadened to include actors who must carry out the assessment (experts, public authorities, stakeholders, but also the public that is affected, in general). Finally, an extended assessment regarding the set of issues to assess, including not only scientific, technical and economic aspects, but also ethical and social issues.

14.6 Conclusions

The main objective of this chapter has been to identify the different issues associated with the development and implementation of a complex, transformative and emerging technological innovation like nanotechnology. Its insistent promises to improve various areas that affect our lives is leading the public to demand that the innovation be carried out responsibly, through democratic assessment. Before surmising what the possible matters of concern might be and the publics affected by them, it was necessary to consider its epistemological status; that is, to clarify if we are discussing a science, a technology or a mixture of both. Here the concept of technoscience sheds light on the ethical and social implications associated with nanotechnology and the possibilities of the public actively participating in decisions regarding its development.

From a pragmatist perspective, based on Dewey's analysis, the publics concerned about the undesired consequences of nanotechnology have not yet emerged. Its issues are still immersed in a cloud of uncertainty. That is why we propose that STS experts in nanotechnology work on a *proactive or anticipatory evaluation* of the different possible scenarios with the collaboration of nanotechnologists to anticipate the consequences of the possible benefits and risks associated with nanotechnology and try to clarify the uncertainty that shrouds them. This will also serve to predict the reaction of publics that may emerge in the foreseeable future. This is not meant to serve as an instrument to cynically avoid negative popular reactions, but rather to provide truly useful guidance.

References

Arnaldi, S. (2014). ¿Qué tan suave debería ser la regulación nano? Identidades sociales y opiniones de los stakeholders italianos. *Mundo Nano*. http://www.revistas.unam.mx/index.php/nano/article/view/48704. Accessed 25 Nov 2017.

Beck, U. (1998). *World risk society*. Cambridge: Polity Press.

de Cózar-Escalante, J. M. (2010). *Nanotecnología, salud y bioética*. Gijón: Junta General del Principado de Asturias (JGPA), Sociedad Internacional de Bioética (SIBI).

Dewey, J. (1927). The public and its problems. In J. A. Boydston (Ed.), (1988) *The later works, 1925–1953* (Vol. 2). Carbondale: Southern Illinois University Press.

Dupuy, J. P. (2007). Some pitfalls in the philosophical foundations of nanoethics. *Journal of Medicine and Philosophy, 32*(3), 237–261. https://doi.org/10.1080/03605310701396992.

European Commission. (2004). *Communication from the commission–towards a European strategy for nanotechnology*. Office for Official Publications of the European Communities, Luxembourg. https://cordis.europa.eu/pub/nanotechnology/docs/nano_com_en_new.pdf. Accessed 25 Nov 2017.

European Commission. (2008). *Communication from the Commission–Regulatory aspects of nanomaterials*. Brussels. http://eurlex.europa.eu/LexUriServ/LexUriServ.do?uri=COM:2008:0366:FIN:en:PDF. Accessed 25 Nov 2017.

European Commission. (2012). *Communication from the Commission–Second regulatory review on nanomaterials*. Brussels. http://eurlex.europa.eu/LexUriServ/LexUriServ.do?uri=COM:2012:0572:FIN:en:PDF. Accessed 25 Nov 2017.

Feenberg, A. (2002). *Transforming technology. A critical theory revisited*. New York: Oxford University Press.

Ferrari, A. (2013). From nanoethics to the normativity of technological visions. *Ethics & Politics*, XV(1): 220–235. http://hdl.handle.net/10077/8898. Accessed 25 Nov 2017.

Foladori, G. (2010). Las nanotecnologías en contexto. Sociología y tecnociencia. *Revista Digital de Sociología del Sistema Tecnocientífico, 0*(2), 35–55. http://dialnet.unirioja.es/servlet/articulo?codigo=3715576. Accessed 25 Nov 2017.

Guston, D., & Sarewitz, D. (2002). Real-time technology assessment. *Technology in Society, 24*(1), 93–109. https://doi.org/10.1016/S0160-791X(01)00047-1.

Hodge, G., Bowman, D., & Maynard, A. (Eds.). (2010). *International handbook on regulating nanotechnologies*. Northampton: Edward Elgar.

Hottois, G. (1984). *Le signe et la technique. La philosophie à l'épreuve de la technique*. Paris: Aubier Montaigne.

Latour, B. (2007). Turning around politics: A note on Gerard de Vries' paper. *Social Studies of Science, 37*(5), 811–820. https://doi.org/10.1177/0306312707081222.

Marres, N. (2007). The issues deserve more credit: Pragmatist contributions to the study of public involvement in controversy. *Social Studies of Science, 37*(5), 759–780. https://doi.org/10.1177/0306312706077367.

Nordmann, A. (2010). Philosophy of technoscience in the regime of vigilance. In G. Hodge, D. Bowman, & A. Maynard (Eds.), *International handbook on regulating nanotechnologies* (pp. 25–45). Northampton: Edward Elgar.

NSTC, National Science and Technology Council. (1999). *Nanotechnology. Shaping the world atom by atom*. http://www.wtec.org/loyola/nano/IWGN.Public.Brochure/. Accessed 25 Nov 2017.

Palm, E., & Hansson, S. O. (2006). The case for ethical technology assessment (eTA). *Technological Forecasting and Social Change, 73*(5), 543–558. https://doi.org/10.1016/j.techfore.2005.06.002.

Sargent J. F. Jr. (2014). *The national nanotechnology initiative: Overview, reauthorization, and appropriations issues*. Congressional Research Service: Library of Congress, Washington, DC. http://www.fas.org/sgp/crs/misc/RL34401.pdf. Accessed 25 Nov 2017.

Wynne, B. (2006). *What could the foundations of NanoBioInfoethics be?* Some lateral thoughts. For Stanford-Paris conference on Social and Ethical Implications of Nano-Bio-Info Convergence, Avignon. http://stanford.edu/dept/france-stanford/Conferences/Ethics/Wynne.pdf. Accessed 25 Nov 2017.

Part V
Case Studies

Chapter 15
Philosophy of Activism and Community Management in Water Systems

Carlos Augusto Osorio Marulanda

Abstract This chapter presents an analysis of the forms of community participation in drinking water production systems, especially in rural areas of Latin American countries, bearing in mind Studies on Science, Technology and Society, also known as Social Studies of Science. The proposed philosophy of activism regarding participatory experiences in water issues is based on the interpretative flexibility that appears in the definition of the technical aspects of the systems and in the ethical-political aspects of water access.

15.1 Introduction

In Studies on Science, Technology and Society (STS) as an interdisciplinary field involving philosophical, sociological and historical approaches, among others, to science and technology, some characteristics may be distinguished for the purpose of analyzing community participation in drinking water systems in Latin America.[1] In recent decades, community participation with respect to water issues has intensified worldwide (Tignino and Sangbana 2015). There have been numerous experiences in Latin America, including the selection of water purification technologies, the construction and maintenance of water supply systems (Gutiérrez et al. 2013; Córdova et al. 2014), as well as more activist expressions linked to the mobilizing of

[1] In its origins, STS presented two separate trends and different aims: an activist STS, which arose from the concern about undesirable effects resulting from technical and scientific development, which was represented by the acronym "Science, Technology and Society"; and another, more academic STS, focused on the analysis of the construction of knowledge based on sociology, history and philosophy of science, which uses the labels "Studies on Science and Technology" and "Social Studies of Science" (González 2004). In either case, the fields of delimitation as well as the "cross fertilizations" between different perspectives are a frequent subject of study (Cozzens 2001, quoted by González 2004).

C. A. Osorio Marulanda (✉)
Escuela de Ingeniería Industrial, Universidad del Valle, Cali, Colombia
e-mail: carlos.osorio@correounivalle.edu.co

B. Laspra, J. A. López Cerezo (eds.), *Spanish Philosophy of Technology*, Philosophy of Engineering and Technology 24,
https://doi.org/10.1007/978-3-319-71958-0_15

communities to achieve access to water services, protests against their privatization and the high costs of services, the recognition of the right to participate in decisions about water source and supply system management, and the demand for quality water resources and their availability, among other issues (Llano 2015; Castro 2007a, b, 2009a, 2012; Miranda 2005; Pérez and Ortiz 2013). In this respect, we will see that, through STS studies, the first types of experience may be understood as a matter of interpretative flexibility in a technical system, while more activist forms of participation are related to ethical-political proposals that are also present in STS studies.

In what follows, some elements of science and technology participation are presented with the aim of defining a reference framework to present the philosophical practices of activism in drinking water systems in Latin America. The characteristics of two types of practices are subsequently presented: participation in technical issues, and participation for the right to water. To end the chapter, some considerations are put forward regarding the importance of participation in water issues and its relation with other approaches involving community participation.

15.2 Participation in Science and Technology

Public participation in matters of science and technology has a background in the protests carried out by social groups and the ecologist movements in the United States in the 1960s and 1970s arising from socially controverted situations linked to science and technology. Ecologist groups also played a significant role in Europe and became new political forces in countries such as Germany, Switzerland, Denmark and Holland. The emergence of these groups was accompanied by new methods of political protest, including demonstrations and blockades against the construction of environmentally controversial projects (e.g. nuclear power plants). Several lessons about direct participation in government decisions have been learned from these experiences. Europe has a long-standing history in testing new public participation methods that include educating citizens, while in the United States, where protests and social movements have been very active, innovations in participation are linked to the creation of alternative methods for conflict resolution (Webler and Renn 1995, p. 19).

An additional social context is related to the debate over the Deficit Model, a model that anticipates the inability of the public to understand and appreciate the achievements of scientific activity. Since 1970s, this deficit was measured in public understanding of science surveys, with the intention of being corrected through scientific literacy. Miller (1983, 1987) pointed out that literacy had to provide comprehension about scientific methods, a positive assessment of science and technology based on their own results, and the rejection of superstitious beliefs. The deficit model has been used to question public participation in decisions regarding science issues and also to lessen the argument about the relationship between science and the public (Buchi and Neresini 2008). Through scientific literacy, information facilitating access to the right knowledge from experts would be guaranteed and hence a favorable attitude towards science, technology and innovation would be achieved. Access to information would thus be a way to involve the public with science. Nevertheless, Rowe and Frewer (2005) have pointed out that access to information does not mean participation in technical and scientific decisions, as participation only occurs when the flow of information goes both ways between the body organizing a participative process and the interested public or whoever represents said public. Participating implies recognizing that the public also has valuable information and knowledge that need to be taken into account.

In line with Renn et al. (1995), Rowe and Frewer (2005), and Bucchi and Neresini (2008), and according to the experiences examined in this chapter, public participation in science and technology will be considered as the set of situations and activities, both spontaneous and organized, in which different kinds of non-expert communities (citizens, other concerned parties, and groups of interest), who each have their own interests and knowledge, participate or become engaged together with experts and other actors with the purpose of evaluating technologies and developing technical and scientific activities, making decisions regarding science policies and projects, and participating in processes of knowledge co-production. Public participation implies a varied presence of participants with their own points of view and interests. If the observations by Nelkin (1984, cited by López and Todt 2004), are taken into account, it could be considered that for disputes, inquiries, evaluation processes and decision making, the engaged public would be the people who are directly affected, those involved or potentially affected, consumers, the aware public, non-governmental organizations, and experts in diverse areas. Public institutions likewise take part at different levels with the function of organizing, coordinating, founding, ensuring citizens rights and even co-opting community participation, as has occurred in certain cases related to water supply (Castro 2012). Other actors may be added to the above list depending on the matter in question, such as international cooperation organizations, or even private companies or trade unions. The latter may possibly represent interests that are questioned by a community, as well as form part of lawsuits and disputes.

Public participation happens to be more relevant when there is a higher intensity of cooperation between social actors of different kinds to produce knowledge (Bucchi and Neresini 2008; Callon et al. 2001). Cooperation can be generated for knowledge transfer processes towards real situations, as part of the stabilization of scientific knowledge itself. It can also be generated when communities contribute with knowledge in order to define and solve research problems. In this regard, there are more sensitive matters for participation, such as the environment, agriculture and health (Hess 2008).

Participation in science and technology becomes a subject of major importance due to the situations that arise in today's societies. Sarewitz (2001) highlights concerns such as the global character of nature (environmental variability, mobility of diseases, invading species) associated with incomplete knowledge of natural phenomena and the increase in uncertainty, and potential risks. Other aspects are related to inequalities between the rich and the poor, which grow in line with scientific and technological development even though, at the same time, science makes it easier for citizens to access social services and participation.

In Latin America, there has been greater public participation in science and technology in environmental issues (Cepal 2013). It is a kind of participation restricted to the field of policies, plans and strategies, with a discretionary scope. Those initiatives that stand out are related to the programs and projects carried out in some countries in Central and South America and more direct participation in countries such as Chile, Colombia, Costa Rica and Ecuador. The challenges represented by these kinds of participation have to do with the problems of policy and program implementation and, to a greater extent, with the protection of the rights of indigenous communities, ethnic groups and minorities (Cepal 2013).

Participation is not exempt from numerous complexities and difficulties, such as issues of public diversity and representation, the motivations for carrying it out, when to participate, the procedures to follow, access to specialized information, the distribution of tasks, the forms of deliberation and decision making, the cost-benefit relationship, the binding nature of decisions, in addition to possible apathy on the part of communities when they are not directly affected by the respective issue (Callon et al. 2001; López and Todt 2004; Bucchi and Neresini 2008).

15.3 Participation in Technical Issues

Water supply systems require an interrelated set of technical, natural and social elements to work in order to provide drinking water in the quantity, quality and continuity required by a community.[2] The organization and setting up by government of a systemic structure of this type, with direct or representational public participation, has been one of the most important accomplishments in Western democracies. In Latin America, where the service of water supply is not available to all citizens, community participation to obtain drinking water tends to be present in technical issues, via initiatives on the part of citizens themselves, as well was through projects that are usually included in development plans or in agendas defined by public and international organizations. Such participation may include water resource management, the design and construction of a supply system, sanitary risk control, and the administration and maintenance of the system, including cleaning and water rate collection actions (Rojas et al. 2013). Some experiences of this kind depend on the role assumed by public organizations regarding the degree of importance in this ways of participation (World Bank 1978; White 1987). This participation can be

[2] For such considerations, water supply systems can be considered technological systems (Hughes 1983, 1987). Constructive technology assessment is a technology generation process that starts from the early, corrective interposition of emerging technologies in order to modulate their evolution, range, and type of impacts. Within this process, orientations by experts and social groups take place in such a way that the assessment process reflects the diversity of values and interests (Rip et al. 1995; García et al. 2001, pp. 65–67).

represented by means of a progression ladder: on the lower rungs, there would be no community participation because the initiative corresponds to government; while on the higher rungs, the community is the one taking the initiative (Van and Boot 2003). In many rural areas and large, poor and marginalized sectors of cities in Latin America, the community often takes the initiative and contributes with its own resources, such as providing the workforce and carrying out specific works related to the water system (Brikké 2000). For instance, on the slopes of the western area of the city of La Paz, local communities took the initiative of deploy the access to drinking water in their neighborhoods, although they also counted on the participation of politicians at that time and, at a certain point, the contribution of a foreign government that supported the community after a landslide destroyed part of the infrastructure that had already been completed. Not only has the community contributed with its own resources to provide the water service throughout a chronicle that has its origins in the 1970s, but also the management and maintenance of the system are currently carried out autonomously (Perales 2014).

Another type of participation in technical issues occurs when interventions by universities and non-governmental organizations establish links with communities in rural areas, indigenous communities or specific ethnic groups. This specificity derives from the way in which communities participate, as well as from the way in which natural resources are managed. For example, in the Ticuna indigenous reserve of Resguardo Nazareth, way upstream of the city of Leticia on the Amazon River (between Colombia and Brazil), a community participation experience was carried out to define water and sanitation technologies in line with the expectations, conditions and world views of the indigenous people (Méndez et al. 2011). By using methodologies that included participatory action research (PAR), a group of professionals in social sciences and engineering worked together with the community until three appropriate technologies where selected, one for water and two for sanitation. PAR is a kind of research which organizes the analysis and the intervention focusing on the removal of privileges and the recognition of the "other", in addition to fostering communication and self-determination on the part of the communities in the projects as well as the relationship between knowledge and action (Borda 1999; Durston and Miranda 2002). The technologies defined in this experience and in many others that have been proposed in Latin America are considered appropriate technologies. They are low-cost, locally-controlled technologies, used with renewable energies and sustainable agriculture (Hess et al. 2008). In the context of water supply, a technologically appropriate solution would be one that is highly efficient yet simple to build, operate and maintain, with a high degree of reliability, flexibility, accessibility and which also uses existing resources (CINARA & IRC 1994, p. 86).

How can this participation in technical issues be understood from the perspective of studies on Science, Technology and Society (STS)? In this respect, the arguments based on the Deficit Model have been recurrent in questioning participation in technical decisions. Communities are excluded for not having scientific knowledge based on data empirically proven by academic communities. Nevertheless, it should be pointed out that this comparison has been refuted when highlighting the fact that the knowledge possessed by communities is qualitatively different. Knowledge based on data is only part of the required knowledge. In an environmental intervention process, the former type of knowledge interacts with others of an extra-epistemic nature, such as value judgments, insights, traditions, perceptions and interests. Knowledge of local conditions can determine what data is sound and relevant and what the related political issues are (Funtowicz and Ravetz 1997).

A way to close the gap between one kind of knowledge and the other, prior to denying their importance and difference, is to ensure the creation of settings in which both kinds of knowledge can come together, as in the Hybrid Forums proposed by Callon et al. (2001), in Constructive Technology Assessment (Rip et al. 1995; García et al. 2001),[3] or in the circumstances created by using participatory action research (PAR) (Borda 1999). In this type of hybrid setting, social groups, experts, politicians and all those who consider themselves to be involved can meet and discuss technical options regarding a situation or problem. Via meetings of this kind, the uncertainty associated with the scientific and technological development and its possible consequences may be better addressed. In participation concerning drinking water, which can be differentiated from the frequent forms of participation in science issues like nuclear energy, transgenic food or environmental justice (Hess 2008), the knowledge possessed by local communities becomes linked in a very diverse way. For example, these communities are the ones that know the stories associated with water in a locality, by preserving the anecdotes and memories related to floods and droughts, they can be in possession of different information regarding the upkeep of water resources and their associated forests, as well as having valuable ideas about how to erect a small aqueduct in the most sustainable way.

Another argument about the exclusion of local communities from technical decisions has to do with technological determinism. Technological determinism refers to the belief that technical forces determine social and cultural changes. From this point of view, all human capacity to control the technological development is thus lost. Technology would be produced in a world of inevitable occurrences, where its own progression marks the direction to be followed by society (Ellul 1960; Winner 1977, 2001). One of the most critical approaches questioning technological determinism highlights the fact that technology contains political properties, because of which it becomes necessary to promote the evaluation of social and material infrastructures within technological systems (Winner 1987). In other words, if we reflect about what kind of society we want, we will be able to decide what technical means we will use. Hence the need to define new political roles in institutions where community participation can have its place (Winner 1987). If these questions about technological determinism are taken into account, participation will contribute to the fact that the aims and requirements of a community with respect to drinking water issues will become the subject of permanent attention. Through the presence

[3] Constructive technology assessment is a technology generation process that commences with the early and corrective interposition of emerging technologies in order to modulate their evolution, range, and type of impacts. This process is guided by experts and social groups take place, in such a way that the assessment process reflects the diversity of values and interests (Rip et al. 1995; García et al. 2001, pp. 65–67).

of the community in the design and management of the system, the control of many performance-related decisions can be achieved; a control that is circumscribed by environmental and social variables and technological changes. In the language of technological systems (Hughes 1983, 1987), this means that the involvement of the social actors in the management of the course of the system and not only at the moment of the design, i.e., when the system is being built, could contribute to technology not being interpreted from the perspective of technological determinism.

15.4 Participation for the Right to Water

Castro (2012) posits the existence of three major forms of community participation in matters concerning drinking water in Latin America. Two of these take the form of technocratic management: the first excludes the population in its condition as citizens and customers, while the second permits or induces highly restricted settings for participation that tend to limit such participation to self-help activities or the exercise of consumer rights. The third form of community participation, however, is bottom up, being fostered by social organizations of different types (workers, users, communities and environmental NGOs). The frameworks are not rigid, as a self-managed initiative may result in a formal project, or even be co-opted by other stakeholders. On the other hand, induced projects and those which are controlled to a greater or lesser extent may produce unexpected results. This third community participation initiative is closely related to protests and social mobilization activities for the right to water, which is an inalienable right of every human being and whose demand has been exacerbated during recent decades in the countries in the region because of the trend in public policies towards privatization (Castro 2007b). Some examples of mobilization are those carried out in Cochabamba (Crespo et al. 2003), Mexico (Castro 2007a) and Tucumán (Crenzel 2003). Promoting a private initiative has not just been limited to yielding the rights and markets of water resources to private companies, it is also about practices and processes based on models from the private sector which state-owned companies have ended up incorporating in their policies through institutional reforms promoted since the 1990s (Castro 2007a, b, 2009a, b, 2011, 2012). Yielding to private companies or individuals the rights to goods such as water, which form part of the public domain, start out from the idea that private investment – and especially foreign investment – is more efficient in providing the service. However, it has not been possible to prove that private management of these markets constitutes a mechanism that is sufficiently useful for the management of water resources, or for solving problems of investment, access or usage.

These forms of political participation coincide with some concerns of an activist philosophy of technology and have been the object of concern by Studies on Science, Technology and Society (STS). Feenberg (1995, 1996) considers that a fairer society is possible, starting with a plurality of technological advances capable of internalizing public policies. He states that, with a few exceptions, public involvement with technology is not carried out through processes like voting; the general public is not interested enough in technical matters. However, at a local level as in water-related cases like that presented here, the public becomes involved in protests about technical developments that concern it, especially in domains such as the environment. According to Feenberg (1996), these movements can be interpreted in terms of democratic activity and from social theory as re-appropriations of technology by its users.

Inspired by American Pragmatism in relation to the application of the experimental method to social problems, Durbin (1978) proposes the interdisciplinary effort that supposes cooperating not only with other experts, but also with other citizens to achieve the progressive solution of some complex contemporary problems. Based on diverse experiences in different fields of knowledge, Durbin (1992) proposes some practices that favor activism: not acting alone, as wide-ranging community support is required; proposing reforms as a road plan; relating specific experiences in activism with the experiences of other activists; fostering public education in regard to technical and scientific matters, as well as interdisciplinarity and social responsibility at universities; and relying on ethics and the society that practices them. Furthermore, according to his experience in the struggles for the defense of the rainforests in Costa Rica, he advocates the importance of taking the specificities of the local culture into account and the need for the commitment of activists not to be reduced to the topic of values, which means working for policies that change things (Durbin 2006). Technology is not a force that cannot be controlled, democratic means and socially responsible activism must be used for acting (Durbin 1992, p. 201).

Experiences in social mobilization related to drinking water are activist experiences, as they defend local actions aimed at promoting basic reforms for water service supply in socially adequate conditions such as: affordable costs, opportunity in the service, water quality, system continuity and sustainability. Likewise, it is about the citizens' own freedom in the sense that Sen (1999) has indicated: the recognition of people's rights to ensure a dignified life for themselves. This means the right to participate so as to remove the main sources of deprivation of freedom, including poverty and tyranny, scarcity of economic opportunities and systematic social deprivations, the state of neglect in which public services may be found, and intolerance and excessive intervention by repressive States.

15.5 To Wind Up

Public participation in drinking water systems in Latin America constitutes an expression of the social struggles for gaining access to an essential basic service, while at the same time requiring the relevance of community knowledge and initiatives in technical issues. Participation aimed at gaining access to drinking water is also an ethical issue, an ethic of that which is public with respect to access to an essential service, regardless of the capacity of the individual or the family unit to pay for it (Castro 2012). However, it is also an environmental ethic, especially visible in those social groups struggling for water under an ecocentric perspective, even though they frequently do not have a clearly defined stance regarding issues like privatization. This stance differs from that of social groups which assume an anthropocentric posture, whose claims are aimed at defending the human right to access to water, though with lesser emphasis on the protection of water resources (Castro 2009b). Such an environmental ethic presupposes the existence of a sort of natural law, a certain pre-existing ecologic order which humans have to respect, and that their actions have to be in harmony with this natural law (Mitcham 1996). This sense of unity between a social and a natural order is a growing trend in diverse social movements in Latin America in the fields of food sovereignty and agroecological production of a non-capitalist type (De Sousa 2011), as well as in environmental and cultural practices for the protection of natural resources (Escobar 2010).

Diverse proposals exist for building up changes in relation to technical/scientific problems. Besides activism, Durbin (1992, 2003) defends the issue of the education of experts through ethics courses, while Mitcham (1994) proposes joint work between engineers, technicians, humanists and citizens to approach the ethical analysis. Castro (2007a) advocates an interdisciplinary holistic approach based on an analysis of the epistemic rationalities that appear in water projects in which community rationalities have their place. Among all these proposals, community participation represents a decisive issue in the comprehension of natural and social processes, as well as in the solution of problems requiring science and technology.

Recognizing community participation from the perspective of Studies on Science, Technology and Society means recognizing that there exists interpretative flexibility in technical decisions about water systems, as it is presented in the STS tradition, which focuses on the analysis of knowledge construction processes. It likewise means recognizing the activist perspective that emphasizes the analysis of the social, ethical, and environmental consequences of science and technology, as revealed in the different types of experiences on water supply systems in Latin America. The philosophy of activism in water-related issues thus refers to this interpretative flexibility and to an ethical-political perspective.

References

Borda, F. (1999). Orígenes universales y retos actuales de la IAP (Investigación Acción Participativa). *Análisis Político, 38*, 71–88.

Brikké, F. (2000). *Operation and maintenance of rural water supply and sanitation systems: A training package for managers and planners*. Delft: IRC International Water and Sanitation Centre.

Bucchi, M., Neresini, F. (2008). Science and public participation. E. Hackett, O. Amsterdamska, M. Lynch, J. Wacjman, The handbook of science and technology studies (449–472). Cambridge: The MIT Press.

Callon, M., Lascoumes, P., & Barthe, Y. (2001). *Agir dans un monde incertain: Essai sur la démocratie technique*. Paris: Editions De Seuil.

Castro, J. E. (2007a). El estudio interdisciplinario de los conflictos por el agua en el medio urbano: Una contribución desde la sociología. *Cuadernos del CENDES, 24*(66), 21–46.

Castro, J. E. (2007b). La privatización de los servicios de agua y saneamiento en América Latina. *Nueva Sociedad, 207*, 93–112.
Castro, J. E. (2009a). Apuntes sobre el proceso de mercantilización del agua: Un examen de la privatización en perspectiva histórica. In J. Delclòs (Ed.), *Agua, un derecho y no una mercancía* (pp. 35–55). Barcelona: Icaria Editorial.
Castro, J. E. (2009b). Presentación. *Anuario de Estudios Americanos, 66*(2), 15–22.
Castro, J. E. (2011). El proceso de democratización de la gestión de servicios públicos esenciales de agua y saneamiento. *Hábitat y Sociedad, 2*, 13–37.
Castro, J. E. (2012). Participación y control social en el saneamiento básico. In L. Heller (Ed.), *Agua y saneamiento: En la búsqueda de nuevos paradigmas para las Américas* (pp. 137–156). Washington, DC: Organización Panamericana de la Salud.
CEPAL. (2013). Acceso a la información, participación y justicia en temas ambientales en América Latina y el Caribe: Situación actual, perspectivas y ejemplos de buenas prácticas. *Serie Medio Ambiente y Desarrollo No. 151.*
CINARA & IRC. (1994). *Informe general. Programa TRANSCOL.* Cali: Universidad del Valle.
Córdova, G., Romo, M., & Romero, L. (2014). Acción pública local y prácticas autogestivas en colonias sin agua entubada ni saneamiento, en el estado de Chihuahua. *Gestión y Política Pública, XXIII*(2), 385–420.
Cozzens, S. (2001). Making disciplines disappear in STS. In S. H. Cutcliffe & C. Mitcham (Eds.), *Visions of STS. Counterpoints in science, technology, and society studies.* Albany: State University of New York Press.
Crenzel, A. (2003). Tucumán–Argentina case study report. In J. E. Castro (Ed.), *Prinwass project.* Oxford: University of Oxford.
Crespo, C., Laurie, N., & Ledo, C. (2003). Cochabamba–Bolivia case study report. In J. E. Castro (Ed.), *Prinwass project.* Oxford: University of Oxford.
De Sousa, B. (Ed.). (2011). *Producir para vivir. Los caminos de la producción no capitalista.* México D. F: F.C.E.
Durbin, P. (1978). Toward a social philosophy of technology. In P. T. Durbin & C. Mitcham (Eds.), *Research in philosophy and technology. An annual compilation of research.* Greenwich: Jai Press.
Durbin, P. (1992). *Social responsability in science, technology, and medicine.* Bethlehem: Lehigh University Press.
Durbin, P. (2003). Ética, o cómo tratar democráticamente los problemas tecnosociales. *Isegoría, 28*, 19–31.
Durbin, P. (2006). *Small steps to save the rainforest: A plea for experiential learning.* Delaware: University of Delaware.
Durston, J., & Miranda, F. (2002). Experiencias y metodología de la investigación participativa. *CEPAL- SERIE Políticas Sociales, 58*, 5–71.
Ellul, J. (1960). *El siglo XX y la técnica.* Barcelona: Labor.
Escobar, A. (2010). *Una minga para el postdesarrollo: Lugar, medio ambiente y movimientos sociales en las transformaciones globales.* Lima: Universidad Nacional Mayor de San Marcos.
Feenberg, A. (1995). *Alternative modernity.* Berkeley: University of California Press.
Feenberg, A. (1996). *Summary remarks on my approach to the philosophical study of technology.* Notes for a presentation at Xerox PARC.
Funtowicz, S., & Ravetz, J. (1997). Problemas ambientales, ciencia post-normal y comunidades de evaluadores extendidas. In M. I. González, J. A. López, & J. L. Luján (Eds.), *Ciencia, tecnología y sociedad, lecturas seleccionadas.* Barcelona: Editorial Ariel, S. A.
García, E., et al. (2001). *Ciencia, tecnología y sociedad: Una aproximación conceptual.* Madrid: OEI.
González, M. (2004). *Principales enfoques y nuevas aportaciones en los estudios en Ciencia, Tecnología y Sociedad. Curso de Especialista en CTS+I.* Madrid: OEI.

Gutiérrez, V., Nazar, A., Zapata, E., Contreras, J., & Salvatierra, B. (2013). Género y participación de las mujeres en la gestión del agua en las subcuencas Río Sabinal y Cañón del Sumidero, Berriozábal, Chiapas. *Revista de Estudios de Género: La ventana, 4*(38), 246–276.
Hess, D., Breyman, S., Campbell, N., & Martin, B. (2008). Science, technology, and social movements. In E. Hackett, O. Amsterdamska, M. Lynch, & J. Wacjman (Eds.), *The handbook of science and technology studies* (pp. 449–472). Cambridge: The MIT Press.
Hughes, T. P. (1983). *Networks of power*. Baltimore: The Johns Hopkins University Press.
Hughes, T. P. (1987). The evolution of large technological systems. In W. E. Bijker, T. P. Hughes, & T. Pinch (Eds.), *The social construction of technological systems*. Cambridge: MIT Press.
Llano, V. (2015). Community knowledge sharing and co-production of water services: Two cases of community aqueduct associations in Colombia. *Water Alternatives, 8*(2), 77–98.
López, J., & Todt, O. (2004). *Participación pública en ciencia y tecnología. Curso de Especialista en CTS+I*. Madrid: OEI.
Méndez, S., Pazo, M., & Romero Pérez, B. (2011). Metodología para la apropiación de tecnologías de saneamiento básico en comunidades indígenas. *Cuadernos de Desarrollo Rural, 8*(66), 153–176.
Miller, J. (1983). Scientific literacy: A conceptual and empirical review. *Daedalus, 112*(2), 29–48.
Miller, J. (1987). Scientific literacy in the United States. In *Communicating science to the public* (pp. 19–40). Chichester: Wiley.
Miranda, N. (2005). Recife, Brazil: Building up water and sanitation services through citizens. In B. Balanyá, B. Brennan, O. Hoedeman, S. Kishimoto, & P. Terhorst (Eds.), *Reclaiming public water: Achievements, struggles and visions from around the world* (pp. 113–119). London: Transnational Institute and Corporate Europe Observatory.
Mitcham, C. (1994). *Thinking through technology*. Chicago: University of Chicago Press.
Mitcham, C. (1996). Cuestiones éticas en ciencia y tecnología: Análisis introductorio y bibliografía. In M. I. González, J. A. López, & J. L. Luján (Eds.), *Ciencia, tecnología y sociedad: Una introducción al estudio social de la ciencia y la tecnología*. Madrid: Tecnos.
Nelkin, D. (1984). Science and technology policy and the democratic process. In J. C. Petersen (Ed.), *Citizen participation in science policy* (pp. 18–39). Amherst: University of Massachusetts Press.
Perales, V. (2014). Agua y saneamiento autoconstruidos en La Paz (Bolivia). *Integra Educativa, VII*(1), 139–156.
Pérez, A., & Ortiz, B. (2013). Participación ciudadana en la transformación del manejo del agua en Puerto Rico. *Revista Puertorriqueña de Psicología, 24*, 1–16.
Renn, O., Webler, T., & Wiedemann, P. (1995). A need for discourse on citizen participation: Objectives and structure of the book. In O. Renn, T. Webler, & P. Wiedemann (Eds.), *Fairness and competence in citizen participation*. Dordrecht: Springer.
Rip, A., Misa, T., & Schot, J. (Eds.). (1995). *Managing technology in society: The approach of constructive technology assessment*. London: Pinter.
Rojas, J. H., et al. (2013). Análisis comparativo de modelos e instrumentos de Gestión Integrada del Recurso Hídrico en Suramérica: Los casos de Brasil y Colombia. *Revista Ambiente e Água, 8*(1), 73–97.
Rowe, G., & Frewer, L. (2005). A typology of public engagement mechanisms. *Science, Technology, & Human Values, 30*(2), 251–290.
Sarewitz, D. (2001). Bienestar humano y ciencia federal, ¿cuál es su conexión? In J. A. López & J. M. Sánchez (Eds.), *Ciencia, tecnología, sociedad y cultura*. Madrid: Biblioteca Nueva-OEI.
Sen, A. (1999). *Desarrollo y libertad*. Barcelona: Planeta.
Tignino, M., & Sangbana, K. (2015). *Public participation and water resources management: Where do we stand in international law?* Paris: UNESCO.
Van Ast, J. A., & Boot, S. P. (2003). Participation in European water policy. *Physics and Chemistry of the Earth, 28*, 555–562.

Webler, T., & Renn, O. (1995). A brief primer on participation: Philosophy and practice. In O. Renn, T. Webler, & P. Wiedemann (Eds.), *Fairness and competence in citizen participation*. Dordrecht: Springer.

White, A. (1987). *Participación comunitaria en abastecimiento de agua y saneamiento, conceptos, estrategias y métodos, Serie documentos técnicos* (Vol. 17). La Haya: IRC.

Winner, L. (1977). *Tecnología autónoma*. Barcelona: Editorial Gustavo Gili.

Winner, L. (1987). *La ballena y el reactor: Una búsqueda de los límites en la era de la alta tecnología*. Barcelona: Gedisa.

Winner, L. (2001). Dos visiones de la civilización tecnológica. In J. A. López & J. M. Sánchez (Eds.), *Ciencia, Tecnología, Sociedad y Cultura*. Madrid: Biblioteca Nueva-OEI.

World Bank. (1978). *Towards a typology of popular participation* (p. 16). Policy planning and program review Dept.

Chapter 16
Localities Facing the Construction of Fossil-Fuel Power Plants. Two Experiences to Address the Hostile Face Electricity Infrastructures

Gloria Baigorrotegui and María Teresa Santander Gana

Abstract This chapter is inspired by interesting experiences concerning electricity infrastructures facing local uprisings in two towns separated by a great distance: to address the hostile face building fossil-fuel power plants in mobilized citizen localities. Hostility levels are presented as different pulses, in which unexpected public demonstrations invade the public arena, possibly resulting from socio-technical closures in environmental impact evaluations. They mirror secret energy alliances, showing the influences and pressures of these transnational businesses. We also consider collective actions, incorporating interpretations and new and unexpected alliances. Our proposal draws on two qualitative, ethnographic studies about local socio-environmental conflicts contrary to the construction of fossil-fuel power plants, one of which arose in the Basque region of Spain, and the other in Coquimbo, Chile. Finally, we maintain that these geographically disperse studies, with their differing results, contribute to the discussion of locality as a framework of resistance against violent infrastructures, which goes beyond normalized processes and unfolds on other levels of action.

16.1 Introduction

As infrastructures, energy projects and local resistances help to track philosophical and practical dimensions that are paradoxical as well as being both latent and explicit (Howe et al. 2016; Ureta 2014; Barry 2013; Pacey 1983). In this respect, STS studies have criticized the policies of experts, engineers, and scientists for not

G. Baigorrotegui (✉)
Institute of Advances Studies, University of Santiago of Chile, Santiago, Chile
e-mail: gloria.baigorrotegui@usach.cl

M. T. Santander Gana
Industrial Engineering Department, University of Santiago of Chile, Santiago, Chile
e-mail: mariateresa.santander@usach.cl

B. Laspra, J. A. López Cerezo (eds.), *Spanish Philosophy of Technology*,
Philosophy of Engineering and Technology 24,
https://doi.org/10.1007/978-3-319-71958-0_16

contributing to the socialization of uncertainties, risks, assumptions, previous political agreements, etc. (Albornoz and López Cerezo 2010; Mitcham 2005; Funtowicz and Ravetz 1993).

When the 'dark side' of infrastructure takes the form of confrontational practices, this may turn them into zero sum types (Collins 1975), thus leading to people resorting to abusive practices (Galtung 1996; Baigorrotegui 2008). Hård (1993) comments on this aspect, referring to the bifocality of technology conflicts, and Stankiewicz (2009), to the difficulties for managers in solving these great challenges.

Two examples will show how notions of hostility, socio-technical closures, and secret energy networks are driven to action, bridging the gap of STS in local energy infrastructure politics that emerged over and above state actors in the European Union and the United States (Barry 2013; Hess 2007). The experiences we present are separated by a considerable distance. The first is the situation facing the construction of a Coal-Fired (CF) power plant in La Higuera, Chile and the other, the fight against the Combined Cycled Gas Turbine (CCGT) power plant in Boroa, in the Autonomous Community of the Basque Country (ACBC), Spain.

This is a cross-sectional exploratory study employing primary (interviews, direct observation) and secondary (bibliographic collections, photographic and audiovisual materials) sources and including six field campaigns, during different periods of public visibility for the respective conflicts.

The public representatives of the mobilization in the locality of Boroa, in the Basque Country, Spain, were contacted. The leaders of the citizen platform were interviewed, as well as representatives of both the local government and the autonomous (regional) government. The head of the thermal power station project and experts supporting the local mobilization were also interviewed. Participants were observed in April 2003, June 2003, and August 2007.

In the locality of La Higuera, in the Coquimbo Region of Chile, ethnographic fieldwork occurred during August 2010, April 2013, and June 2013. Citizen platform leaders were contacted, as well as representatives from the local and regional governments, and experts supporting local mobilizations.

16.2 Socio-technical Closures in Electricity Grid Infrastructures and Energy Secrecy for Siting Project

Summerton (1992) materially recognizes the technology dependencies involved in the high voltage network in other pre-existing utility distribution grids, such as water, gas and carbon. She therefore tones down the radical, possibly confrontational tone of the controversies and prefers to focus on mediation and negotiation.

Material practices in infrastructures tend toward stabilization and entrenchment, which Hughes has called *momentum* (Hughes 1993), while Summerton (1992, p. 880) has recognized as "invisible" associations involved in them. These are located on the "hard side" of negotiations, where "some presumptive subscribers proved recalcitrant; in this sense as well, the physical artefacts were essentially 'soft' while actors sometimes proved 'hard'" (Summerton 1994, p. 252). State regulators mainly participate in these invisible associations at different levels in legislative and political bodies (parliaments, national ministries of energy, city halls, and political parties), financial institutions (banks, stockbrokers), and professional interest groups (disciplinary organizations).

As all of these actors and agencies face up to one another, novel confrontations will clearly arise; e.g., between the public and the private, the local and the transnational.

Trust has plummeted in systems-builders responsible for siting power plants, especially after the trans-nationalization and privatization of energy infrastructures. Unequal access to the administrative, knowledge and judicial processes of localities have been proof of the contradictions or obscure maneuvers that can emerge from the relationships between policy representatives and transnational energy business directors (Howe et al. 2016; Aalto and Korkmaz 2014).

The process of Environmental Impact Assessment (EIA) in the siting of energy facilities could be considered a mechanism of conflict management. Affected areas feel the need to go beyond technocratic and non-participatory EIA (Rajan 2005; Barandiarán 2015; Sepúlveda 2000) in the name of human rights and Environmental Justice (Walker 2010).

16.3 Local Hostile Behavior Contrary to the Siting of Energy Projects

Reaching consensus is becoming increasingly difficult. In Latin America, these energy projects have been denounced as part of a neo-extractivist model, lacking in consideration for human rights, ancestral community rights, local knowledge, biodiversity, and environmental governance (Urkidi and Walter 2011; Saguier 2012; Martinez Alier 2003; Baigorrotegui et al. 2014). In southern Europe, particularly in the Basque Country, Citizen Platforms[1] have orchestrated struggles for municipal self-determination in energy policies, (Larrinaga 2006; Zubiaga 1995) with differing degrees of manifest hostility.

[1] By Citizen Platforms we refer to agencies that include individuals, and groups who gather thematically, usually to support or reject a project or local plan (Estévez and González 2005; Devine-Wright 2009). We link citizen platforms closer to grassroots groups than to national or international NGO movements and campaigns, although they are also served by citizen platforms. Moreover, these citizen platforms are constituted by emerged identities in different time and spaces thanks to the agency of social media platforms. They are closer to hybridized collective actions than stabilized social movements (Rodríguez-Giralt 2011).

The politicization of facts in socio-environmental controversies[2] and engagement between experts and activists (McCormick 2009) contribute to lobbying in favor of local resistances (Baigorrotegui 2015; Barandiarán 2015) as forms of counter-expertise attuned to their demands (Hess 2015; Frickel et al. 2010; McCormick 2007).

16.4 The "Pulses" of Socio-technical Closures, Energy Secrecy and Hostile Infrastructures in Siting Fossil-Fuel Power Plants

While both conflicts are incommensurable experiences, we propose to represent them in terms of 'pulses' to pinpoint the relationship/s between: (i) socio-technical closures, (ii) secret energy relations, and (iii) the levels of hostility in such confrontations, which are all linked in particular ways.

Socio-technical Closures This dimension represents pulses to evaluate the scientific and technical opinions of energy project siting and certifications presented by authorities in the EIA process. The considered time period in both representations is marked by the timing of legal opinions, the delivery of reports and studies, addenda requests, and the extension, or not, of time periods related to institutionally mandated time schedules and projects.

Invisible Energy Alliances This axis represents all the relationships between regulators, business people, and technicians that are not transparently presented to citizens, including secret relationships. Although these relationships are not presented as being directly linked to the conflict in question, they mark attempts to influence public opinion in favor of those in charge of the project. Diplomatic energy actions and concealing risks fall into this dimension.

Intensity of Hostile Behavior This represents manifest evidence of hostile behavior within socio-technical practices. It does refer to some feelings and emotional spheres of practices, but we only represent the manifest part of the resistance practice.

This dimension is plotted intuitively and is contingent and interactive. For example, in creating the EJOLT database,[3] five scales are defined in terms of: (1) High (widespread, mass mobilization, violence, arrests, etc.), (2) Medium (street protests, visible mobilization), (3) Low (some local organizing), (3) Latent (no visible

[2] Pellizoni (2011) recognizes that politicization refers to making an issue public, which is different from bringing it into a private environment, where the events assume reduced interest and where not all of the actors would have the legitimacy to polemicize them.

[3] Known as Environmental Justice Organizations, Liabilities and Trade (EJOLT) this database represents a widely range of conflicts about Ecological Distribution around the world. For more details, visit: http://www.ejolt.org/maps/. Accessed August 3, 2016.

organizing at the moment), and (4) Unknown. Another example is the inventory of the Environmental NGO Terram in Chile,[4] in which a conflict is deemed as such starting with complaints by citizens in the community and newspaper articles: in short, with its visibility in mass media and in EIA reports.

16.5 Barrancones and Boroa as Specifically Hostile Electricity Infrastructure Sitings

The liberalization of the energy market first occurred in Chile in 1980 and then in Spain in 1982. In Chile, the emphasis was on the free market, where direct foreign investments predominate over governmental investment. In Spain, however, national and state interests create tensions between regional autonomy and energy sovereignty in the European and global energy markets. These specific pathways of each country toward energy liberalization over the last 40 years are concurrent with the technology transfer of fossil-fuel power plants and the rise of local socio-environmental conflicts contrary to their construction. Diplomatic efforts on the part of energy sector members, either to justify a 540 MW Coal-Fired (CF) power plant in Chile (ME 2012), or in favor of the construction of an 800 MW Combined Cycle Gas Turbine (CCGT) power plant in the Autonomous Community of Basque Country (ACBC) (PE3E 2010) show the interests of their system-builders to verticalize the business of installing these infrastructures.

16.5.1 The Explosive Construction of the Combined Cycle Gas Turbine (CCGT) Power Plant in Boroa, Basque Country, Spain

In 1998, some of the residents of the town of Amorebieta-Echano in Biscay, Spain, became aware via their informal networks that the construction of a CCGT power plant was being planned in the rural neighborhood of Boroa, on the same site where the construction of an automotive assembly plant had once been planned (Baigorrotegui 2015).

In response to the central government's environmental and land regulation offices authorizing the energy generation project, with an installed capacity of 800 MW, some one hundred and twenty residents pledged in 2009 to work to prevent the advent of what they identified as "a monster".

[4]For more details, visit: http://www.terram.cl/quienes_somos/historia-de-terram/. Accessed August 4, 2016.

The citizen platform called Zornotza Alive, later known as Zornotza Bizirik (ZB),[5] was officially formed in 2000 by a group of residents of different political leanings and ages, which included housewives, local business owners, urbanists, students, members of the ecological movement, along with Basque nationalist activists as their consultants.

On the one hand, the Basque Government (BG) was promoting land urbanization to build an industrial park, where the Basques would also be given access to local work. On the other, organized residents spoke out against serious hazards to the ecosystem and to their health and quality of life. While the energy authorities stated that the plant would enable energy independence and energy generation using CCGT technology, previously unknown in the Basque Country, resistance groups argued that the plant would be dependent on purchasing natural gas to generate energy and that the construction would be illegal according to municipal bylaws.

16.5.1.1 First Move: EIA as an Instance of Public-Private Redesign

ZB highlighted the rashness of constructing the plant in the aforementioned location, which was still rural, and not planned to accommodate chimneys above 10 m in height. Meanwhile, governmental technical teams assessed the magnitude of the impacts of the construction, but also suggested changes to the refrigeration system in order to protect the ecological flow of the River Ibaizabal. The EIA, supposedly a mechanism for public assessment of a private project, acted as a mechanism for redesigning private-public energy plans. In fact, the CCGT power plant was to function as a "roaring power plant" in the ears of its neighbors, due to its proposed use of aero-generators instead of river water (Baigorrotegui 2015). The public- and private-sector technicians worked in alliance with the CCGT to gain environmental approval.

In response to complaints, the municipality commissioned a report from the local consultancy firm European Ecological Consultancy (EKOS) S.L. The report suggested not building the plant, in the absence of joint discussion and a comprehensive analysis of alternative options. However, the authorities ignored this alternative report, which was used by ZB to reaffirm its claims.

16.5.1.2 Second Move: In Updating Development Regulations and Their Technical Justifications

The EIA of the power plant was approved given the favorable reception of the related facilities, the environmental permits and local land arrangements. Local resistance strategies were put into action: mass protests and claims of violation of active municipal bylaws. ZB urban planning experts presented numerous official allegations to local and district authorities, in the hope that the municipal council

[5] For its meaning in Basque.

would deny the plant's license to operate as it had done previously in 2000, when construction of the plant was not authorized due to its exceeding the 10-m height limit (Larrinaga 2003). In this context, ZB presented fifteen thousand signatures to lobby the authorities. Thus, the diplomatic visit by Minister of the Irish Republic Mary O'Rourke to the Basque authorities in September 2000, which highlighted the importance of co-developing energy projects, had begun to pay off. The intensity of the discussion in the municipal ordinary sessions escalate to such an extent that the mayor herself and her team found it impossible to complete the desperately needed referendum, according to the ZB. After pressure from the mayor's political party, the Basque Nationalist Party (PNV in Spanish), the council approved the modifications of the industrial park plans.

Without a quorum at its council meeting, the mayor angered the ZB platform, who in response decided to stage their own alternative referendum. So, on April 14, 2002, a guarantee committee of close to 50 people certified that 43% of population of Amorebieta-Echano sensed in 2002 showing a rejection rate of 98% to the plant's construction.

Neither the publicly presented alternative referendum results nor the claims of inadmissibility of the operating and siting licenses stopped the plant from being built. The updated development bylaws allowed heights over 10 m, thanks to technical justifications. In order to do so, the municipal council's legal team reinterpreted the context of the regulation, thereby favoring the individual liberty of the applicants to build and utilize public utilities in the construction.[6]

16.5.1.3 Third Move: The Spill and Explosive Attack on the Works

With all the legal applications and regulatory bylaws resolved in favor of the plant's construction, ZB intensified its protests and calls, focusing more on PNV. Throughout 2003 and into January 2004, human chains, mass street protests, sit-ins, and even a hunger strike continued to take place.

In January 2005, in the process of cleaning the boilers already installed on the ground, there was a spill of 12,000 l of washing fluid next to the San Pedro stream. This infuriated the ZB citizen platform, who claimed that the amount spilled was 17 times higher still.

Six months later, ETA[7] announced the installation of four explosive devices around the plant which exploded without causing harm to any victims. This opened

---

[6]The municipal council's urban planning and legal teams had to justify the construction, which they had previously not authorized. The primacy of private interest and individual liberty was highlighted in a new interpretation of urban bylaws (Amorebieta-Etchano Municipal Council Urban Planning Department Report 11/07/2002. Case File UR-5033. Bizkaia Energía S.L.).

[7]ETA, anacronym for *Euskadi Ta Askatasuna* (Basque Country and Freedom in English translation) is an armed Basque nationalist and separatist organization in northern Spain and southwestern France. Given its policy of armed force and acts of violence, the organization has been catalogued as a terrorist organization. On October 20, 2011, ETA announced the definitive cessation of its armed activity. It is currently acknowledged as being inactive.

Table 16.1 Moves to situate a combined gas turbine power plant in Boroa

MOVE I	EIA as an instance of public/private redesign
SCT	Re-assessment of the cooling system. EIA's approved non-synchronized
IEA	Rejection of the EKOS consultancy report
IHB	Low. ZB's formalization of their questioning of the Energy Plan. EKOS report as a reference document
MOVE II	In amending development regulations and their technical justifications
SCT	Report to justify urban rules changes. Atmospheric monitoring. Landscape adaptation
IEA	Upgrading of local regulations with private criteria preference. Irish Minister's visit
IHB	Medium. Alternative citizen referendum. Numerous urban planning allegations
MOVE III	The spill and attack with explosives at the work site
SCT	Approval of the construction
IEA	Inauguration of the plant. Contradictory evidence of spill
IHB	High. Street demonstrations. Sit-ins, hunger strike

up the discussion to transition from this local environmental conflict to the broader Basque conflict.

In the following table, we present the conflict for Boroa as a function of Socio-Technical Closures (STC), Invisible Energy Alliances (IEA), and the Intensity of Hostile Behavior (IHB). Throughout the course of these events, the principal collective actions of the ZB citizen platform in socio-technical terms are highlighted with the intensity of this flow (Table 16.1).

This table shows the three moves that shaped the Boroas CCGT power plant. The politicization of the facts was small-scale initially, being shaped without citizen influence. The redesign of the plant's refrigeration system was suggested by governmental experts and was presented as a requirement to increase information within the EIA. After the company "responded" to this suggestion, each EIA approval for the plant was reached in different administrative levels. The suggestions made in the EKOS S.L. report, requested by the municipal council, were to be rejected and ZB organized marches and protests. Between the second and third moves, resistance actions concentrated on the pronouncement of the municipal council. Despite efforts by ZB's technical teams to challenge the violation of urban bylaws and territorial planning, the urban planning regulations were amended and re-interpreted by authorities based on criteria of private interests and individual liberty.

Furthermore, the Irish Minister's visit to the Basque authorities created a backdrop to justify development plans, as well as landscape adaptation and atmospheric monitoring. These presented the results of the alternative citizen referendum as completely delegitimized by the autonomous authorities and the company itself. The final move represents the dismissal of the administrative appeals filed by ZB, along with the contradictory evidence regarding the magnitude of the spill in the local stream, and the installation of four explosive devices. These events followed the inauguration of work at the plant by former President of the Basque Government Juan José Ibarratxe.

During this course of events, presented in Table 16.1, increased levels of hostility became evident on both sides of the conflict. This occurred in the form of socio-technical closures, which were fiercely supported by politicians and the autonomous government. This situation shows the confluence of socio-technical closures regarding secret energy alliances. Here, urban planning experts and the collective action of ZB were excluded from the moves.

When the regulations were amended and the technical justification plans were applied for, with the corresponding approvals, information regarding the CCGT power plant and its environmental impacts on the land were once more *blackboxed* for the local population,[8] hostility increased outside the actual construction grounds, and complaints shifted, placing blame on the local government, which was most closely related with the Basque conflict.

16.5.2 VI.II The Sudden Cessation of the Building of the Coal-Fired Power Plant in Barrancones, Chile

Just like in Boroa, the local residents in the towns of Choros and Punta de Choros informally (Spoerer 2013) found out about the plans to build three coal-fired power plants, right in front of the Humboldt Penguin National Park in Norte Chico, Chile. The scientists that routinely drove on the only dirt road from Los Choros to Punta de Choros to observe the penguins, whales, and bottlenose dolphins, among other animals, were alerted by local residents of the national energy plans for the zone.

The people born in Punta de Choros practise traditional, artisanal fishing, as well as intensely focusing on agricultural activities and tourism which is dependent on nature and the beauty of local scenery, so they did not hesitate to mobilize against what they considered a dire threat.

Although the authorities recognized the energy needs of mining activities in the region and consultants tried to placate residents by reminding them that there are small environmental impacts for every project, people were not satisfied. Thus, the energy authorities at the time, who requested the drawing up of a scientific report to measure the synergistic effects of installing the CF power plant, became nervous.

16.5.2.1 First Move: Environmental Impacts on the Eco-system Are Impossible to Simulate, But Can Be Monitored

The risks of installing three coal-fired power plants overlooking marine reserves and the National Park seemed difficult to measure. The damage from extracting and releasing water with temperatures of at least 9 °C at the rate of 27 l/s into coastal

[8] The blackboxing metaphor illustrates the invisibility of the internal mechanisms of an infrastructure or object, which can mainly be understood by its inputs and outputs. According to Bruno Latour, this occurs when science and technology is qualified as successful, noncontroversial, and a matter of fact.

waters seemed especially irreversible. Without clear-cut answers, approximately 200 inhabitants decided to protest in front of a nearby artisan fishery town called Caleta Hornos. When they decided to block the Pan-American route, the national police responded employing a great deal of repression, especially against women.

Although the environmental authorities of the region discussed the results of the scientific report in a public meeting, it was far from scientific consensus. To the contrary, participating scientists were confronted with unreasonable interpretations of the uncertainty of the data for the coastal zone. On the one hand, the project director, responsible for the introduction and conclusions of the study, acknowledged the difficulty in simulating the impacts on the eco-system, but he did suggest that the building of the thermoelectric plant would be vigilantly monitored for adverse effects on local flora and fauna. On the other hand, scientists responsible for certain chapters of the report, rejected this point of view and advised against the construction of the Coal-Fired (CF) power plant due to the lack of evidence ensuring against irreversible damages to this unique ecosystem. Finally, after a controversy involving local politicians and fishermen, the National Forestry Corporation's study at the time concluded as follows:

> No foreseeable compensatory measures exist on the Humboldt National Penguin reserve against the possible impacts of thermoelectric projects. This reserve is susceptible to being affected by thermoelectric projects; and there are no conservation agreements regarding the inconvenience of installing thermoelectric projects in sites in the Region. (Session Act 3/2009, COREMA Coquimbo: 7)

Although these conclusions probably contributed to the rejection of two of the three projects by their own promoters, the third coal-fired power plant leader for the transnational company GDF Suez Energy continued its EIA, even using some of the evidence from the study in question in its favor. Members of the Environmental Defense Movement,—*Movimiento de Defensa del Medioambiente*, MODEMA in Spanish—reported to the court having received anonymous death threats via texts messages and phone calls.

16.5.2.2 Second Move: Environmental Impact Assessment Approved Without a Marine Permit

The citizen platform MODEMA was aware that the plant could not be built without a marine permit, so they did everything they could to assert their influence in political lobbies to ensure the withholding of the permit. The regional sub-secretary for marine affairs at the time resisted pressure from his colleagues in the Party for Democracy—*Partido Por la Democracy* (PPD) in Spanish—and decided not to grant the permit.

However, the possibility for the company to delay submitting their responses to the administrative authorities in the process of an EIA is an example of the leeway allowed to companies so that they can redesign original projects, in addition to the possibility for business people to await the political climate that would be more favorable to their projects.

Along with the change in governmental authorities in 2010, there was also a change in the authorities in the regional ministry. Thus, some of the technical reports that rejected the granting of the permit, later supported the construction of the plant.

After dealing with the aftermath of the earthquake of February 2010, the new authorities were visited by the Suez Energy manager for Latin America, Gerard Mestrallet. On behalf of the largest energy consortium in the world, he promised to continue supporting Chile's development through the country's choice of energy project mix. Representatives from NGO's like Oceana Chile criticized this stance, claiming that the zone had enough scientific evidence to be considered a multi-use protected zone, and that for 7 years the environmental NGO Terram had required that thermoelectric regulations be left to congress.

The Chilean EIA process and the submission of technical reports resulted in a consolidated technical report compiled by the director of the relevant administrative sector. This report was given to political policy makers, who also reworked the information to fit their priorities. Thus, the change in authorities, particularly in the National Forestry Commission, was crucial for receiving EIA approval. This proves just how closely socio-technical closure is intertwined with political opinions and regulations throughout the EIA process.

16.5.2.3 Third Move: The Dissemination of a Public Official's Text Message and Massive Spontaneous Demonstrations Throughout Chile

While the previous moves relate to the *Salvemos Punta de Choros* (SPC) campaign, they became a visible network in coordination with citizen platform MODEMA members. Interest in bird and marine plant diversity in the area attracted public figures such as Celine Cousteau. The presidential candidate at the time and future president, Mr. Sebastián Piñera, was also attracted to the issue and came to the area to scuba dive and appreciate the beauty of the local scenery. Similarly, a group of surfers and young people, some of whom were knowledgeable in marketing and advertising techniques, joined in the resistance to the building of the plant construction. This group compiled abundant audiovisual material and created a documentary called "Chao Pescao", used to effectively disseminate the harmful effects of coal-fired power plant technologies on the area. The members of SPC, always with their video cameras in hand, were involved in campaigning for years, acquiring expertise in publicity and marketing for group action. One particular scene that they disseminated captured one of the officials of the Regional Environmental Commissions referring to the activists as "*hippies de mierda*" ("shitty hippies") in a text message, which specifically sparked public outrage. Mass media journalists, who had previously shown little interest in the issue, were now also calling themselves "shitty hippies".

By this time, people all over the country were not hesitating to protest publically. It was the largest simultaneous mass protest that Chile had seen to date and was achieved through cyber-activism. The term "thermoelectric" became a

Table 16.2 Moves to situate a coal-fired power plant in Barrancones

MOVE I	Environmental impacts on the eco-system are impossible to simulate but can be monitored
STC	Controversial presentation of scientific report
IEA	Scientific controversy. Unconsidered risks in EIA. Secrecy of authorities
IHB	High. MODEMA alerts scientists of risks. Strong protests and fierce police repression of protests. Death threats to MODEMA members
MOVE II	EIA approved without marine permit
SCT	EIA approval of the CF power plant
IEA	Political lobbying of sub-secretary, visit by Suez energy CEO to representatives
IHB	Low. SPC campaign. MODEMA lobby
MOVE III	The dissemination of a public official's text message and spontaneous mass demonstrations throughout Chile
SCT	Abandonment of the construction of the CF power plant
IEA	Presidential phone call to suggest re-location of CF power plant
IHB	Medium high. Campaign on-line and offline. Thermal power plant as a "trending topic." Simultaneous mass street protests. "Shitty hippies". Text messages

"trending topic" on Twitter on August 26, 2010. The reaction was so strong that the authorities were given no choice but to react. Even former President Piñera called Gerard Mastrellet to suggest modifying the location of the plant. He later published on his Twitter account, "we have achieved a great solution enabling us to protect the nature sanctuary of Punta Choros, Isla Damas and Gaviota for future generations" (Table 16.2).[9]

The peak in Move I corresponds with the controversial presentation of UCN's scientific report, along with the subsequent development of the Barrancones project without considering the risks that the report had highlighted. This rise evened out as soon as there was a change in government in response to the regional authorities' secrecy, under the influence of the visit from the Suez Energy CEO and the environmental approval of the project, still without being granted permission by the marine authorities. In Move III, hostility rises again when, because of a public official's carelessness, the public learned of the "shitty hippies" text, which resulted in simultaneous mass protests and media intervention. At this time, former President Piñera called the CEO of Suez Energy Latin America to suggest re-locating the plant, a request which was ultimately rejected by its owner.

The project of this CF power plant shows, on the one hand, limited politicization of facts and, on the other hand, how levels of hostility and nonviolent actions were controlled by MODEMA and its SPC campaign in order to maintain the conflict in the public arena, thus avoiding its criminalization. Invisible energy alliances were broken down following this presidential decision, which paved the way for criticism and calling for improved environmental and energy institutions.

[9] Sebastián Piñera's personal twitter account. Available at: https://twitter.com/sebastianpinera/status/22189173783. Accessed March 10, 2015.

16.6 Discussion of Hostile Situated Infrastructures

Public rejection of both projects was driven by the negative local effects of building these types of thermal power plants. The sources of information regarding the power plant for concerned local residents were informal, namely through their networks of personal contacts and local environmental groups. Therefore, local inhabitants were surprised and were not invited to participate in modifying the plans for local development necessary for the projects to come to fruition. Thus, the system builders appeared to be working behind their backs, which paved the way for resistance to the project and its leaders.

The socio-technical complexity of ACBC regulation enabled the EIAs to be approved separately at different administrative levels, some corresponding with the Central Government of Spain, and some with the ACBC, roughly over a 3-year period under the same president. The EIA approval in Barrancones took about the same amount of time, in contested administrative stages of local, national and Regional Environmental Commission, with two presidents in charge and prolonged by Suez Energy's late submission of their *addendum*. After the approval of the EIAs for the plant, ZB continued with its strong allegations directed to the local government, which requested three plans that camouflaged the building of the plant (Baigorrotegui 2015). On the other hand, for MODEMA, the local council of Higuera only represented a docile institution that would abide by regional and hence national opinion. The national courts of justice were the next arena to continue the fight. ZB did not take mobilization beyond the opinion of the nationalist authority.

When facing both national and autonomous citizen platforms, governmental agencies and institutions demonstrated different mechanisms to generate hostility and different levels of unfriendliness in the confrontation. However, the contingency between the events and the arrogant actions and lack of care on the part of certain public officials added to collective public actions. In Boroa, their way of rejecting the options of a referendum in Amorebieta was crucial, while in Barrancones the change in verdicts in the technical reports was linked to the change in authorities, and the contempt with which public officials referred to MODEMA demonstrated how aligned the invisible energy alliances really were with the idea of progress by means of these technologies.

16.7 Conclusions

Local opposition in Chile and the Basque Country, Spain, was not expected. The definition of who, when and which regulations were considered controversial arenas rather than preset data. Struggles for these definitions re-signified the meaning of local, going beyond geographical boundaries, in a hostile manner in both experiences, but with different outcomes.

In general, we could recognize that the evidence of the so-called secret energy alliances influences socio-technical closures. This, in part, corroborates the assumptions of post-normal science (Functowicz and Ravetz 1993), especially when there are high stakes and the environmental risks are evaded in EIAs.

The level of hostility is linked to hidden risks in official socio-technical closures Although the scale of hostility differs from case to case, in both, the type of communication of the authorities with the citizen platforms involved was decisive in increasing its intensity.

We found that hostility levels in both confrontations were lowest while waiting for the final EIA verdicts. This proves that citizen platforms articulate their collective actions with great attention to these processes. From that point onward, formal institutional actions continue to be important in managing degrees of hostility and, for platforms to work, they must be flooded with alternative mechanisms of visibility. The results of these actions were to decrease the speed of official socio-technical closures.

Similarly, when alternative technical reports postponed official socio-technical closures, the authorities did not facilitate their dissemination (Boroa) or accept biased opinions (Barrancones).

The arguments of conservation scientists and some governmental technical experts criticizing final EIA reports were visible on the street and were also used in EIA citizen allegations. They became dissident (Delborne 2008) or activist-scientists (Flood et al. 2013) due to their disagreement with the official reports. The international dimension of these associations, through personal contacts or NGO representations, was more relevant in Barrancones than in Boroa.

EIA mechanisms may lead to transparency processes, while opening the window to new information and socio-technical justifications (Barry 2013). The fact that local demands and resistance were distanced from EIA procedures influenced the increased hostility and declining socio-technical arguments about the projects.

Finally, we present in which ways buildings of energy infrastructures become hostile infrastructure sittings. In both experiences across non manifest alliances between energy policy members and transnational energy business actors acting as the hard side of socio-technical negotiations (Summerton 1994, 1992), particularly evident for citizen platforms and grassroots groups.

References

Aalto, P., & Korkmaz, T. (2014). European energy security: Natural gas and the integration process. *Journal of Common Market Studies, 52*(4), 758–774.

Albornoz, M., & López Cerezo, J. (Eds.). (2010). *Ciencia, Tecnología y Universidad en Iberoamérica*. Buenos Aires: Eudeba.

Baigorrotegui, G. (2008). Gobernanza y participación en energía. Modos alternativos para la mediación y la deliberación socio-técnica. In F. Gómez & J. A. López Cerezo (Eds.), *Apropiación Social de la Ciencia* (pp. 215–235). Madrid: Biblioteca Nueva.

Baigorrotegui, G. (2015). Socio-environmental conflict in the construction of a power plant in Spain's Basque Country: Strategies, expertise and certifications, local environment. *The International Journal of Justice and Sustainability, 20*(1), 114–129.

Baigorrotegui, G., Parker, C., & Estenssoro, F. (2014). Visiones sobre los tránsitos socio-técnicos hacia patrones de consumo sustentable en agua y energía en la minería sudamericana – dos controversias argentinas. *Sociologías, 16*(37), 72–111.

Barandiarán, J. (2015). Chile's environmental assessments: Contested knowledge in an emerging democracy. *Science and Culture*. https://doi.org/10.1080/09505431.2014.992332.

Barry, A. (2013). *Material politics. Disputes along the pipeline*. Chichester: Wiley.

Collins, R. (1975). *Conflict sociology. Towards an explanatory science*. London: Academic.

Delborne, J. (2008). Transgenes and transgressions: Scientific dissent as heterogeneous practice. *Social Studies of Science, 38*(4), 509–541.

Devine-Wright, P. (2009). Rethinking NIMBYism: The role of place attachment and place identity in explaining place-protective action. *Journal of Community and Applied Social Psycology, 19*(6), 426–441.

Estévez, B., & González, M. (2005, September). Participación, comunicación y negociación en conflictos ambientales: Energía eólica marina en el mar de Trafalgar. *Arbor, 715*, 377–392.

Flood, M., Martin, B., & Dreher, T. (2013). Combining academia and activism. Common obstacles and useful tools. *Australian Universities Review, 55*(1), 17–26.

Frickel, S., Gibbon, S., Howard, J., Kempner, J., Ottinger, G., & Hess, D. (2010). Undone science: Charting social movement and civil society challenges to research agenda setting. *Science, Technology and Human Values, 35*(4), 444–473.

Functowicz, S., & Ravetz, J. (1993). Science for the post-normal age. *Futures, 25*(7), 739–755.

Galtung, J. (1996). *Peace by peaceful means. Peace and conflict, development and civilization, International Peace Research, Oslo*. London: Sage Publications.

Hård, M. (1993). Beyond harmony and consensus: A social conflict approach to technology. *Science, Technology and Human Values, 18*(4), 408–432.

Hess, D. (2007). *Alternative pathways in science and industry. Activism, innovation, and the environment in an era of globalization*. Cambridge: MIT Press.

Hess, D. (2015). Public as treats? Integrating science and technology studies and social movement studies. *Science as Culture, 24*(1), 69–82.

Howe, C., Lockrem, J., Appel, H., Hackett, E., Boyer, D., Hall, R., Schneider-Mayerson, M., Pope, A., Gupta, A., Rodwell, E., Ballestero, A., Durbin, T., el-Dahdah, F., Long, E., & Mody, C. (2016). Paradoxical infraestructures: Ruins, retrofit and risk. *Science, Technology and Human Values, 41*(3), 547–565.

Hughes, T. (1993). *Networks of power. Electrification in western society 1880–1930*. London: Johns Hopkins University Press.

Larrinaga, J. (2003). Zornotza Bizirik: Ez Gure Etxe Globalaren Baratze Lokalean. No en el huerto local de nuestra casa global. In I. Bárcena (Ed.), *Globalizazioa eta Inguramena: Euskal begirada bat*. Bilbao: UEU.

Larrinaga, J. (2006). Nimby power: el referéndum sobre la central térmica de Amorebieta y la potencia de la negación. In J. Encina & I. Bárcena (Coord.), *Democracia ecológica: Formas y experiencias de participación en la crisis ambiental* (pp. 217–237). Sevilla: UNILCO.

Martínez Alier, J. (2003). *Deuda ecológica: ¿Quién debe a quién?* Barcelona: Icaria.

McCormick, S. (2007). The governance of hydro-electric dams in Brazil. *Journal of Latin American Studies, 39*, 227–261.

McCormick, S. (2009). *Mobilizing science. Movements, participation, and the remarking of knowledge*. Philadelphia: Temple University Press.

ME, Ministerio de Energía. (2012). *Estrategia Nacional de Energía 2012–2030*. Energía para el futuro: Limpia, segura y económica del Gobierno de Chile. http://portal.mma.gob.cl/wp-content/uploads/2014/10/3_Estrategia-Nacional-de-Energia-2012-2030_Energia-para-el-Futuro.pdf. Accessed 3 Aug 2016.

Mitcham, C. (Ed.). (2005). *Encyclopedia of science. Technology and ethics*. Detroit: Macmillan Reference USA.
Pacey, A. (1983). *The culture of technology*. Cambridge, MA: MIT Press.
PE3E. (2010). *Hacia un desarrollo energético sostenible*. Estrategia Energética Euskadi 2010: Política Energética Vasca, Ente Vasco de la Energía -EVE-. http://www.eve.eus/Publicaciones/Planes-Energeticos/Estrategia-Energetica-Euskadi-2010-Hacia-un-Desarr.aspx. Accessed 3 Aug 2016.
Pellizoni, L. (2011). The politics of facts: Local environmental conflicts and expertise. *Environmental Politics, 20*(6), 765–785.
Rajan, R. (2005). Science, state and violence: An Indian critique reconsidered. *Science as Culture, 14*(3), 1–13.
Rodríguez-Giralt, I. (2011). Social movements as actor-networks: Prospects for a symmentrical approach to Dolana's environmentalist protests. *Convergencia, 56*, 13–35.
Saguier, M. (2012). Socio-environmental regionalism in South America: Tensions in new development models. In P. Riggionizzi & D. Tuddie (Eds.), *The rise of post-hegemonic regionalism* (pp. 125–145). United Nations University Series on Regionalism 4. Springer. https://doi.org/10.1007/978-94-007-2694-9_7
Sepúlveda, C. (2000). El proyecto Celulosa Valdivia: La legitimidad del Sistema de Evaluación de Impacto Ambiental puesta en juego. In F. Sabatini, C. Sepúlveda, & H. Blanco (Eds.), *Participación ciudadana para enfrentar conflictos ambientales –Desafíos para el Sistema de Evaluación de Impacto Ambiental* (pp. 2–23). Santiago: Ediciones LOM.
Spoerer, M. (2013). Participación ciudadana e incidencia política: Estudio del caso Barrancones. *Persona y Sociedad, XXVII*(1), 17–44.
Stankiewicz, P. (2009). The role of risks and uncertainties in technological conflicts: Three strategies of constructing ignorance innovation. *The European Journal of Social Science Research, 22*(1), 105–124.
Summerton, J. (1992). *District heating comes to town. The social shaping of an energy system*. Linköping: Department of Technology and Social Change.
Summerton, J. (1994). *Changing large technical systems*. Boulder: Westview Press.
Ureta, S. (2014). Normalizing Transantiago: On the challenges (and limits) of repairing infrastructures. *Social Studies of Science, 44*(3), 368–392.
Urkidi, L., & Walter, M. (2011). Environmental justice dimensions in anti-gold mining movements in Latin-America. *Geoforum, 42*, 683–695.
Walker, G. (2010). Environmental justice, impact assessment and the politics of knowledge: The implications of assessing the social distribution of environmental outcomes. *Environmental Impact Assessment Review, 30*(5), 312–318.
Zubiaga, M. (1995). Euskal herriko mugimendu ekologistaren ingurune politikoa. A modo de extracto soluble. In I. Bárcena, P. Ibarra, & M. Zubiaga (Eds.), *Nacionalismo y ecología. Conflicto e institucionalización en el movimiento ecologista vasco* (pp. 178–186). Madrid: Los Libros de la Catarata.

Chapter 17
The Challenge of Transversal Education Through Teaching Ethics in Engineering: From Hubris to Hybrid

Emilio Muñoz, Ana García-Laso, and Domingo Alfonso Martín-Sánchez

Abstract Throughout the twentieth and twenty-first centuries, Spain has remained backward as regards the modern scientific policy initiatives developed in Europe and the United States. One of the challenges has been to introduce the ethical perspective in studies of a technological nature within a framework within which a culture of hubris evolved in the scientific and technological domain. In this context, this chapter is structured in two parts. The first develops the theoretical framework, i.e., scientific policy in Europe, with special focus on Spain. Based on the innovations developed by the authors at the School of Mining and Energy, Technical University of Madrid, the second part describes the experimental exercise of an integrative learning process carried out by introducing ethics and transversal skills in Engineering Studies.

17.1 Philosophy of Scientific Policy

17.1.1 The Origin and Evolution of Science Policy

Some thinkers and analysts attribute the origin of modern scientific policy to the initiative resulting from the collaboration between US President Roosevelt and his scientific advisor, the engineer Vannevar Bush, which gave rise to the historical

E. Muñoz (✉)
Unit of Scientific Culture Research, CIEMAT, Madrid, Spain
e-mail: emilio.munoz@ciemat.es

A. García-Laso · D. A. Martín-Sánchez
UESEVI, E.T.S.I. of Mines and Energy, Polytechnic University of Madrid, Madrid, Spain
e-mail: ana.garcia.laso@upm.es; domingoalfonso.martin@upm.es

B. Laspra, J. A. López Cerezo (eds.), *Spanish Philosophy of Technology*, Philosophy of Engineering and Technology 24,
https://doi.org/10.1007/978-3-319-71958-0_17

document *Science, the Endless Frontier*, reprinted by the National Science Foundation (Bush 1960). However, Emilio Muñoz has put forward the thesis of a European origin for this policy. This origin was linked to World War I (the Great War, 1914–1918), especially the postwar period; i.e., he places the beginnings of modern science policy in the first third of the past century. This thesis has been advocated in a number of publications (Muñoz 2014a; Red CTI 2008) and has its roots in the book *Pour une politique scientifique* (Piganiol and Villecourt 1963).

In this period, essentially following World War I, Europe recognized that the industrial revolution and industrial society were the drivers of the evolution of science and technology. Therefore, that century and half ago, Europeans were living within a scientific civilization which expressed the need to establish a science policy or scientific policy in terms of competitive strategy. This political strategy acknowledged from the very start the convenience of a commitment to science as a basic element for social and economic development. Besides this commitment, an essential component of the political discourse was the understanding of the private sector as an entrepreneurial driver and executor, as well as that of policy decision makers as agencies of promotion and management.

After World War II, the United States assumed the leadership of modern scientific policy through the aforementioned report, *Science, the Endless Frontier*, whose author, Vannevar Bush, presented it to President Truman as a program for postwar scientific research. The report has been considered a sort of foundational act of modern scientific policy that has attained historical value and has merited a great number of recognitions and related papers. The report recognizes the essential role of science and scientists in winning the war and proposes a "policy for science"; i.e., a policy which considers public investment as a critical instrument for the development of science that should be managed by the scientific community itself.

From this view of the promotion of scientific development was to emerge the Mertonian perspective of the sociology of science (Merton 1985). In addition to this strong commitment to the promotion and autonomy of scientific activity, the Bush report also considered that the applications and uses of science and advances in technology would gain value through the manufacturing world. Some analysts have seen in this declaration the origin of what was later to be called – at a time of European concern regarding the utility of science – the "linear model of innovation". This model follows the sequence: scientific knowledge → technological development → innovative products (placement on the market). This model was successful and continued to gain ground until achieving broad social recognition, bringing about statistical analytical approaches with the creation of the concepts of R&D and their entry into the laboratories of big firms.

In the international arena, the United States led the creation of the Organization for Economic Cooperation and Development (OECD). In this period, Europe main aim was to rebuild and, while pursuing this goal, it incorporated the North American view of scientific policy. During this process, the distinction between three elements of the concept of R&D became widespread in the sphere of knowledge production: (1) basic or fundamental science, the object of which is to contribute to the increase of knowledge; (2) applied science, aimed at acting with and forming part of the technical and business world; and (3) technology, which may serve as a driver for knowledge production or be the result of scientific development.

This strategy of division is closer to European mainstream thinking rather than to that of the US and yielded positive effects both in the organizing and structuring of mechanisms for fostering research and infrastructures. It also had side effects, however; problems triggered by its tendency towards fragmentation and compartmentalization. This took place when it was necessary to bring together and integrate knowledge production agents and knowledge users, when the strategic goal was to link academia and business.

However, this model of scientific policy, this social contract on science, began to be questioned as a project of the European Union in the 1970s as a consequence of the energy crisis and Europe's constant concern about the failure of European firms to profit from scientific knowledge. The reaction was an economic viewpoint that developed a lack of trust in science and the scientific community as a factor of economic development. This resulted in a period of change in European scientific policies throughout this period of hesitancy and preoccupations, eventually giving rise to the diffuse concept of innovation, which rapidly gained relevance and was supported via substantial investment (Muñoz 2009).

17.1.2 Science and Technology: Their Characteristics, Pathways and Dynamics

Scientific policy has thus traditionally focused on the promotion of science and only under complex and critical situations has it evolved to seek out the link between science and technology. In current contexts, there is recognition of the close relationship between science and technology, but this has not been the common pattern throughout recent human history.

The natural sciences were initially associated with philosophy, subsequently evolving through bottom-up strategies fostered by elites interested in the search for knowledge about nature. This led to a process of diversification, similar to that of speciation in evolutionary biology. Scientific disciplines such as physics, chemistry, and biology emerged and also underwent processes of diversification that generated sub-disciplines or novel disciplines, as well as others generated via hybridization.[1] On the other hand, those scientific areas associated with the survival of humankind through adaptation to the environment, such as the agricultural and biomedical sciences, followed evolutionary pathways linked to civilization, to cultures, as was also the case of mathematics. Running parallel to the development of pathways that seek forms of production based on diversified scientific knowledge, the scientific disciplines encouraged political and socio-economic recognition. Perhaps the critical point was the creation of the Nobel Prizes in 1900, pursuant to the last will and

[1] Hubris is exaggerated pride or self-confidence; a hybrid is something of mixed origin and combined properties, formed or composed of heterogeneous elements.

testament of the Swedish scientist and inventor Alfred Nobel. The culture of hubris of the scientific community thus reached its consecration with the Nobel Prize.

Technologies may share bottom-up origins with experimental sciences, although their evolution has been different. They co-evolved with civilizations as a reaction to the needs and environments for social developments made by adaptive mechanisms for survival, thus emerging in association with particular cultures of each period of development of humankind and its corresponding societies. Whatever the case may be, technologies have developed according to auto-poietic processes profiting from flow rates of evolution and the hybridization potential in order to achieve implantation and survival. Today, the Information and Communications Technologies (ICTs) and biotechnologies are the visible technological manifestation of our own society (Muñoz 2016).

17.1.3 Spain and Its Pursuit of a Modern Science Policy

For centuries, Spain has pursued scientific and technical development as a coveted trophy to incorporate it into its cultural heritage and national identity. This aim has proved unattainable for several reasons. One of these may be because its industrial and political revolutions did not take place at the key historical moments of neighboring countries. In any case, what sums up and goes beyond all these reasons is the fact that the foreign spark required to light the evolutionary fuse from outside has failed. For instance, we learned from evolutionary biologists in the summer of 2015 that the upsurge of oxygen was the cause of the Cambrian explosion. Those who frequently use evolutionary biology as a tool for socio-political analysis have a moral obligation to dismantle the wild Darwinian postulates arising from the neoliberal doctrine about the belief that evolution is driven to the selection of the strongest (Desmond and Moore 1991) These postulates are so often mistakenly used by some of our politicians and economists that there is a demand to change their mind sets regarding current ideas that the 'winners' in evolution are those better able to adapt: cooperation, altruism and culture being positive factors for such a process. As Spanish Nobel Laureate Ramón y Cajal (1889, p. 52) wrote when referring to scientists, "In a favorable environment, even the dissident feels his forces are growing; a hostile or indifferent environment knocks down the best tempered mood". Spain today needs a peremptory and urgent upsurge of oxygen in order to implement the pressing scientific revolution so that it may become a mature, modern and competitive country. However, the budgets and economic criteria implemented after the Civil War in 1936, throughout the twentieth century and since the beginning of the twenty-first century have followed a continuous process of up and downs, a kind of sine curve. Dramatically, the government that came into power in 2011 in the midst of the economic crisis seemed, on the contrary, to be inclined towards suffocating the R&D and Innovation system. Perhaps they thought that the Cambrian explosion can wait.

17.2 Ethics in the Transversal Curriculum in Engineering Studies

The aim of our strategy to develop an innovative transversal teaching curriculum at the School of Mining and Energy (Spanish acronym, ETSIME), Polytechnic University of Madrid (Spanish acronym, UPM), was to contribute to counteracting certain trends depicted in the previous reflection. The interdisciplinary and intergenerational team formed by the authors was the fruit of a challenge faced by Emilio Muñoz when he was invited to head a unit to incorporate these studies at the ETSIME. This invitation took place in 2006, within the framework of an ill-defined science policy in Spain and in a paradoxical socio-economic context. At the time, Spaniards lived under the illusion of a huge bubble generated by speculation in the building sector (easy money, lots of work) financed by expensive credit provided by banks to the housing sector. Meanwhile, a new crisis was detected in the highly developed countries precisely caused by the errors of financial institutions. This fact coincided with the concerned interest of the (renowned) usually privately-owned Business Schools in Spain to introduce ethics in their emergent curricula, following the recommendations of North American experts from highly reputable institutions such as the Massachusetts Institute of Technology. These gurus began disembarking in Spain – on the eve of the US crisis – to lecture, give interviews and be the subject of reports in economic newspapers and in the business supplements of mainstream newspapers.

Therefore our approach had to be different. We sought an approach based on a scientific perspective on the issue and sustained by the scaffolding of an innovative philosophy of science. To be innovative, the philosophy of science should focus on sub-areas and/or exploratory lines such as the philosophy of scientific policy and must grow on the theoretical and quasi-experimental model offered by the theory of evolution in its most integrative and current version. To embark on this intellectual adventure, it was deemed necessary to abandon the hubris resulting both from the prevailing neoliberal economy applied in recent decades, from those engineering graduates who had benefited from economic and social recognition during the crisis, and from those scientists belonging to science disciplines seeking to impose their trends and patterns of excellence throughout the field of knowledge production.

This change of course was conceived to lead to a pathway providing strategies for hybridization and cross-breeding. By chance, the three authors belonged to different fields of study (chemical and pharmaceutical sciences, telecommunications engineering, and Earth sciences), coming together in a project that may be qualified as a synthesis of educational innovation and participation with a commitment to social development. This exciting intellectual experience in the field of education enabled us to achieve a new synthesis that can be summarized in the comparison between two options: one is that of the strategic type of policy to confront top-down and bottom-up views; the other is that of applied moral philosophy to contrast the cultures of hubris and hybridization.

17.2.1 The Evolutionary Course of the Educational Program

Our proposal to introduce ethics in transversal educational activities at an engineering school emerged following a top-down approach commissioned by the head of the ETSIME. The historical and cultural tradition of ethics rendered the evolutionary viewpoint a commendable way to explore its incorporation in an innovative educational program. Subsequently, the proposal was enhanced by taking into account the history of the subject as well as the backgrounds of those who were to develop and implement the teaching program.

As none of those involved in the proposal held a degree in philosophy, it was considered appropriate that the first theoretical aspects of the ethics courses should be imparted by philosophers who had taught and researched in the field of studies on Science, Technology and Society (STS studies). Concepts related to practical issues incorporated in the curricula of Business or Management School, such as corporate social responsibility, the social economy and economics, the base of the pyramid, and social innovation initiatives, were also programmed as lectures to be given by external experts. The option was thus to hybridize scientific disciplines with social and economic issues and a mix of lecturers belonging to different fields in the tutorial action plan.

As the theory of evolution serves as our guiding principle, we were convinced from the beginning of the teaching program that it should evolve under the influence of the reactions and appraisals of the students involved. The program thus moved progressively from a top-down to a bottom-up approach, in transit from a scenario of hubris triggered by isolation to a hybrid ‘space’ resulting from cooperation.

17.2.2 The Teaching Program and Its Ethical Dimensions

In 2006, the ETSIME received international funding to introduce ethical dimensions throughout the teaching program in professional degree courses. This commitment led to the creation of a Chair in Ethics and Values as a first objective, subsequently followed by the creation of an “Ethics in Science and Engineering” course as a curricular experience to test the impact of this academic discipline among undergraduates and even among graduate students. The course was built around the criteria and strategies listed below.

First, the basic concepts were borrowed from Aristotle and his book *The Nicomachean Ethics,* with special focus on his sentence, “Each art and each question and in a similar manner each action and each pursuit are designed to get some good; and for this reason, the good has been recognized, in the correct form, as the goal towards which all things are directed” (Book I, Chapter I – in Bartlett and Collins 2011). The selection of the primary readings and contents to serve as the foundations for our course was performed by examining the works and theories of a number of philosophers of technology and education such as Ian Barbour, Carl

Mitcham, John Dewey and, especially, Hans Jonas and his ethics of responsibility (Jonas 1995, 2000).

Second, entry into the field of applied ethics took place via the question: Why embed ethics in science, technology and engineering? An initial answer arose through the awareness of the amazing scope of developments in science and technology. A second answer came from the STS studies approach, as a way of understanding social attitudes and engagement regarding science and technology. Finally, a third answer emerged from the comparison of the social challenges derived from the concepts of the "risk society" and "governance", as applied to the field of science and technology (Muñoz 2006).

These three answers enabled the development of an integrative framework for introducing new ethical dimensions and concepts such as *interethics*[2] and *spaces*[3] (Muñoz 2006, 2008). The concept of interethics relies on a consequentialist view of ethics, a more focused view than that of the broader viewpoint of utilitarianism. Interethics attempts to determine the good or evil resulting from the consequences of an act by exploring case-by-case values, costs, and benefits and is preferentially based on possessive values (i.e., values that may be quantified). Nevertheless, we went one step further, taking into account the 'spaces' in which the ethical dimensions and options are to be applied or even appraised. In a global and highly complex socio-economic space, such as the one established by the global world where a great variety of actors operate, the concept of 'interethics' should be studied and applied within a milieu of selected and predetermined values according to the issues or problems being addressed. Taking into account the current socio-economic situation, we considered the following values: the right to know the truth, regulation in and for the exercise of freedom, empathic capacity, the pursuit of justice, reactions to the generation of poverty and inequality, and the recovery of equity.

17.3 The Pathway to Teaching Social Skills in Engineering Studies

A major issue raised by the new course was the development of ethically-based social skills. Two concerns were taken into account in relation to this issue. First, surveys conducted both by the Polytechnic University of Madrid and professional associations showed that graduates who had followed recent curricula reached an adequate level of knowledge in technical skills, but showed significant gaps in terms of social skills (Martín-Sánchez and García-Laso 2014). Second, former approaches

[2] The notion of interethics, or interrelated ethics, is a concept that goes beyond the more traditional concept of deontology in the sense of overcoming a rigid ethics based on fixed principles, as found in modern version of Luddism or extreme utilitarianism (Muñoz 2008, 2014b; Muñoz et al. 2011).

[3] 'Spaces' is a concept that contrasts with and extends the notion of systems in relation to the actors connecting processes, in terms of more openness and flexibilities for the integration of different elements.

to the teaching of ethics in professional degree courses usually taught deontological ethics, an approach presenting difficulties in complying with the requirements of complex and highly interactive processes characterizing current global, multipolar societies and their socio-economic issues. Becoming aware that the limitations to be overcome, we developed the interethics approach.

Meanwhile, with the onset of the global socio-economic crisis, the Chair in Ethics and Values lost its original funding, which resulted in an interesting organizational innovation. It was transformed into the Social Entrepreneurship, Ethics and Values in Engineering Unit (Polytechnic University of Madrid), within the Quality and Social Responsibility Division. Furthermore, there was a profound educational change at the university level, as our university took measures to better adjust to the European Higher Education Area (The Bologna Process), transforming the classic concept of teaching by lecturers to one of interactive learning in which student capabilities and performance acquire a greater role in the learning process (Martín-Sánchez et al. 2016).

The results of the experience are fairly good. The number of students has increased from less than ten to twenty as a result of increased acceptance of the subject. Moreover, the educational activity rapidly achieved very good results in student satisfaction surveys (4.5/5). The major success of this course was primarily achieved by word of mouth, reaching its maximum with the introduction and development of a second course entitled "Towards Social Entrepreneurship: Service-Learning Practice". This initiative is based on the methodology of Service-Learning (S-L), which abandons the view of students as mere recipients of the teaching process in order to understand them as protagonists of the learning process (see below). A bottom-up approach was progressively adopted in the continuous evolution towards the acquisition of social skills by engineering graduates.

17.3.1 Doing While Learning and Teaching: An Evolutionary Dream?

With the advent of the new methodology advocated by the Bologna process, academic disciplines (courses) also came to be understood as training activities, which, in our case, are not specific to the ETSIME, but are implemented at the UPM level. These training initiatives are complementary in our case and form part of a pathway, although the curriculum designed for each of them can be considered independently and may be pursued in different courses.

The first of the training activities is called "Ethics and Values in Engineering", which is divided into three modules. In the first, the theoretical foundations of ethics are presented to students. In the second module, students receive an innovation and creativity course, with particular emphasis on the concept of social innovation. This concept integrates various innovative solutions to address the most pressing problems in our society, such as poverty, unemployment or the environment (Innovation for Social Change 2014). In the third module, which follows a bottom-up approach, students are expected to develop a Social Entrepreneurship (SE) project in which they have to detect and propose solutions for critical issues in their immediate environment. Recent projects have been submitted for evaluation either to the directorial board of the ETSIME or to external experts in social innovation from Hub Madrid.[4] The most likely option is that students will develop projects that are related to their immediate university environment, thus covering issues and demands of the ETSIME to improve student life at the university.

The feasibility of these projects is assessed by the directorial board of the ETSIME, which may also provide the necessary funds for their development if the ideas outlined in the proposals are valued positively. Some examples of these SE projects have been: "Engineer, A worker for society?", in which students proposed a greater number of social activities with credit recognition to the directorial board of the ETSIME; and "Energy Efficiency House", which involves the creation of a sustainable space that serves as a classroom demonstration of different alternative energy solutions. The most relevant example of social innovation proposed and developed by students in recent years has been the design of a training activity which has been integrated as a specific product of our university center called the "Monitor Project".

This project constitutes an addition and supplement to the activity of the Internal Quality Assurance Systems in Higher Education called the "Mentor Project", which consists of the mentoring of new students by more experienced learners. While the Mentor Project is responsible for reducing the social impact which may affect some students on entering the university, the Monitor Project focuses primarily on the leveling up of skills acquired in high school which seem necessary for their appropriate development in the first year of academic disciplines. These Mentor and Monitor projects are based on a successful methodology consisting of the peer-to-peer transmission of information and basic knowledge, once again making students the true protagonists of the educational process.

17.3.2 Service Learning: Goal and Point of Departure on the Pathway from Hubris to Hybrid Cultures

The educational methodology called S-L (Service Learning) has constituted a decisive framework for our responsible engineering students to recognize and appraise the practical dimensions of being trained in ethics and values. The structure of this activity fits our theoretical framework (Tapia 2006) and consists of three main stages:

1. *Preparation.* In this stage, students seek to detect the needs and demands of people in risk of social exclusion. Up to now, our particular targets have been unemployed immigrant women, young people dealing with prior failure at

[4] Hub Madrid is an ecosystem of social innovation present in more than 85 cities of the world made up of 15,000 entrepreneurs. It constitutes an infrastructure to support new projects that is based on three pillars: collaborative spaces, a community of professionals with a common approach, and a series of training and mentoring services to increase the impact of projects.

school, and women who are victims of gender violence. This process is carried out by combining contacts with institutions involved in solving these problems with personal interviews of those willing to participate in the exercise.
2. *Implementation.* This stage consists in the preparation and imparting of training courses that may facilitate a reduction in the social inequality found in the aforementioned groups of excluded individuals. The proposed actions take into account the facilities made available to students by the ETSIME: infrastructure and support services. Some examples are basic computer courses or cookery courses. In addition, we have sought out the cooperation of external companies which may add professional value to the project.
3. *Evaluation.* This exercise was considered necessary to monitor the outcomes of our teaching program and the performance of the trainees involved. Due to the evident complexities of assessing social skills, we developed a combination of quantitative[5] and qualitative[6] mechanisms.

Interestingly enough, the core values most consistently established by our students in these activities are responsibility, honesty, commitment, sensitivity and initiative. Here are some excerpts from their diaries as a mirror image of the values mentioned above.

> In the course of this academic activity I learnt that something that starts out very small or as just an idea can lead to great results and even change what is around us.

> People are the ones who create, change and make anything possible. Much of the success of our projects is associated with the life, strength and soul they might have.

The authors are aware of the success attained in their innovative educational and research adventure, but also recognize the existing limitations in terms of academic recognition and project funding support. In this respect, the scientific journal *Nature* published an article on June 29, 2016, that reveals that interdisciplinary projects are not funded as positively as proposals within narrow fields of study. *The Scientist* echoed the same study in its issue of the same day, stating that "An analysis of Australian Research Council data reveals grant proposals that integrate a broad array of academic fields are less likely to be funded". In our view and from our experience, this points out an important flaw in contemporary science policy influenced by a traditional, bureaucratic viewpoint. The challenge remains open.

Acknowledgments Funding for this study by Fundación Caja de Ingenieros and Fundación Gómez Pardo is gratefully acknowledged.

---

[5] As to quantitative evaluation, students are rated according to the following weighted distribution: 10% for classroom activities, 20% for a proactive attitude in class, 20% for the level of knowledge acquired while attending the course, and 50% for the SE project or for involvement in the S-L.

[6] Concerning qualitative evaluation, Ethics and S-L are valued from field diaries recording individual perceptions of their performance. This evaluation of the mandatory field diaries allows the students' own experiences to emerge in a free form. This tool has proved to provide a major insight, as it allows us to precisely determine the values that have evolved in the learning process.

References

Bartlett, R. C. &, Collins, S. D. (Trad.). (2011). *Aristotle's nicomachean ethics*. Chicago: University of Chicago Press.

Bush, V. (1960). *Science, the endless frontier*. Washington, DC: National Science Foundation.

Desmond, A., & Moore, J. (1991). *Darwin*. London: Michael Joseph Ltd.

Innovation for Social Change. (2014). http://innovationforsocialchange.org/innovacion-para-el-cambio-social/. Accessed 27 Aug 2016.

Jonas, H. (1995). *El principio de responsabilidad: Ensayo de una ética para la civilización tecnológica*. Barcelona: Herder.

Jonas, H. (2000). *El principio vida: Hacia una biología filosófica*. Madrid: Trotta.

Martín-Sánchez, D. A., & García Laso, A. (2014). Experiences in social innovation: A platform for ethics through a school of engineering studies. *Journal of Cases on Information Technology, 16*(3), 4–17.

Martín-Sánchez, D. A., García Laso, A., Muñoz van den Eynde, A., Lopera Pareja, E., Cornejo Cañamares, M., & Muñoz, E. (2016). Integrative learning toward social responsibility in teaching engineering. In W. Nuninger & J. Châtelet (Eds.), *Handbook of research on quality assurance and value management in higher education* (pp. 279–302). Hershey: IGI Global. https://doi.org/10.4018/978-1-5225-0024-7.ch011.

Merton, R. K. (1985). *La sociología de la ciencia, investigaciones, teóricas y empíricas* (Vol. 2). Madrid: Alianza.

Muñoz, E. (2006). Gobernanza, ciencia, tecnología y política: Trayectoria y evolución. *Arbor, 181*(715), 287–300.

Muñoz, E. (2008). Dinámica y dimensiones de la ética en la investigación científica y técnica. *Arbor, 184*(730), 197–206.

Muñoz, E. (2009). La crisis de la política científica: Patologías degenerativas y terapias regenerativas. A modo de epílogo. *Arbor, 185*(738), 837–850.

Muñoz, E. (2014a). La apuesta por la ciencia: Imperiosa necesidad para afrontar las crisis. *Por un nuevo contrato social. Gaceta Sindical, 22*, 269–281.

Muñoz, E. (2014b). Desde una evolución convergente hacia la emergente economía biológica: Un itinerario por la biología evolutiva, la interdisciplinariedad y las interéticas. *Dilemata, 6*(14), 165–187.

Muñoz, E. (2016). *¿Estamos ya en el siglo de la biología? Un análisis sobre sus impactos sociales y económicos*. Encuentro Tendencias Científico-Tecnológicas: Retos, potencialidades y problemas sociales. Segundo panel: Tendencias en Biotecnología y sus impactos sociales, Facultad de Ciencias Políticas y Sociología de la UNED, Departamento Sociología III (Tendencias Sociales), April, 8–9. (in press).

Muñoz, E., Flor, B., García-Laso, A., & Martin-Sánchez, D. A. (2011). Experiencias cruzadas sobre innovación social: El valor de las interéticas. In I. Díaz García & A. Muñoz van den Eynde (Eds.), *Participación y cultura científica en perspectiva internacional* (pp. 225–250). Madrid: CSIC-La Catarata.

Piganiol, P., & Villecourt, L. (1963). *Pour une politique scientifique*. Paris: Flammarion.

Ramón y Cajal, S. F. (1889). *Reglas y consejos sobre investigación biológica* (2nd ed.). Madrid: Fortanet.

Red CTI (2008). *¿Hacia dónde va la política científica (y tecnológica) en España?* (J. Sebastián, I. Ramos Vielba, & M. Fernández Esquinas, Eds.). Madrid: CSIC. http://digital.csic.es/bitstream/10261/25204/1/view.cor. Accessed 21 June 2016.

Tapia, M. N. (2006). *Aprendizaje y servicio solidario: En el sistema educativo y las organizaciones juveniles*. Buenos Aires: Ciudad Nueva.

Chapter 18
Technique and Technology in the Practice of Distance Running

Marta I. González García

Abstract Recent accounts of the development of the philosophy of technology identify a distinct empirical turn in the 1980s and 1990s characterized by a focus on specific technologies and an understanding of technology as a contingent product of a heterogeneous array of factors. In this contribution, I explore the overlaps of this so-called empirical turn with other research fields such as science studies, practice studies or even consumer research via a case study on sports technology, that of the running shoe.

18.1 Introduction

In his well-known book *Thinking through technology* (1994), Carl Mitcham divided philosophical research on technology into two broad traditions: engineering philosophy of technology and humanities philosophy of technology. Over the following decades, the field has developed and spread, rendering it difficult to sustain the distinction between a pro-technology and analytic attempt at a technological philosophy and a critical and interpretative reflection on technology itself. Bringing order to such diverse and divided research area is not an easy task, even more so because the philosophy of technology cannot be neatly separated from other analyses of technology.

Recent accounts of the development of the philosophy of technology identify a distinct empirical turn in the 1980s and 1990s (Mitcham et al. 2001; Brey 2010). While classical philosophy of technology focused largely on technology in general, new research following the empirical turn focuses on specific technologies, understanding technology as a contingent product of a heterogeneous array of factors including social aspects. This narrower perspective, centred on particular case studies, departs from traditional philosophical approaches to technology as a phenomenon. It thus avoids global deterministic stances regarding technology as a whole,

M. I. González García (✉)
Department of Philosophy, University of Oviedo, Oviedo, Spain
e-mail: martaig@uniovi.es

B. Laspra, J. A. López Cerezo (eds.), *Spanish Philosophy of Technology*,
Philosophy of Engineering and Technology 24,
https://doi.org/10.1007/978-3-319-71958-0_18

as well as both indiscriminate technophilia and technophobia. This so-called empirical turn shares a great deal of cross-fertilization with other research fields such as science studies, practice studies or even consumer research (Ingram et al. 2007).

Where these research traditions overlap, it is possible to track the evolution of technologies understood as complex systems of materiality, practices, and meanings in context. In this chapter, I will explore these overlaps via a case study on sports technology, namely that of the running shoe. The birth of the modern running shoe can be related to the social boom in running culture that commenced in the 1960s and 1970s and underwent a second revolution around the turn of the century. Since then, as a technological artefact, the running shoe has been transformed from a rather simple protection of the foot to a high-tech tool. This case study has an additional element of interest. The recent emergence in recreational running of an "anti-technology" movement that defends the benefits of barefoot running may be understood as the advent of a new social group that has destabilized the technological trajectory of running shoes, promoting the introduction of new lines of research and technological innovation.

Running cultures and running shoes are being simultaneously produced in this case study, in a process that challenges extensionist views of technology, displays the co-evolution of humans and non-humans, and shows the interchanging roles of experts and lay users. The history of the running shoe could thus provide some insights into the evolution of practices, technologies, and users in "context", bringing together different approaches involved in the empirical turn in research on technology.

18.2 The Empirical Turn and Its Consequences

Bicycles may be a good starting point for discussing the empirical turn and its consequences. In 1984, they were the case study chosen by Wiebe Bijker and Trevor Pinch to present their Programme on the Social Construction of Technology (SCOT). Stabilized technologies, they argued, are the contingent product of complex processes of variation and selection, in which social groups and their interests shape artefacts by selecting among competing designs for a given technology in construction. In their seminal paper, Bijker and Pinch (1984) took the air tyre as an example of the power of dominant social groups to impose their concerns and definitions of the artefact, i.e., their power to shape technology.

However, bicycles were also used as a case study to point out some limitations of the original SCOT programme. In their insistence on relevant social groups, Pinch and Bijker not only forgot about "irrelevant" social groups (Winner 1993), but also about social and cultural meanings at large. A decade after the appearance of bicycles on the science studies map, Paul Rosen (1993) returned to them, arguing that the "invention" of mountain bikes around 1980 reveals some shortcomings of the SCOT approach as regards the nature and agency of social groups and the wider social context. His main problem with Pinch and Bijker's approach is their simplistic account of society, made up of underanalysed "relevant social groups" and a never reached "third stage" of wider social factors. In Rosen's reconstruction of the history of the mountain bike – its origin, development, and current precarious stabilization–, he maintains that the "wider social context" is essential to understand both the shaping of technology and the nature and dynamics of the "relevant social groups" involved. Social context, Rosen argues, cannot be a final and secondary step, because social groups and their problems cannot be accounted for without it. Somehow paradoxically, according to Rosen, SCOT analyses tend to focus on the inner workings of technology, thereby losing the crucial information that its social context can provide.

In Rosen's view, social and cultural meanings, marketing strategies and economic forces show up as relevant factors in the modelling of technologies. Born out of the drive and creativity of a bunch of Californian young bikers in the 1970s, mountain bikes have developed from an artisanal artefact made from junk parts of old bicycles for their mountain adventures into a complex technological product made up of parts manufactured all over the world. In modern mountain bikes, meanings of nostalgia and adventure are combined with a post-Fordist industry that promotes decentralization and constant innovation, making stabilization improbable. In this new context, marketing strategies gain relevance. The mountain bike industry creates at one and the same time both products for markets and markets for their products, selling nature – what Rosen (1993, p. 499) calls "de-modernizing themes of frontiers, pioneers, adventure, wilderness…" – through technology. These strategies give rise to new social groups of mountain bikes enthusiasts attracted by appeals to nostalgia and youth. Mountain bikers, including those who use their bikes mostly to move around the city, subverting the original technological script, are constructed in the same processes as the artefact itself.

While the original SCOT approach focused on stabilization, modern mountain bikes are only partially stabilized artefacts. Their design is continuously changing in order to attract consumers and produce increasingly specialized bikes. As new type of bikes for new social groups emerge at each step, diversity rather than stabilization becomes the norm (Rosen 1993). Moreover, following post-Fordist trends, bicycle production is nowadays radically marked by delocalization. Design, frames, and other components are manufactured and assembled at different geographical locations, giving rise to political and economic issues well beyond the original SCOT approach.

The "seamless web" (Hughes 1986) of mountain bikes and their social, cultural, and economic context presented by Rosen becomes even broader-ranging as later analyses explore the co-production of bike technology and bike culture (Rosen 2002). A rather simple technological artefact, like a bicycle, is embedded in complex "sociotechnical ensembles" (Bijker 1995),[1] composed of social and material

[1] According to Bijker, the "sociotechnical ensemble" as a unit of analysis refers to Callon's principle of general symmetry, "society is not determined by technology, nor is technology determined by society. Both emerge as two sides of the sociotechnical coin during the construction processes of artifacts, facts, and relevant social groups" (Bijker 1995, p. 274).

relationships between industry, users and their diverse interests and needs, sport authorities, environmental problems, and policies. These interactions are inscribed in the artefact bicycle, which evolves as a result of them. Rosen narrates the rise of different "bike countercultures" resisting the stabilization of the artefact brought about by the cycling industry and sports regulations. These appropriations of the bicycle reflect positions of resistance: to the pervasive "car culture", the constraints on innovation imposed by the interests of the industry, and the invisibilization of female riders in bicycle design. Through widening his focus, Rosen enters the field of the politics of technology.

As a paradigmatic case study of contemporary social-oriented analysis of technology, bicycles exemplify a number of common trends in a family of approaches emerging from the social constructivism of the 1980s. Technologies and societies are mutually shaped in processes of co-production, transcending both social and technological determinisms. Technology, economics, politics, and culture (material and social elements) develop together in the construction of technological systems. Of course, together with their design, technologies carry "scripts" that guide their use and impacts (Akrich 1992), reach "momentum" (Hughes 1987) or present "obduracy" (Bijker 1995), making them difficult to revert. Technology is a powerful modelling force of our societies. However, technology is not autonomous in the strong sense of the term. Human agency plays an active role in maintaining stabilized technologies and has the capacity to destabilize, resist and rewrite "scripts". Technology, even simple artefacts, is political and has politics in a variety of senses. A growing number of voices are now calling for what Brey (2014) terms a constructive philosophy of technology, implied in the solution of social problems. Although Brey does not think this constructive analysis of technology should also be necessarily constructivist, 10 years earlier, Wiebe Bijker (2003) had argued that STS scholars may be the answer to the need for public intellectuals committed to technology problems.

Bicycles also reveal the interest and complexity of everyday technologies. "Mundane technologies", as opposed to exotic or novel technologies, are those so integrated in our lives that they remain mostly unnoticed, i.e., they are taken for granted (Michael 2000a, b). While many analyses of the social impact of technology tend to focus on technological systems linked to radical social changes, such as information technologies and biotechnology, mundane technologies permeate our everyday life, serving the "production and reproduction of local social configurations" (Michael 2000a, p. 3). We might likewise add, they are also a potential locus for resistance and reconfiguration. Mundane technologies carry scripts that guide our lives in conjunction with them, but they are also appropriated by their users in unexpected ways. The 'context' of a technology (if such a term may still be applied) is constituted both by the conditions of its design and production and the practices of its users.

Bicycles as a case study show, all in all, that, through the empirical turn, social-oriented philosophy of technology in a sense weaves its own "seamless web" with social studies to build an image of technology as a process of production of unstable sociotechnical ensembles in which material, social, organizational, and symbolic elements converge. Research on technology has come all the way from macro to micro analyses and back again; although this time against the backdrop of well-developed case studies. Recent attention to "mundane technologies" and the particularities of their interaction with their users points to yet another common space that the philosophy of technology may share with a related "turn" in science studies: the turn to practice (Pickering 1995; Schatzki et al. 2001; Shove and Pantzar 2005).

I will follow the thread from the paradigmatic example of bicycles to present sport as a field where co-production can be seen in action. Sport offers a wide range of opportunities to explore "the coevolving relation between human and nonhuman actors (objects) jointly implicated in the process of 'doing'" (Ingram et al. 2007). Practice-oriented approaches, focused on the interlocking between human competences or skills and materialities, are well suited to analyse these relations and may offer a more integrated image of sociotechnical ensembles.

18.3 Technology and Sport

Technology and sport have a long history of fruitful relationships. Science and technology play a leading role in the huge progress in terms of records in every discipline. Lighter bicycles, swimsuits that improve performance, oxygen equipment for Himalaya climbers… the list is unending, as is the controversy surrounding most of these technological innovations applied to competitive sport (Dyer 2015). Full body polyurethane computer-designed swimsuits, for example, were central to 98% of Beijing Olympic Games medals and were banned by the International Swimming Federation the following year. Now forbidden as a sort of 'technological doping', these innovative swimsuits show that the boundary between what is legitimate and what is cheating, between what the human body is able to do on its own (whatever that means) and what technological mediation allows it to do, is continuously being negotiated and reconstructed in each episode of technoscientific development and human ambition.

Classic extensionist accounts of technology assume a transparent relationship between artefacts and bodies (Brey 2000). Technologies may boost speed, endurance, or any other ability humans need to excel in sport. Moreover, they do so by extending human capacities. While Paralympic athletes compete with artificial legs substituting their missing natural limbs, able-bodied competitors use technology to enhance the way their natural bodies perform. In both cases, however, this supposed transparency of technological mediation may be put under scrutiny. Technology does not substitute or enhance human organs and functions innocently. Its alliance with human bodies creates difference and transformations (Kiran and Verbeek 2010).

German long jumper and amputee Markus Rehm is a good example. The Paralympic champion did not succeed in his attempt to compete in the Olympic Games held in Río de Janeiro (2016) alongside able-bodied athletes. Experts were not able to agree on the advantage his prosthetic leg could give him in the take-off of the jump because of the springiness of his carbon-fibre limb, while allowing that this same limb makes his approach run less efficient. The long jump with a prosthetic leg seems to be quite different than jumping with two natural legs. Substitution is not a simple act.[2]

In the realm of sport, the examples of how new technologies do not result in improved performance, but in a different kind of performance, abound. The artificial does not simply adds to the natural, but reconfigures it. The introduction of klapskate, for instance, signalled a revolution in speed skating (van Hilvoorde et al. 2007). In modern klapskates introduced at the end of twentieth century, the boot is attached to the blade by a hinge at the front. This way, the movement of the foot is smoother and allows the skater to reach a greater speed than when using traditional rigid skates. The new technology also implied that skaters should re-learn how to perform, i.e., new skating technology called for new skating technique, leaving behind those skaters unable to adapt. The way new artefacts combine with athletes' bodies and skills requires turning the focus to practice (jumping, skating, etc.), where, as active agents, users of the technology become key elements in its understanding. Humans and non-humans together constitute the practice of the sport.

Applying practice theory to innovation and consumer research, Shove and Pantzar (2005, 2010) conceptualized practices as active combinations of materiality, meanings, and skills.[3] They present the invention of Nordic walking, their case study, as a novel integration of pre-existing elements: the skill of walking, the stick as an artefact, and images that relate to wellbeing, health, sport, and nature. In order to stabilize Nordic walking as a sport, the practice of walking with sticks should be separated from the disability arena to associate it with the practice of sport as a way to attain health and wellbeing. Walking with sticks should be something sportier than a simple walk, but nothing as hard and strenuous as other sports in which sticks are used, such us mountaineering or cross-country skiing. It also requires learning the correct technique that characterizes this new sport and differentiates it from the unconscious customary walk to transform it into an expert act with a specific end in itself. Moreover, these combinations of elements are typical of each geographical and cultural context, so that transferring Nordic walking from Finland to other countries consists, according to Shove and Pantzar, in reinventing rather than exporting.

Running for running's sake, not for survival or competition, also had to be invented as a cultural practice. Meanings of wellbeing, health, achievement, and community have developed at the same time as material commodities (apparel, shoes, technologies such as GPS watches, heart monitors, and hydration systems) and the needed skills to perform the action. Professional and amateur athletes,

[2] Bionic limb expert Hugh Herr, recent Princess of Asturias Award for Technical and Scientific Research (2016), is another excellent example. A gifted rock climber from his youth, he had both legs amputated after a mountain accident and directed his efforts to fight disability. His smart prostheses are much more than simple 'artificial legs', and his mission to end disability is continuous with the dream of human enhancement through the merging of body and technology.

[3] Their emphasis on material mediation as an essential component of practices is based upon Schatzki (2001), as opposed to alternative analyses of practices such as Bourdieu's, which focuses on routine and repetition.

coaches, industry, retailers, and race organizers, among other social groups, have shaped the contemporary practice of running in interaction with all these elements, constructing it as a hybrid space in which the boundaries between science and culture, lay users and experts, become blurred. The final result, recreational running, is precariously stabilized in that space, in which users play a creative and active role through their shared practices, experiences, and knowledge (González García 2012).

18.4 Running as Practice and Culture

Long distance running is becoming a growing religion. Portrayed as a community of practice, a social world, a culture, a market, a boom or a hype, the Western practice of running has experienced an amazing boom over the last 50 years, the time lapse in which the more laid back practice of jogging has been transformed into the serious running, an activity that deserves academic analysis from a number of different approaches in social sciences and humanities, from sociology and psychology to geography, philosophy, and consumer research (Bale 2004; Bridel et al. 2015; Latham 2015).

In our post-everything society, running serves as a way to get fit, socialize, seek out achievable challenges, or improve one's self-esteem. Recreational running is much more than a simple leisure activity. It shapes not only bodies and minds, but also landscapes, wardrobes, and timetables. Running is now routine, a habit, a normalized activity in Western societies, where runners of all ages and aptitudes have become taken for granted as occupants of streets, parks, mountains and pathways.

Running as a sports activity for the general population emerged in New Zealand in the 1960s (Latham 2015). Track and field coach Bill Bowerman, the founder of Nike, exported the practice to the USA in an attempt to solve health problems linked to sedentary urban life. The history of the running boom is thus relatively independent from professional athletics, although there are obvious overlaps. The first running boom arose as part of a *recreational (r)evolution* that normalized the public display of physical activity (Scheerder et al. 2015), previously confined to specialized sports facilities such as gymnasiums and play courts. Desinstitutionalized and desportified, leisure running parted ways with competitive running, changing its scenario from the track and field court to roads, parks and other public spaces.

A second running boom took place around the turn of the century, this time characterized by the fragmentation and diversification of its practitioners. Running became a more democratic activity, spreading to previously excluded groups, paradigmatically women and the elderly. The social group of recreational runners is nowadays composed of a diverse array of people. As running became more popular, business around it flourished (Scheerder et al. 2011). Consumer segments may thus be an appropriate way to classify social groups of runners according with their willingness to acquire running-related goods (such as a pair of running shoes), experiences (such as entering to run a marathon), or expertise (such as a coach or a physiotherapy service).

Bowerman devised training logs for joggers based on the sophisticated training schedules of his athletes. Since then, the gap between professional track and field and recreational running has widened. Nowadays, however, long distance running has become a peculiar sport in which amateurs and professionals compete alongside one another in many events. In major world marathons, such as those of New York, Boston, Berlin and London, elite runners pursuing records or economic prizes run together with thousands of more or less dedicated amateur runners. Almost 50,000 runners finished the most famous marathon in the world, the New York City Marathon, in 2015. In Spain, Barcelona holds the most popular marathon, in which more than 15,000 runners took part in 2016. Professionals and amateurs share the roads in competitions and also share running-related technologies, although running may constitute a very different practice for each group.

18.5 The Technology of Running

Running seems a simple matter, the most natural sport, based on an ability we all acquire very early in our childhood. However, contemporary runners are equipped with a growing number of artefacts: shoes, clothes, heart rate monitors, GPS devices, smart watches, hydration systems, compression garments, sport drinks, and so on. Even complex training schedules and nutrition planning may be considered part of the technologies of running. Among all these, shoes are the most obvious and apparently unavoidable tool for safe and efficient running. "A pair of comfortable shoes" is all the jogger needs to go out on the streets, according to the pioneering 1963 pamphlet *The Joggers Manual* (Latham 2015).

The advice of looking for a "pair of comfortable shoes" was indeed a great tip for those running novices pounding the pavement in their first miles. Nowadays, long distance runners, both amateurs and professionals, carefully choose the shoes they will be training and racing with. Especially in the case of marathoners and ultramarathoners, a wise selection of shoes is key to success. In a marathon, faster runners will be running for something more of a couple of hours. Runners at the end of the pack will need 5 or 6 h to complete the 26.2 miles. Such a prolonged effort requires careful attention to one's feet. A bad choice may make what is already an extreme experience unbearable. The best pair of running shoes is in fact that which goes unnoticed, a perfect mundane technology that is taken for granted. However, not even running shoes are compulsory. At the same time as the first jogging boom began to emerge in America, Abebe Bikila ran and won the Olympic marathon in Rome barefoot (1960).

The emergence of the running shoe has in fact been described as a consequence of the spread of the practice of jogging (Tenner 2004). From the "comfortable pair of shoes" of the 1960s to modern and highly specialized trainers and racing flats, the history of the running shoe may be written alongside the history of the running craze. This most basic piece of running equipment has experienced a slow, yet evident, evolution since the advent of the jogging boom. Running shoes have progressively incorporated new technologies at the same time as running has gained in popularity among the Western urban population (Shorten 2000), constituting a growing market for athletic footwear companies.

Converted into a complex artefact that needs expert knowledge for its selection, the running shoe has taken on new meaning as a tool to avoid injury, a crucial goal in the social world of runners. Although the general population was advised to take up running as a healthy activity, an excessive amount of recreational runners get injured every year. Muscles, tendons and joints suffer the repetitive impact of running and modern trainers promise to help elude these side effects: Achilles tendinitis, plantar fasciitis, runner's knee, shin splits, IT band syndrome and the like.

The trainer that has become consolidated over the last 30 years is a well-cushioned shoe, with a drop of approximately 12 mm from heel to toe, and with arch support. Cushioning developed in order to decrease the impact of each stride, under the assumption that reduced impact will prevent injuries. High heel-to-toe drop was meant to reduce stress on the Achilles tendon, while arch support is supposed to help ovepronating runners.[4] The modern running shoe thus evolved from the 1970s and 1980s to the beginning of twenty-first century, gaining weight through systems of cushioning, stability and control, guiding foot movement to reach an ideal stride, and elevating it from the terrain to absorb impact.

Choosing running shoes is a complex decision. Factors like the weight and sex of the runner, footstrike pattern, training speed, weekly mileage, and usual running surface are relevant factors to consider. Expert advice, usually from retailers or podiatrists, has become the recommended practice for the responsible runner.

18.6 From Technological to Natural Running

However, several studies published from the late 1980s state that the aforementioned allegations are not backed up by scientific evidence (Frederick 1986). What evidence does seem to show is that, at best, high-tech running shoes do not prevent injuries; and, at worst, they may be a risk factor for injury. Steve Robbins, an enthusiastic advocate of 'freeing' feet, published a series of papers as early as the late 1980s arguing that contemporary running shoes were mainly to blame for the running injury epidemic (Robbins and Hanna 1987; Robbins and Gouw 1990, 1991). According to Robbins and his co-authors, as footwear reduces sensory feedback, runners on modern shoes unconsciously underestimate the impact of each step, overstriding and pounding the pavement heavily with their heels. Barefoot runners, however, who get direct feedback from their naked feet, absorb the shock more efficiently by landing on the forefront or mid-part of their feet and shortening their

[4] A runner overpronates when the inward rotation of his or her foot is excessive during running, flattening out the arch foot. Overpronation is related to a number of injuries (Hintermann and Nigg 1998).

strides. Long-term use of shoes also weakens natural foot structures, well adapted for barefoot running, making them more prone to injury.

Scientific warnings regarding the difficult relationship between running shoes technology and injuries have developed over the years hand in hand with the increasing reliance on the technological fix prescribed by shoe makers, retailers, and podiatrists. However, the growing dissatisfaction of some groups of runners eventually gave rise to a new understanding of the practice of running and the role of technology. The bestseller *Born to run* (McDougall 2009) constituted an important turning point in the social perception of running shoes. Mixing narrative and essay, journalist and amateur runner Christopher McDougall relates the story of his redemption from persistent injuries through barefoot running. Scientific literature on human evolution and running biomechanics, together with the experiences of traditional running communities such as the Tarahumara from Mexico, make up an allegation against the technologization of running. McDougall denounces shoe companies for selling science and technology without evidence. According to the new credo, the human species is naturally adapted to running, though not with those heavy, cushioned, motion-control shoes sold at speciality stores, but rather… barefoot.

Anthropology Professor at Harvard University Daniel Lieberman stands out as the scientific authority backing up McDougall arguments. His *Nature* paper (Lieberman et al. 2010), in which he compared the footstrike patterns of shod and barefoot runners, put the debate on the scientific map. Lieberman argued that humans have run barefoot or almost barefoot during all their evolutionary history and that human anatomy is well adapted to long distance running. By comparing footstrike patterns, he showed that barefoot runners tend to land on their forefoot or midfoot, while runners shod with modern trainers do so on their heels. Heelstriking is only made possible by bulky, well- cushioned shoes, because otherwise it is too painful. With their cushioned sole and high drop, running shoes change the way we humans are adapted to run and are responsible for a good number of common injuries. The technological fix does not protect runners from injuries; it causes them. It was marketing, not science, what has got runners to buy extremely expensive artefacts that offer them no benefit (Clinghan et al. 2008; Richards et al. 2009).

McDougall's book was illuminating for thousands of desperate runners. Tired of biomechanical analysis, expensive running shoes, and orthopaedic insoles, they found a simple and hopeful alternative. To solve their problems, they should do just the opposite they were told to do: re-learn to run barefoot, as we are adapted to do. From the USA to the rest of Western countries, 'natural running' became a fast growing trend. The barefoot running movement developed counter to accepted expert knowledge, while new information and research opened up new perspectives on running and technology. Until then, to be an informed runner meant following the advice of podiatrists, physiotherapists, and experts. To explore the possibilities of barefoot running, however, means looking for alternative sources of information and relying on peer advice and experiences (Rothschild 2012). Now, scientific evidence for barefoot running, still scarce and fragmented, is a question of debate and does not appear as a corpus of hard facts, ready to adopt by the responsible runner. The black box is open (Jenkins and Cauthon 2011; Bowles et al. 2012).

Barefoot running is not a stabilized practice, with a well-defined, fixed combination of technical and cultural aspects. Its definition is in progress and material elements (shoes), skills (running technique), and meanings (its 'natural' character) are being negotiated through both expert knowledge and shared experience. The Internet and social networks enthusiastically received the new boom and opened up a good number of spaces in which users can share knowledge and experiences.[5]

Running has undergone some important changes with the emergence of barefoot movement. Barefoot running is, literally, a new way of running for Western practitioners. From a biomechanical point of view, it clearly differs from shod running. In fact, the biomechanical difference between barefoot and shod running is the main argument for discarding modern shoes. Going barefoot thus requires forgetting old habits acquired through the use of shoes and re-learning the 'natural' way of running.

An appropriate "transition" is, in fact, key to successfully adopting the new practice. While recreational runners took to the streets with their expensive shoes, entrusting their structural and functional problems to experts and technology, returning to natural running requires awareness and attention. Runners should now focus on their body, posture, and movement, replacing running technology by running technique. After years of disciplining our feet, suddenly going barefoot will be traumatic and may cause new injuries. Paradoxically, 'natural running' requires an important learning curve.

Over-enthusiastic natural runners have in fact populated physiotherapy, podiatry, and orthopaedic clinics. Muscles, tendons, and joints suffer when accommodating to what should be natural. Running drills, earlier the exclusive domain of professional athletes, have begun to be practised by recreational runners in order to improve their running technique. Landing style as well as cadence, stride length, and body alignment are all elements to take into account in order to become a competent barefoot runner. Natural running courses and instructors flourish everywhere. Now, when a novice barefoot runner suffers any injury, the cause is not poor technology, but poor technique.

18.6.1 From Barefoot to Minimalist Running

From the benefits of barefoot running, we can also subtract a good number of problems. Learning the new ability requires time and effort. There is also a risk of injury if the adaptation is not slow enough. Moreover, contemporary running surfaces are not as friendly on runner's feet as the savannah of early hunters. The negative

[5] *Barefootrunning.com* (accessed October 25, 2016), online since 1997, claims to be the original barefoot site. *Thebarefootrunners.org* is the website of the Barefoot Runners Society, founded in 2009 (accessed October 25, 2016). *Barefootrunning.fas.harvard.edu* gathers scientific information on barefoot running following Daniel Lieberman's research and is partially funded by the Vibram company. *Naturalrunningcenter.com* is led by Dr. Marc Cucuzzella and runner Bill Katovsky, a curious combination of authority and experience, representing the hybrid spirit of the new trend (accessed October 25, 2016). *Correrdescalzos.es* and *correrdescalzo.com* are two Spanish websites on natural running (accessed October 25, 2016).

consequences of the spread of the barefoot movement have also been especially worrying for shoe manufacturers. Their business was built upon obedient runners who bought a new pair of expensive running trainers every 1000 km, as advised by experts.

Shoe companies, however, have been readier to adapt than many runners and have sought out new marketing strategies in tune with the new trend. As early as 2004, Nike had already launched its new shoe line "Nike Free", designed to allow the natural movement of the feet (Davis 2014). According to the company, they had observed that some coaches have their athletes run barefoot on grass in order to strength their feet and joints. Apparently, doing some of their mileage barefoot helped prevent injuries. Promoting barefoot running was not an appropriate strategy, though, for a company whose business is selling shoes, so the *Nike Innovation Kitchen* created the first shoe designed to run 'as if' barefoot. "Nike Free" appeared as the first of a new kind: the minimalist running shoe.

Nowadays, the market offers a wide variety of minimalist shoes. Some models are extremely radical, such as Vibram FiveFingers,[6] (with no cushion, zero drop, and individual toe sections) and Huarache sandals (imitating those of the Tarahumara featured in McDougall's best seller). However, most runners seduced by the arguments of 'less is more' have been more willing to try less radical alternatives, at least at the beginning. 'Transition shoes', a tamed version of these radical minimalist variations, thus became the main bet of shoe companies and represented a huge market success. With a bit of drop and a bit of cushion, they offer the 'natural running' experience without most of its risks. Even heel strike runners could moderately use these minimalist shoes to help them change their running style, train their technique, and strength their feet and joints.

Today, each shoe manufacturing company has its 'minimalist line' (Adidas, New Balance, Saucony, ASICS, Brooks, etc.). They do so, however, avoiding taking sides in the controversy and abandoning their traditional buyers. They sell technology and its rival at the same time. Some of them have adopted features such us less drop or weight in all their shoes. Generally, however, they have simply extended their offer by adding a minimalist series to their conventional cushioned and guiding shoes collections.

18.6.2 The Co-constitution of Running Practices and Running Cultures

Barefoot or minimalist running emerges as a new combination between experiences and abilities, artefacts, science, technology, and marketing strategies. A new meaning of running related to its 'natural' character, associated with a new technology (the minimalist running shoe), and the appropriate running technique are some of the elements that define this unstable, open practice.

[6]Originally designed for sailing and water sports, its potential use for 'almost barefoot' running was introduced later as a suggestion of an enthusiastic barefoot runner and instructor (McDougall 2009). Vibram received a lawsuit in 2012 for the unproven health claims related to the benefits of its FiveFingers. The company accepted a settlement agreement and moved on.

What it means to be a barefoot or minimalist runner is a question without a single answer. Running barefoot is not the same for the Tarahumara or the Kalenjin, who run for a variety of purposes: from ceremonial races to transportation. Western runners also attach very different meanings to the practice. For some, it means resistance to big companies and technologies, a return to a more authentic experience far from the mass leisure activity which recreational running has turned into. For others, runners in pain, it is yet another attempt to escape from injury and keep running. Some runners have even found new uses for which minimalist shoes were not designed nor marketed, such as racing shoes or speed training, a niche previously occupied by traditional racing flats. For these, minimalist shoes may be a means to improve their performance.[7] Nor does the minimalist running shoe have a closed identity: transition, race, training or strengthening, etc.: its promotion and use comprise a diversity of runners and aims.

Going from barefoot natural runners to minimalist urban runners presents a number of problems mediated by science, marketing, and experiences. The risk of injury, the difficulties of changing one's running technique, and the contradictory evidence found in scientific research are hindering the stabilization of the practice. In fact, many runners remain forever in the 'transition' phase, in an intermediate place between the old and the new practice, showing that stabilization may not be a necessary end point. Runners live actively in this space of open possibilities and unstable relationships in which they had to construct their own practice of running.

Science and technology are at the service of sport. However, their role is not simply that of providing understanding and solutions to basic problems that worry recreational runners, i.e., how to run better and faster without getting injured. Science informs, analyses, and tests the allegations of the parts in conflict, but it is not the "obligatory passage point". Its work runs in parallel with what is happening on Internet forums, social networks and the streets. Technology, on the other hand, has found in minimalist running shoes a way to reintroduce itself in a practice that came into being as an anti-technological movement. Science and technology are thus yet another element in the complex ensemble of interactions from which good running is produced and practised.

The process of the co-constitution of running technologies and running cultures brings together different traditions in the analysis of technology under the empirical turn. The common theme in philosophy of technology of the human/artefact interaction may be situated in the analysis of technologically mediated practices. Practices, understood as "embodied, materially mediated arrays of human activity centrally organized around shared practical understanding" (Schatzki 2001, p. 11), constitute the focus for the production and reproduction of "sociotechnical ensembles" composed of material and social relations between human and non-human actors. The inner workings of technology and its interaction with its users – such as the effects

[7] Contemporary minimalist running shoes share a good number of features with the traditional racing flats athletes use for competition and fast workouts: they are lightweight, low drop and thin-soled (Davis 2014).

of different shoe construction on runners' biomechanics – and the symbolic value shoes have in the constitution of identities (Hockey et al. 2015) are both examples of these relations, which also include knowledge-sharing among practitioners, marketing and manufacturing strategies, and the multiple meanings individually and collectively attached to running as a contemporary lifestyle. The minimalist runner and the minimalist shoe are both being simultaneously produced.

While early social constructivist views described the forces and factors that result in the stabilization of artefacts, understanding artefacts as elements of technologically mediated practices, both individual and social, allows us to give account of how stabilization may be a never reached end, without implying an unsuccessful sociotechnical ensembling. Practices are also an appropriate place from which to analyse the different scripting and resistance strategies of designers and users. In the context of consumption, these may include the joint creation of products for markets and markets for products, as well as the activist or civic practices in which citizens engage in order to resist or re-script industry devised associations.

Acknowledgments This contribution has been possible thanks to the financial support provided by the Spanish National R&D&I Plan (MINECO-15-FFI2014-58269-P) and by the FICYT (FC-15-GRUPIN14-128).

References

Akrich, M. (1992). The de-scription of technical objects. In W. Bijker & J. Law (Eds.), *Shaping technology/building society: Studies in sociotechnical change* (pp. 205–224). Cambridge, MA: MIT Press.

Bale, J. (2004). *Running cultures. Racing in time and space*. London/New York: Routledge.

Bijker, W. E. (1995). *Of bicycles, bakelites and bulbs. Toward a theory of sociotechnical change*. Cambridge, MA: MIT Press.

Bijker, W. (2003). The need for public intellectuals: A space for STS. *Science, Technology and Human Values, 28*(4), 443–450.

Bijker, W. E., & Pinch, T. (1984). The social construction of facts and artifacts: Or how the sociology of science and the sociology of technology might benefit each other. *Social Studies of Science, 14*(3), 399–441.

Bowles, C., Ambegaonkar, J., Cortes, N., & Caswell, S. (2012). Footwear for distance runners: The minimalism trend. *International Journal of Athletic Therapy & Training, 17*(6), 14–18.

Brey, P. (2000). Technology as extension of human faculties. In C. Mitcham (Ed.), *Metaphysics, epistemology, and technology. Research in philosophy and technology* (Vol. 19). London: Elsevier/JAI Press.

Brey, P. (2010). Philosophy of technology after the empirical turn. *Techné: Research in Philosophy and Technology, 14*(1), 36–48.

Brey, P. (2014). From reflective to constructive philosophy of technology. *Journal of Engineering Studies, 6*(2), 129–136. (preprint version in: https://www.utwente.nl/bms/wijsb/organization/brey/Publicaties_Brey/brey-2014-constructive-philosophy.pdf).

Bridel, W., Markula, P., & Denison, J. (Eds.). (2015). *Endurance running: A socio-cultural examination*. London: Routledge.

Clinghan, R. J., Arnold, G. P., Cochrane, L. A., & Abboud, R. J. (2008). Do you get value for money when you buy an expensive pair of running shoes? *British Journal of Sports Medicine, 42*, 189–193.

Davis, I. S. (2014). The re-emergence of the minimal running shoe. *The Journal of Orthopaedic and Sports Physical Therapy, 44*(10), 775–784.

Dyer, B. (2015). The controversy of sports technology: A systematic review. *Springer Plus, 4*, 524. https://doi.org/10.1186/s40064-015-1331-x.

Frederick, E. C. (1986). Kinematically mediated effects of sports shoe design: A review. *Journal of Sports Sciences, 4*, 169–184.

González García, M. I. (2012). Correr descalzos. Entre la ciencia y la cultura. In M. I. González García, J. A. López Cerezo, & J. A. (Eds.), *Fronteras de la ciencia. Hibridaciones* (pp. 185–198). Madrid: Biblioteca Nueva.

Hintermann, B., & Nigg, B. M. (1998). Pronation in runners. *Sports Medicine, 26*(3), 169–176.

Hockey, J., Dilley, R., Robinson, V., & Sherlock, A. (2015). 'There's not just trainers or non-trainers. There's like degrees of trainers': Commodisation, singularisation and identity. *Journal of Material Culture, 20*(1), 21–42.

Hughes, T. (1986). The seamless web: Technology, science, etcetera, etcetera. *Social Studies of Science, 16*(2), 281–292.

Hughes, T. (1987). The evolution of large technological systems. In W. Bijker, T. Hughes, & T. Pinch (Eds.), *The social construction of technological systems* (pp. 51–82). Cambridge, MA: MIT Press.

Ingram, J., Shove, E., & Watson, M. (2007). Products and practices: Selected concepts from science and technology studies and from social theories of consumption and practice. *Design Issues, 23*(2), 3–16.

Jenkins, D., & Cauthon, D. (2011). Barefoot running claims and controversies. A review of the literature. *Journal of the American Podiatric Medical Association, 101*(3), 231–246.

Kiran, A. H., & Verbeek, P. P. (2010). Trusting our selves to technology. *Knowledge, Technology, & Policy, 23*(3), 409–427.

Latham, A. (2015). The history of a habit: Jogging as a palliative to sedentariness in 1960s America. *Cultural Geographies, 22*(1), 103–126.

Lieberman, D., et al. (2010). Foot strike patterns and collision forces in habitually barefoot versus shod runners. *Nature, 463*, 531–535.

McDougall, C. (2009). *Born to run*. New York: Knopf.

Michael, M. (2000a). *Reconnecting technology, culture and nature*. London/New York: Taylor and Francis.

Michael, M. (2000b). These boots are made for walking…: Mundane technology, the body and human-environment relations. *Body & Society, 6*(3–4), 107–126.

Mitcham, C. (1994). *Thinking through technology*. Chicago: The University of Chicago Press.

Mitcham, C., Kroes, P., & Meijers, A. (Eds.). (2001). *The empirical turn in the philosophy of technology, Research in Philosophy & Technology* (Vol. 20). Greenwich: Jai Press.

Pickering, A. (1995). *The mangle of practice*. Chicago: University of Chicago Press.

Richards, C. S., Magin, P. J., & Callister, R. (2009). Is your prescription of distance running shoes evidence-based? *British Journal of Sports Medicine, 43*, 159–162.

Robbins, S. E., & Gouw, G. J. (1990). Athletic footwear and chronic overloading. A brief summary. *Sports Medicine, 9*(2), 76–85.

Robbins, S. E., & Gouw, G. J. (1991). Athletic footwear: Unsafe due to perceptual illusions. *Medicine and Science in Sports and Exercise, 23*(2), 217–224.

Robbins, S. E., & Hanna, A. M. (1987). Running-related injury prevention through barefoot adaptations. *Medicine and Science in Sports and Exercise, 19*(2), 148–156.

Rosen, P. (1993). The social construction of mountain bikes: Technology and postmodernity in the cycle industry. *Social Studies of Science, 23*(3), 479–513.

Rosen, P. (2002). Up the Vélorution: Appropriating the bicycle and the politics of technology. In R. Eglash et al. (Eds.), *Appropriating technology*. Minneapolis: University of Minnesota Press.
Rothschild, C. E. (2012). Primitive running: A survey analysis of runner's interest, participation, and implementation. *Journal of Strength and Conditioning Research, 26*(8), 2021–2026.
Schatzki, T. (2001). *Social practices*. Cambridge: Cambridge University Press.
Schatzki, T., Knorr-Cetina, K., & von Savigny, E. (Eds.). (2001). *The practice turn in contemporary theory*. London/New York: Routledge.
Scheerder, J., Vos, S., & Taks, M. (2011). Expenditures on sport apparel: Creating consumer profiles through interval regression modelling. *European Sport Management Quarterly, 11*(3), 251–274.
Scheerder, J., Breedveld, K., & Borgers, J. (2015). Who is doing a run with the running boom? In J. Scheerder, K. Breedveld, & A. Danchev (Eds.), *Running across Europe* (pp. 1–27). London: Palgrave McMillan.
Shorten, M. R. (2000). Running shoe design: Protection and performance. In D. Tunstall-Pedoe (Ed.), *Marathon medicine*. London: Royal Society of Medicine.
Shove, E., & Pantzar, M. (2005). Consumers, producers and practices. Understanding the invention and reinvention of Nordic walking. *Journal of Consumer Culture, 5*(1), 43–64.
Tenner, E. (2004). *Our own devices: How technology remakes humanity*. New York: Vintage.
van Hilvoorde, I., Vos, R., & Wert, G. (2007). Flopping, klapping and gene doping: Dichotomies between 'natural' and 'artificial' in elite sport. *Social Studies of Science, 37*(2), 173–200.
Winner, L. (1993). Upon opening the black-box and finding it empty: Social constructivism and the philosophy of technology. *Science, Technology and Human Values, 18*(3), 362–378.

Chapter 19
Public Participation in Science and Technology and Social Conflict: The Case of Aerial Spraying with Glyphosate in the Fight Against Drugs in Colombia

Mónica Lozano

Abstract In October 2015, the government of Colombia suspended aerial spraying with glyphosate as a means of controlling illicit crops in the country. That decision marked the beginning of a new stage in the over 30-year history of aerial spraying in the fight against drugs. This article analyzes the role that science and technology has played in shaping the social conflict generated by the use of glyphosate fumigation and the forms of public participation that subsequently emerged. There are three mechanisms whereby actors have participated in that conflict, each of which involved scientific-technological representation in the dispute: (1) definition of the conflict; (2) establishment of alliances; and (3) performance, or placing the conflict in the spotlight. I propose the term 'Rashomon effect for science and technology' to describe the phenomenon characterized by the fact that, in situations of social conflict, science and technology are appropriated by the actors involved in the conflict. This leads to a multiplicity of perspectives depending on the actors' interests, values, and beliefs and are based on different representations of science. The analysis of these mechanisms reveals an undesired effect in terms of democratic participation: the loss of direct dialogue with the affected communities, including considerations of their social and political perspectives. Instead of helping to establish a basis for better political decision making, scientific and technological representation ends up contributing to the exclusion of the communities directly affected by the spraying from the decision-making process.

M. Lozano (✉)
Psychology Program, Universidad del Rosario, Bogotá, Colombia
e-mail: monica.lozano@urosario.edu.co

B. Laspra, J. A. López Cerezo (eds.), *Spanish Philosophy of Technology*, Philosophy of Engineering and Technology 24,
https://doi.org/10.1007/978-3-319-71958-0_19

19.1 Public Participation in Science and Technology and Social Conflict

In democratic societies, social conflicts – particularly those whose their definition or solution involves science and technology – pose important challenges for understanding public participation in science and technology. How do scientific knowledge and other types of knowledge relate to definitions of conflict and the search for political solutions? What are the mechanisms of public participation in science and technology in these processes? What role does public participation in science and technology play in the decision making?

From a technocratic perspective, decision making has traditionally been based on the distinction between two spheres: one related to policy, the other to knowledge. Politics is supposedly capable of interpreting the interests, needs, and expectations of citizens through representative processes, i.e., representative democracy. Conversely, the field of knowledge – represented by experts – is assumed to be able to produce objective, valid, reliable knowledge that can be used to design possible options and scenarios, all of which are important in decision making. Trust in decision-making expertise is based on three assumptions: (1) the idea of a linear sequence that ranges from problem definition (*policy*) to (*expert*) advice to decision (*policy*); (2) the value neutrality of scientific knowledge (knowledge is neither good nor bad per se); and (3) the disinterestedness or political neutrality of scientists (Weingart 2001).

This line of thinking has been questioned by perspectives based on a stronger vision of democracy, i.e., deliberative democracy, which raises the need to extend the sphere of political decision making to citizens. In this view, citizens are regarded as an active part of the processes of deliberation and decision making. Public participation in science and technology is also promoted and budgets that justify the separation between the spheres of politics and knowledge are called into question.

In a broad sense, public participation in science and technology is understood to be a diverse set of more or less spontaneous, organized, structured activities or situations. Via these activities or situations, non-experts become involved in and provide their own contributions to processes such as agenda setting, decision making, and education policy, as well as to knowledge production in science and technology (Callon et al. 2001; Rowe and Frewer 2005; cited by Bucchi and Neresini 2008) and in evaluating the results and impacts on science and technology (Joss 1999). A series of formal initiatives is currently being developed that seeks to strengthen public participation in science and technology; such initiatives are justified by arguments that support democratization in the contemporary world.

However, academic understanding of public participation in science and technology is still poor; being weighed down by the conventional model of cognitive deficit and instrumental conception of participation (López Cerezo 2005). Discussions on this topic have focused on participatory mechanisms, the qualification of the public to participate in scientific and technological discussions, the relationships between experts and laypeople, and the validity of non-scientific knowledge in decision making. However, as noted by Wynne (2007), discussions have failed to highlight the far-reaching political implications of participatory processes; they have failed to address the fundamental question of what is meant by "science and technology" in these processes. Wynne makes a critique of viewpoints that focus on discussions about the public's qualification to be involved in expert technical debate and suggests the need to distinguish this from the public's qualification to be involved in *public issues involving* technical expertise. He also draws attention to uninvited public participation that occurs outside the sponsored channels of participation. Wynne notes how such perspectives address new challenges in understanding the public in relation to science and identifies new criteria for public participation.

Bucchi and Neresini (2008) made the distinction between *sponsored public participation* (such as public opinion surveys, participatory technology assessments, and deliberative democracy initiatives, i.e., consensus conferences, citizens' juries or panels and referenda) and *spontaneous public participation*. The latter is characterized by not being deliberately promoted by a sponsor and includes such processes as local protests, social movements, and hybrid forums in which laypeople and experts participate (as in the case of community research or patient associations).

The study of social conflict involving science and technology (or their representations) offers new possibilities of understanding public participation in issues related to science and technology in social contexts. The present chapter examines the social conflict generated by aerial glyphosate fumigation aimed at controlling illicit crops in Colombia, examining the role of science and technology in the configuration and development of this conflict. Furthermore, it analyzes the social representations of science and technology that have emerged in the press, scientific and technical reports, and interviews with some of the actors involved. The representations are analyzed from three different perspectives: (1) the ideas that emerged during that social conflict regarding science, technology, and scientists; (2) the scientific and technical developments (graphs, data, discursive productions) that occurred during the conflict; and (3) the legal and political understanding of social representations in representative democracies, which permits an analysis of how scientists deal with the interests of the various groups involved.

In the case studied here, public involvement in science and technology is one means of achieving the objectives pursued by the social actors in the conflict; science and technology constitute one scenario in which the confrontation takes place. In this case, speaking of *spontaneous participation* carries the risk of suggesting that the forms of participation in science and technology that occur during conflict (and which are not structured by a sponsor) arise without the participants being obliged to engage in them. The analysis shows how public participation by the affected groups and their allies in science and technology is a strategy for translating a series of demands of the groups affected by the fumigation. The analysis does so in a context in which internationalization of the conflict has restricted the groups' possibility of being direct participants in the national political debate.

The demands of communities or groups are translated by virtue of the possibility of scientific representation into studies, data, and reports. This material competes with other studies, data, and reports produced by other actors that generally have greater political power. In the process, a multiplicity of accounts and perspectives on the conflict develops. The result is what I call a 'Rashomon effect for science and technology' in situations of social conflict.

The mechanisms of participation employed by the actors (definition of conflict, establishment of alliances, and performance or placing the conflict in the spotlight) are forms via which they develop arguments in the face of conflict and engage in the process. However, these mechanisms have an undesired effect, in which the possibility of the affected communities directly communicating their social and political perspectives is lost.

19.2 Colombia: Illicit Crops and Social Conflict

The UN *World Drug Report 2017* ranked Colombia as the most important producer of cocaine, followed by Peru and Bolivia. Between 2000 and 2013, the area of coca bush cultivated in Colombia declined from 163,300 to 48,000 ha due to a series of combined strategies, including alternative development programs and the eradication of illicit crops. By 2016, in the midst of the peace process involving the FARC (Revolutionary Armed Forces of Colombia), the cultivated area showed a twofold increase over the 2013 figure, reaching 96,000 ha (Fig. 19.1). Unlike in the case of coca plants, Colombia does not possess a systematic means of determining the area planted with cannabis plants and opium poppies, although the aforementioned report registered an increase in cultivated areas of opium poppies compared with 2013 from 298 to 462 ha. In 2016, however, there was a 22% decrease in the cultivated

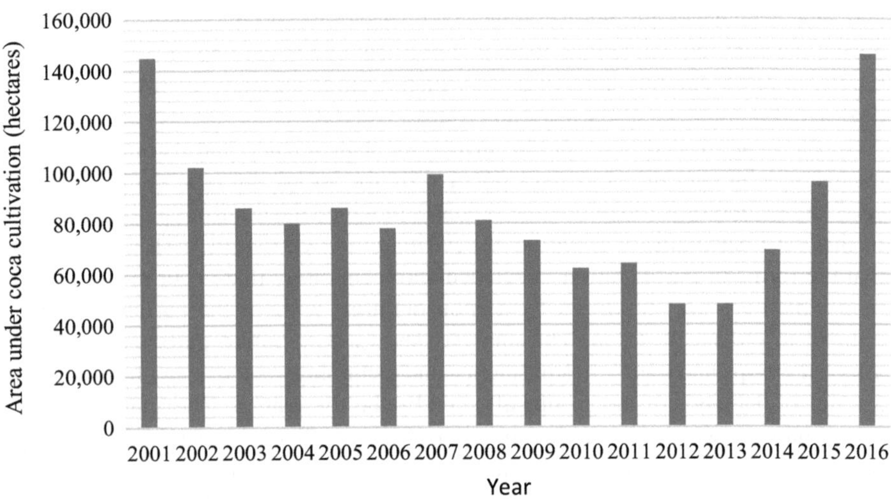

Fig. 19.1 Global coca cultivation in Colombia, 2001–2016 (Compiled by the author based on data from the UNODC Coca cultivation survey in Colombia 2017 and the previous year)

area of opium poppies compared with 2015. As regards cannabis plants, there were 201 cultivated hectares in 2016 (United Nations Office on Drugs and Crime 2017).

Until the mid-1970s, drug production in Colombia centered around marijuana. In the 1980s, opium poppies were introduced to the country in some marginal agricultural areas. Cocaine trafficking began in the mid-1970s. Initially, Colombia processed the coca leaf produced in Peru and Bolivia; subsequently, trafficking routes were established between the producing countries and Mexico and the United States. Intensive coca cultivation began in Colombia in the 1980s. The first crops were grown in the settlement areas of the eastern plains and the Amazon basin. The early 1990s were characterized by the expansion of crops. By the end of the decade, this expansion became concentrated in the south of the country, mainly in the departments of Putumayo, Caquetá, Guaviare, Nariño, Cauca, Huila, and Tolima. These territories, which form part of the Colombian Massif, are located in the southeast of the country, which includes the Amazon basin.

Initially, the trafficking and production of cocaine was in the hands of the so-called Medellin, Cali, and North Coast drug cartels. However, in the 1990s, following state persecution of these groups and the dismantling of the strongest, much of the control of the illicit crop production passed into the hands of guerrilla and self-defense groups, constituting one of their main sources of funding. Initially, the guerrillas' participation in the cocaine business was the tax they levied on peasants and drug traffickers. In exchange for protecting properties such as laboratories, trucks, and shipments of illegal drugs, peasants paid a tax of 10–15% of their coca leaf production or cocaine to the drug traffickers. In the 1990s, the guerrillas' participation increased until each group had established its own system of production, transportation, and commercialization (Díaz and Sánchez 2004). However, it is a question of debate as to whether the guerrillas managed to establish international networks for the traffic of narcotics at any point. For the guerrillas, their involvement in drug trafficking was a gradual process and was contrary to their ideological foundations. The self-defense groups, however, established a relationship with drug-trafficking organizations right from the outset and were funded by drug cartels as part of an expansion strategy when combatting the guerrillas to gain dominance in the coca-producing regions.

The relationship between social conflict and drug production has been widely studied in Colombia. There are different interpretations regarding the links and correlations between them (Guaquetá 2002). The social, political, and economic motivations that gave rise to conflict in the 1960s are widely recognized. This conflict emerged as peasant protest against the political and socioeconomic exclusion of a large proportion of the country's population and took the form of an armed left-wing movement. However, explanations about the maintenance of this movement until its dissolution in 2017 are less clear. Likewise unclear is the role played by drug trafficking in maintaining and escalating the conflict at the national level. Studies have identified a spatial correlation between coca production and armed activity: one of the main causes of the expansion of the coca economy was armed conflict (Díaz and Sánchez 2004). Perhaps more significant for the present study, however, is the fact that areas that produce coca, cannabis, and opium poppies are also the most excluded areas at the national level. These areas lack basic sanitation, education, electrification, and roads, and the state exerts precarious control over them. Under these conditions, the local populations find the possibility of basic sustenance in the form of illicit crops.

19.3 Conflict and Public Participation in Aerial Glyphosate Spraying of Illicit Crops

From 1978 until the end of 2015, aerial fumigation was one of the most important weapons in the fight against drugs in Colombia. The first herbicide used was paraquat. The use of glyphosate was introduced in 1984 based on the recommendation of the National Council of Narcotic Drugs (Consejo Nacional de Estupefacientes [CNE]). Produced by the Monsanto Company, glyphosate is a broad-spectrum herbicide and is the most widely used agricultural herbicide worldwide.

The Colombian government did not open up the decision regarding the use of glyphosate in its aerial eradication policy to public debate. However, since fumigation began, local as well as international social actors (from the civil, political, and even state sectors) have tried to influence the direction of this policy via different strategies. Among others, these have included peasant mobilization and marches, political pressure, media statements, the production and dissemination of impact studies, and constitutional demands. However, they have also included the use of violence in the form of armed strikes in rural populations and shooting at the aircraft involved in fumigation.

From the perspective of public participation of the actors directly affected by fumigation, there are two major stages in the social conflict related to glyphosate spraying to control illicit crops. The first is characterized by greater direct participation of groups of farmers and peasants affected by the fumigation. This stage reached its high point in the major coca marches of 1994–1996. These marches put pressure on the government to agree on a series of negotiated proposals that took into account farmer and peasant perspectives on the problem and ways to solve it. The peasant leaders brought to the discussion table issues such as economic, social, and health issues for the populations affected by fumigation. Agreements were reached with government representatives, including a reduction in aerial glyphosate spraying, the gradual disappearance of illicit crops within 10 years, and investment in roads, electrical grids, health, and education. The 1994 march ended with agreements being signed by the government, which were subsequently breached. The 1996 march ended with specific agreements with some peasant groups. However, there was also a weakening of the movement and military repression: paramilitary groups killed and wounded a number of peasant leaders, and several of them disappeared.

The second stage began in 1999 with the implementation of Plan Colombia – a bilateral United States and Colombian approach to combating the production and trafficking of drugs – and ended in 2016 with the suspension of fumigation. This stage was characterized by the dominance of scientific and technical arguments used by the actors, or their representatives, to support different stances. It was also marked by reduced opportunity for directly affected groups to participate in the discussions.

Protests, strikes, and peasant mobilizations against fumigation continued during the second stage. However, they did not succeed in becoming an articulated national movement capable of engaging in dialogue with the government. Public participation in the debate – particularly that of the directly affected communities – was influenced by intensification of the war against the guerrillas, which directly affected the producing areas, and by international rhetoric about the fight against terrorism. As in the first stage, in the second stage the participation of those directly affected extended beyond the formal channels established for political discussion. However, this stage was characterized by greater media support and internationalization of the conflict. At the same time, a number of mediators emerged among the farmers and there were instances of the mediators being able to affect government decisions. In this process, scientific and technological discourse became increasingly dominant.

In 1999, the government of President Andrés Pastrana (1998–2002) implemented Plan Colombia, which involved the receipt of international aid to help address drug trafficking. The plan was based on the principle of shared responsibility: it recognized that responsibility for the drug problem was shared by the international community as a whole and that it required a comprehensive, balanced approach to address the demand and supply of illicit drugs. At the same time, the first major effort was made to secure a negotiated peace with the FARC, although this eventually ended in failure.

As already noted, Plan Colombia aimed to fight drug trafficking and included the so-called Program for the Eradication of Illicit Crops. One of its basic approaches was extensive aerial spraying of coca, opium poppy, and cannabis crops using the herbicide Roundup Ultra (also produced by Monsanto), whose active principle is glyphosate. This program included a series of strategies aimed at institutional strengthening and the economic and social revival of affected areas. In practice, these actions were less important than the military aid and fumigation. The use of aerial spraying was justified by the increase in the area of crops, the difficulty of accessing the areas in question, and problems in implementing other types of eradication (such as manual eradication, due to safety concerns for eradicators[1] and the need for police and military forces to protect them).

Plan Colombia is still operative, although its approaches and priorities have changed with the different governments. During this process, Colombia has become one of the main allies of the USA in the fight against drugs and one of the largest beneficiaries of US foreign aid. From 2000 to 2015, the United States invested $9.6 billion in the program.[2]

---

[1] From 1990 to March 2017, there were 11,481 victims of anti-personnel mines in Colombia; 40% of these were civilians. Among these civilians, 25% were manual eradicators (Centro Nacional de Memoria Histórica 2017).

[2] Data from the National Planning Department (DNP 2016).

When Álvaro Uribe assumed the presidency in 2002 and following the failure of Pastrana's government to negotiate a peace deal with the FARC, Plan Colombia took on a new agenda. This agenda was reinforced by the effect of the attacks of September 11, 2001 in the United States, and it became a more focused strategy on the fight against drug trafficking and terrorism (DNP 2003).

During the Uribe governments (2002–2006 and 2006–2010), the main objective of foreign policy regarding drugs was to secure material support and logistic, military, and economic cooperation in the fight against them. This objective was addressed strategically so as to place the drug issue within the same sphere as such major international problems as terrorism, the environment, and human rights (Borda 2011). The Colombian government's political stance in this period was based on denying the existence of an armed conflict in the country and on regarding guerrillas (historically of peasant and liberal origin) as terrorist groups. This position allowed Plan Colombia resources to be used in the military struggle against these groups while at the same time intensifying the policy of fumigation.

When Juan Manuel Santos assumed the presidency (2010–2014 and 2014–2018) and peace negotiations resumed with the FARC, the fumigation policy underwent revision. During this stage, there has been a reduction in the resources contributed by the United States to Plan Colombia, which has directly affected those allocated for fumigation. Nevertheless, fumigation continued until 2015. Subsequently, the CNE suspended glyphosate fumigation as a precautionary measure based on recommendations from the Ministry of Health. This measure was backed by the World Health Organization's report published in March of that year, in which glyphosate was classified as a probable human carcinogen (International Agency for Research on Cancer 2015).

Since the announcement was made in May 2015, the decision to suspend glyphosate fumigation has not been devoid of controversy. As at other times, discussions on the use of other herbicides have re-emerged. Conversely, the decision has been seen by right-wing objectors to the peace process as a secret concession to the FARC in the agreements reached in Havana, Cuba. At the time of writing, the debate has taken a new turn owing to increased production of illicit crops, reviving the specter of fumigation in Colombia.

19.4 Science, Technology, and Public Participation in the Social Conflict Generated by Fumigation from 1999 to 2010

During the period of focus of this article (1999–2010), the conflict generated by fumigation has undergone change, both in its social configuration and in the types of actors involved. Despite these changes, the issues in the conflict can be categorized in four areas: risk of fumigation for human health; risk to the environment; effectiveness of fumigation in eradicating illicit crops; and the constitutionality of aerial fumigation (involving complaints about possible violation of the rights of affected populations). Regarding the first three areas,[3] opinions can be polarized. On the one hand, there are those who advocate the safety of glyphosate – both in terms of health and the environment – and defend the effectiveness of fumigation as a measure for controlling the proliferation of illicit crops. On the other hand, there are those who regard glyphosate as detrimental to both health and the environment and believe it to be ineffective.

The first standpoint in its entirety, or at least aspects of it, has been defended by the Colombian presidency (governments headed by Andrés Pastrana and Álvaro Úribe) and agencies responsible for anti-drug control, including the CNE and the national police force. It has also been supported by the Bureau of International Narcotics and Law Enforcement Affairs and United States Drug Enforcement Agency (DEA) and has been backed by international agencies, such as the United Nations through the Office on Drugs and Crime (UNODC), and the Organization of American States through the Inter-American Commission for the Control of Drug Abuse (CICAD). At the other extreme are to be found environmentalist groups, organizations for protecting human rights, some sectors of academia, the Ecuadorian government (affected by fumigation on border areas between the two countries), peasant and indigenous groups affected by fumigation, and certain state bodies, such as that of the Ombudsman's Office during some presidencies.

During the intensification of the fumigation policy, the issue of environmental and human health effects and the social impact constituted recurring elements in the complaints made by peasant or indigenous groups, non-government organizations (NGOs), and some state control agencies, such as the Attorney General's Office and the Inspector General's Office.[4] However, very few specific allegations were reported in the media. Out of 1474[5] news items published in *El Tiempo*[6] related to glyphosate fumigation from 1990 to 2011, only 179 referred to complaints from villagers about the effects of fumigation. Of these, merely in 51 cases were the affected population and type of impact specified, while only in a few cases was there any follow-up to the complaints.

These figures highlight the low relevance in the press of complaints by people and groups directly affected in public discussions; in this case, in *El Tiempo* on the

[3] The fourth is not addressed in this article.

[4] The stance of these bodies has not been constant in the history of fumigation. Although the Attorney General's Office and the Inspector General's Office have played important roles in some governments in denouncing the impacts of fumigation, there have also been periods during which they remained on the sidelines or clearly in favor of this policy. This was the case with the latter body during the presidency of Juan Manuel Santos, when Inspector General Alejandro Ordóñez was the only member of the National Council on Narcotic Drugs to vote against the suspension of glyphosate fumigation.

[5] Search conducted using the key words "denuncias" + "fumigaciones" + "glifosato".

[6] *El Tiempo* is one of the most important newspapers in Colombia. It is currently the newspaper with the largest circulation. For 7 years, it was practically the only one with a national circulation owing to a crisis with its main competitor, *El Espectador*, which in 2001 became a weekly newspaper, but returned to being a daily in 2008.

subject of glyphosate. This relative invisibility in the press of the communities directly affected is related to their exclusion from the debate and to the emergence of the representation processes, in which NGOs, the Inspector General, or other actors assumed these people's complaints and demands.

The analysis of this case allows me to identify at least three mechanisms via which representation is constructed through science and technology in the conflict:

1. Definition of the conflict: the Rashomon effect for science and technology,
2. Search and establishment of alliances; and
3. Performance or staging.

19.4.1 Definition of the Conflict: The Rashomon Effect for Science and Technology

As previously demonstrated (Lozano 2007, 2012), science and technology played a fundamental role in contributing to the definition of conflict from the perspectives, values, and interests of the different actors involved in the social conflict arising from glyphosate fumigation. Science and technology supported the arguments put forward to convince the public about the different stances and produced data for assessing the health and environmental risks of aerial spraying.

The conflict varied depending on the actor and their position in the conflict. At different times, the conflict was presented as a world health problem with links to armed struggles, a manifestation of terrorism, the continuation of a US imperialist strategy, a health problem created by economic and political interests concerned with displacing peasants in fumigated areas, a health and environmental problem, a problem of food safety, a border breach, or a threat to national sovereignty. With their ability to create diverse types of data, reports, and statistics, science and technology support different perspectives on the problems, or at least different aspects of these.

At the same time, scientific discourse supports the arguments formulated by different actors to convince national and international public opinion about the legitimacy of their stances; it contributes via the production of data and reports on the results of a strategy, as well as on its possible risks. This leads to conflicting results, depending on who produces the reports, the United Nations, the United States, the Colombian government, or scientists, who in turn engage different actors to carry out the risk assessment.

This function of science and technology to contribute to the construction of different perspectives on the conflict leads to the Rashomon effect for science and technology in situations of conflict. The effect is expressed in situations of social conflict, in which science and technology are appropriated by the diverse actors, producing a multiplicity of stories and perspectives regarding the conflict that respond to their own interests, visions, histories, situations, and values.

As in Ryunosuke Akutagawa's 1915 and 1922 stories (*Rashomon* and *In a Grove*, respectively) and the film version directed by Akira Kurosawa in 1950,[7] with the development of the conflict generated by glyphosate fumigation, we find ourselves, as in *Rashomon*, faced with stories told by different actors who elaborate their own version of what is happening. What is the actual problem? What do the actors involved in the conflict seek? Is glyphosate harmful or harmless? Does fumigation serve to avoid a much greater problem? The story is created and re-created depending on who is telling it and their interests, perspectives, and stance regarding the conflict. Science and technology – or the complex system of practices, institutions, scientific and technical knowledge, subsidies, and articles – create and contribute to the development of these stories and foster the notion of factuality and impartiality as regards each version of the story. One consequence of this multiplicity of perspectives on the conflict and the scientific arguments surrounding them is that it generates uncertainty regarding the situations that need to be assessed and distrust with respect to political decisions.

19.4.2 Establishment of Alliances

Related to the previous point, the other mechanism I have identified for representations of science and technology is the establishment of alliances among different actors. The expected role of science and technology in the conflict (providing reliable answers to support decision making) requires the establishment of alliances among different actors. In the search for useful knowledge to support different standpoints, governments select experts, while NGOs work with one another, with universities, and with peasant groups.

In this arena, there are excellent opportunities for participation by social movements. However, they only do so to the extent that they can build alliances and networks with NGOs and other movements, which provides visibility across borders.

[7] In 1915 the Japanese writer Ryunosuke Akutagawa wrote the story *Rashomon*, and 6 years later *In a Grove*. Both stories focus on the decline of feudal Japan and pose a series of questions to the reader: "What is the truth?", "What is really happening?", "What story is being told?", and "What justifies the action of the characters?" Yet the facts and the stories change depending on who tells the story, their intentions, and their perspective on the subject. In *Rashomon*, the servant who has been fired and shelters in the castle converted into a morgue, garbage dump, and den of thieves is stuck between starving or becoming a thief. As the story advances, his convictions about what he has to do radically change, depending on the stories he hears and on the different perspectives he develops regarding his life. In the story *In a Grove*, readers have access to different versions of the death of a young man, as told by a woodcutter, a Buddhist monk, an old woman, a snitch, a murderer, the young wife, and the very spirit of the dead man, through a witch. The accounts of the death, of who the murderer is, and what motivates him are different and antagonistic. The reader never knows what really happened, despite the evidence of the body of the dead man and hearing the confessions of three of the people involved who claim to have committed the murder. Perhaps the most well-known version of these stories is Akira Kurosawa's 1950 film *Rashomon*, based on both Akutagawa stories.

This is a very important point in understanding the dynamics of the conflict. It is only when national actors manage to take the discussion to the international arena that they are able to find allies who provide reinforcement through non-invited participation and hence their own actions can achieve greater legitimacy.

For example, right after the attacks of September 11, 2001, it was through its translation into the language of terrorism that the Colombian government managed to maintain and transform the Plan Colombia strategy to its own benefit by positioning the drug debate as a central issue of the problem. The peasants were transformed into narco-terrorists, the groups that supported them were helpers of narco-terrorism, and the fight against coca became a global crusade against terrorism. This provided a new standing for Plan Colombia and made Colombia one of the 'allies' in the fight against terrorism alongside the United States. Terrorists were no longer just distant figures hiding in Afghanistan's caves; they could also be found in Colombia's jungles.

This interpretation is also valid for other actors, such as peasant groups. The international diffusion of the account of aerial fumigation of an indigenous village and its effects appeared in a European newspaper. The publicizing of the incident gave rise to the first study in Colombia on the impact on health of fumigation as part of the strategy for eradicating illicit crops[8] and contributed to the development of a system for gathering and addressing residents' complaints.

In December 2000, Dutch journalist Marjon van Royen, from NRC Handelsblad, published the article *Driven mad by itch*, in which she reported the intoxication of 80% of the children in the Aponte indigenous reservation in the Tablón de Gómez municipality (department of Nariño, located in the south of Colombia). Based on the testimony of a local doctor, the children suffered as a result of aerial glyphosate fumigation.[9] Among other effects, publication of the article resulted in a request by the US Congress for an environmental risk assessment of the fumigation being carried out,[10] with Congress provisionally suspending the use of the resources of Plan Colombia until evidence was produced regarding the safety of glyphosate. The involvement of the Dutch journalist was part of the strategies developed by peasant groups and NGOs who were seeking support: they needed international allies in dealing with the problems created by fumigation. When such a story is circulated in this manner, it transcends national borders, allowing the claims of the villagers to be heard by a wider audience. This development also implies redefining and translating the prob-

[8] Up until 2001, the Colombian government used numerous specialized articles about the use of glyphosate and its possible effects on health and the environment; many of these articles were financed or disseminated by the manufacturer, Monsanto.

[9] An English version of the story is available on the journalist's website: http://www.marjonvanroyen.nl/index.php?option=com_flexicontent&view=items&cid=147:about-marjon-van-royen&id=525:driven-mad-by-itch&Itemid=85

[10] The studies entitled "Final report. Study of the complaints of the damages to the health related to the aerial eradication in Colombia. Department of Nariño, Municipality El Tablón de Gómez" and "Alleged effects of glyphosate on human health" appeared in 2001. Both were conducted by Dr. Camilo Uribe Granja, from the Uribe Cualla Toxicology Clinic, and contracted by the Narcotics Affairs Section of the Embassy of the United States in Colombia (Clínica Uribe Cualla 2001; Uribe et al. 2001).

lem in terms of the interests of other groups. In this case, the voice of the doctor was used to reveal the impact of fumigation on the health of indigenous children. But it also allowed the journalist and photographer to place the issue in the spotlight and open the discussion in other areas, resulting in more effective participation by the communities involved in the decision-making process regarding fumigation.

The establishment of alliances passes through a negotiation process among the actors. Communities may view some NGOs, journalists, and researchers with suspicion; alliances are established based on agreements and negotiations. For example, a commitment regarding the form of participation occurred with Acción Ecológica. This NGO pledged to not seek compensation for its work supporting Ecuadorian border communities in their litigation against Colombia, although its work was important for the Ecuadorian government's claim against Colombia at the International Court of The Hague. From my own experience, the possibility of working with the Cofán community in Putumayo arose because I managed to transform research on social studies of science and technology, adapting it to the requirements of the community. To be able to conduct research in the area, I had to show in advance that the study could be useful to the community's interests.

However, the establishment of alliances also implies an active role for the allies; at least in some aspects of their joint work. The Colombian government and US embassy worked together to produce a response to a news article denouncing the effects of fumigation on indigenous havens. The United Nations and the government of Colombia worked together in monitoring the spread of illegal crops. Border communities are working with Acción Ecológica in collecting blood samples from remote communities as part of studies on the mutagenic effects of glyphosate and its environmental and social effects.

19.4.3 Performance or Staging

Finally, the last mechanism is performance. In an artistic scene, this is understood as a staged performance of short duration, in which the body and various objects are involved. It is designed to create an audience response and provoke a reaction in the public. In the conflict, performance is a mechanism through which actors construct a representation of the conflict (or some of its constituent topics), while at the same time helping to create links between representatives.

Here are two examples of performance. In January 1995, amid the intensification of fumigation during the administration of Ernesto Samper, a French delegation consisting of the President of the Congress and some members of parliament was officially received in Colombia. The purpose was to demonstrate fumigation in the field. During the trip, the delegation was accompanied by Minister of Defense Fernando Botero. During the fumigation demonstration in a field in Huila, the delegation was soaked with glyphosate. A journalist reported the event as follows:

> Botero knew about the plan. The plane flew over them three times and sprayed them all. The minister gave the order that the Turbo Thrush, one of the five small planes that fumigate in Colombia, had to spray the chemical over him and the other 30 or so members of the

delegation. They were being sprayed with a chemical that was considered dangerous for people, animals, and the environment.

Botero told the surprised French delegates, who had never imagined they would be subjected to such a procedure, that glyphosate was not harmful to health. The French had simply requested a visit so that they could see the hectares of opium poppies being cultivated, the glyphosate, the planes used for the fumigation, and the peasants and their plantations, whether licit or illicit. Those who came – a foreign policy and defense commission – can have no doubts that they had become thoroughly immersed in the subject.[11]

This was not the only time the Colombian government adopted such an approach (or tried to explain the move as premeditated). During the Uribe administration, the Colombian ambassador to the United States led a delegation of officials from that country on a fact-finding mission into the fumigation process. During the trip, they were likewise doused with glyphosate. The press release tackled the subject in a humorous fashion, in which the ambassador reassured the bewildered delegation by telling them that it had happened to him before: "I have been doused in glyphosate before. And so far, nothing has happened to me; except that I stopped growing!" To understand the ambassador's joke, it is necessary to realize that he was short in height compared to the American politicians.

Affected farmer groups have also used such performances on several occasions. During some protests in Ecuador in 2007, one peasant sprayed glyphosate onto the garden of the Colombian embassy there. When I visited Sucumbios in Ecuador, one of the communal houses of the peasants featured a pile of maize damaged by glyphosate in the center of a meeting room, while emaciated livestock had been tied to nearby fences to demonstrate the ravages of the chemical (Fig. 19.2). The effects of fumigation on the community were documented in the form of tangible evidence presented to me. Villagers used photographs and medical documents and told stories about the sufferings they had endured.

19.5 Discussion

In the conflict related to spraying illicit crops with glyphosate, no consultation process took place at the national level aimed at publicly deciding whether or not glyphosate should be used. Fundamentally, this was because the decision was never open to democratic debate and for a long time, I suspect, it was not even a sovereign decision on the part of Colombia. However, the handling of the conflict and its translation into techno-scientific argumentation as a way of legitimizing the spraying provides useful lessons (not only for Colombia, but also for sectors in the United States and Europe) when considering issues of public participation in science and technology in situations of social conflict.

[11] *El Tiempo*, January 28 1995, *Mindefensa y franceses se empaparon de glifosato*. Avaible via *El Tiempo*: http://www.eltiempo.com/archivo/documento/MAM-300420. Accessed 11 Sept 2011.

Fig. 19.2 Focal group in Sucumbios Province–Ecuador (Photo by Mónica Lozano)

In Colombia, no formal institutional processes were promoted by the government. However, throughout the history of fumigation, there has been non-invited, active public participation by different social actors that has succeeded in influencing political decisions on the subject. A core element and one that led to a breakdown in the forms of participation occurred with the establishment of Plan Colombia. Before this plan, social mobilization and protests succeeded in forcing the government to sit down and discuss different possibilities with the crop-producing communities. However, the implementation of this strategy to fight drugs led to a preponderance of a scientific and technical discourse in defining the social conflict involved. This situation resulted in a loss of political agency on the part of the affected communities in participating in political decision making. It also led to the emergence of numerous representatives, who translated problems into techno-scientific, environmental, political, or terrorist arguments. Finally, the situation created multiple mechanisms for representation: defining the conflict from different perspectives, establishing alliances, and performance.

It is important to understand that the issues discussed here do not hinge on controversies about science and technology: they are fundamentally about visions of society, human and other forms of life, economics, and social justice. These questions are irreducible to the traditional, hegemonic vision of science. From this perspective, it should be pointed out that the challenge Colombia faces is that of constructing new ways of understanding public participation in science and technology; a challenge that goes beyond the debate about the distinction between experts and laypeople. As to the distinction between civil and political society, it is necessary to recognize that, in participatory exercises (especially ones that take place in non-formal types of participation), there are many fruitful ways of approaching and understanding conflicts… and perhaps also of providing solutions to them.

Acknowledgments I am grateful to Zabrina Welter and the Edanz Group (www.edanzediting.com/ac) for editing a draft of this manuscript.

References

Borda, S. (2011). La política multilateral colombiana de drogas durante las dos administraciones Uribe: Hacia el activismo prohibicionista en una era de distención. In A. Gaviria & D. Mejía (Eds.), *Políticas antidroga en Colombia: Éxitos, fracasos y extravíos* (pp. 149–176). Bogotá: Universidad de los Andes.

Bucchi, M., & Neresini, F. (2008). Science and public participation. In E. Hackett, O. Amsterdamska, M. Lynch, & J. Wacjman (Eds.), *The handbook of science and technology studies* (pp. 449–472). Cambridge, MA: MIT Press.

Centro Nacional de Memoria Histórica. (2017). *La guerra escondida*. Bogotá: Centro Nacional de Memoria Histórica.

Clínica Uribe Cualla. (2001, September). *Informe final. Estudio de las denuncias de daños a la salud relacionadas con la erradicación aérea en Colombia*. Bogotá: Clinica Uribe Cualla.

Díaz, A., & Sánchez, F. (2004). *Geografía de los cultivos ilícitos y conflicto armado en Colombia*. Universidad de los Andes. Available via Economia Uniandes: https://economia.uniandes.edu.co/assets/archivos/Documentos_CEDE/D2004-18.pdf. Accessed 22 Aug 2017.

DNP. (2003). *Balance del Plan Colombia*. Bogotá: DNP.

DNP. (2016). *Plan Colombia DNP 15 años*. Departamento Nacional de Planeación. https://www.dnp.gov.co/Paginas/Plan-Colombia-DNP-15-a%C3%B1os.aspx. Accessed 15 Aug 2017.

Guaquetá, A. (2002). Dimensiones políticas y económicas del conflicto armado en Colombia: Anotaciones teóricas y empíricas. *Colombia Internacional, 55*, 19–36.

International Agency for Research on Cancer. (2015, March 3–10). *Volume 112: Some organophosphate insecticides and herbicides: Tetrachlorvinphos, parathion, malathion, diazinon and glyphosate*. Lyon: IARC Working Group. http://monographs.iarc.fr/ENG/Monographs/vol112/mono112.pdf. Accessed 3 Sept 2017

Joss, S. (1999). Public participation in science and technology policy- and decision- making – Ephemeral phenomenon or lasting change? *Science and Public Policy, 26*(5), 290–293.

López Cerezo, J. A. (2005). Participación ciudadana y cultura científica. *Arbor, CLXXXI*(715), 351–362.

Lozano, M. (2007). El síndrome de Rashomon o la comunicación de la ciencia y la tecnología en situaciones de conflicto. *Redes, 13*(26), 235–260.

Lozano, M. (2012). *Ciencia en conflicto: Representaciones de la ciencia, conflicto social y democracia*. Doctoral dissertation, Universidad Nacional Autónoma de México.

United Nations Office on Drugs and Crime. (2017). *World Drug Report 2017*. United Nations Publications.

Uribe, C., Melo, O., Salcedo, A., Cespedes, J., Santodomingo, T., Gastelbondo, Y., & Hernandez, B. (2001, December). *Supuestos efectos del glifosato en la salud humana. Informe final*. Bogotá: Clínica Uribe Cualla.

Weingart, P. (2001). Paradoxes of scientific advice to politics. In OECD (Ed.), *Social science for knowledge and decision making* (pp. 79–94). Paris: OECD.

Wynne, B. (2007). Public participation in science and technology: Performing and obscuring a political–conceptual category mistake. *East Asian Science, Technology and Society: An International Journal, 1*, 99–110.

MIX
Papier aus verantwortungsvollen Quellen
Paper from responsible sources
FSC® C105338

If you have any concerns about our products,
you can contact us on
ProductSafety@springernature.com

In case Publisher is established outside the EU,
the EU authorized representative is:
Springer Nature Customer Service Center GmbH
Europaplatz 3, 69115 Heidelberg, Germany

Printed by Libri Plureos GmbH
in Hamburg, Germany